ANALYSIS AND DESIGN OF PRESTRESSED CONCRETE

ANALYSIS AND DESIGN OF PRESTRESSED CONCRETE

DI HU

Associate Professor, School of Civil Engineering,
Central South University, China

ELSEVIER

Elsevier
Radarweg 29, PO Box 211, 1000 AE Amsterdam, Netherlands
The Boulevard, Langford Lane, Kidlington, Oxford OX5 1GB, United Kingdom
50 Hampshire Street, 5th Floor, Cambridge, MA 02139, United States

Notices

Knowledge and best practice in this field are constantly changing. As new research and experience broaden our understanding, changes in research methods, professional practices, or medical treatment may become necessary.

Practitioners and researchers must always rely on their own experience and knowledge in evaluating and using any information, methods, compounds, or experiments described herein. In using such information or methods they should be mindful of their own safety and the safety of others, including parties for whom they have a professional responsibility.

To the fullest extent of the law, neither the Publisher nor the authors, contributors, or editors, assume any liability for any injury and/or damage to persons or property as a matter of products liability, negligence or otherwise, or from any use or operation of any methods, products, instructions, or ideas contained in the material herein.

ISBN: 978-0-12-824425-8

For information on all Elsevier publications visit our website at
https://www.elsevier.com/books-and-journals

Publisher: Matthew Deans
Acquisitions Editor: Glyn Jones
Editorial Project Manager: Naomi Robertson
Production Project Manager: Surya Narayanan Jayachandran
Cover Designer: Greg Harris

Working together
to grow libraries in
developing countries

www.elsevier.com • www.bookaid.org

Typeset by TNQ Technologies

Contents

9. Ultimate bearing capacity of flexural members 259

Preface

Prestressed concrete has been one of the most widely used structural materials in railway bridges, highway bridges, and civil and industrial buildings since its first trial application over 100 years ago. A prime example of its application is prestressed concrete beams, which are almost the strongest competitive superstructures in bridges with spans from 30 to 250 m. In view of the excessive deflection and cracking of some built long-span bridges that are still in service and the strict requirements for high-speed railway bridges for long-term camber or deflection, combined with the development of material science, construction technology, and structural analysis, the following trends are emerging: prestressing materials with higher strength and mechanical properties will be developed and used, the construction technologies will be more intelligent and standardized, the performance-based design will draw increased attention, and most of the analysis and design of prestressed concrete structures will be completed by computers through specific software rather than by hand. Accordingly, the textbooks or reference books for the analysis and design of prestressed concrete must keep up with new developments.

This book systematically introduces the basic concepts of prestressed concrete and fundamental principles of analysis and design for prestressed concrete structures subjected to bending, shear, tension, compression, and torsion based on the ultimate limit state method.

The most distinct feature of this book is that the calculation of prestress losses in a step-by-step manner from the construction to the service period is expounded comprehensively, and an effective-stress-based analysis system as well as the calculation formulas for design parameters at the serviceability limit state are constructed. The accurate calculation of effective stress in prestressing tendons is the basis of stress analysis, crack control, and long-term deflection prediction.

Another distinct feature of this book is that it builds a bridge between general analysis methods and the simplified approaches provided in the relevant codes. For a strictly performance-based design with the premise of sufficient strength and durability, when the prestress losses due to friction, creep and shrinkage of concrete, and relaxation of prestressing tendons are considered comprehensively, it is very complicated and tedious to carry out stress and deflection analysis of a prestressed concrete member with a large number of different prestressing tendons. For statically indeterminate prestressed concrete beams with multi spans, it is almost impossible to complete the analysis using a simplified approach. Therefore, it is necessary to elaborate the precise analysis approaches. However, the associated codes and technical standards guiding the design often specify the simplified methods and approximate formulas for many key design parameters.

This book first expounds the train of thought of general analysis, and then introduces simplified approaches to the codes. By reading this book, readers will be able to easily master the general analysis and simplified calculation strategies for pretensioned and post-tensioned concrete structures, and cultivate their design ability.

The author sincerely thanks Prof. Teng Wu from University at Buffalo, Prof. Tony Yang from University of British Columbia, Prof. Wenyu Ji from Beijing Jiaotong University, and Prof. Xiaojun Wei and Menggang Yang from Central South University for their tremendous support and valuable suggestions throughout the whole process of writing this book. Acknowledgments are also due to the author's former students, senior engineers Jiping Guo from Guizhou Communications Construction Group Co., Ltd. and Yong Tang from Guangxi Communications Design Group Co., Ltd. for providing the figures, and current graduate students Jingwei Wu, Yujie Li, and Dangnan Jwakdak for contributing to typing and drawing of the figures. Special thanks are due to the editors, Surya Narayanan Jayachandran and Yingwei Liu, for their excellence in editing and proofreading.

This book can be used as a textbook for graduates and undergraduates majoring in civil engineering and relevant fields, as well as a reference for researchers, engineers, technicians, and managers engaged in prestressed concrete-related projects.

Author biography

Dr. Di Hu is currently an associate professor at the School of Civil Engineering, Central South University (CSU), Changsha, China. He received a Ph.D. degree in Civil Engineering from CSU in 2003, and fulfilled his post-doctoral research on transportation engineering in CSU during 2009—2010. He was invited to study at the University of British Columbia. Canada in 2018. Dr. Hu's main research includes nonlinear analysis of bridge structures, theory of creep effect on concrete structures, assessment and strengthening design of existing bridges, and novel joints of concrete-filled steel tubular columns in high-rise buildings. In particular, he has proposed a refined method for calculating the stress loss in prestressing steels due to anchorage set considering the reverse-friction effect, a refined method (automatically step-up method) and a simplified method (steel restraint influence coefficient method) for analyzing the creep effect on prestressed concrete structures. Dr. Hu has published over 40 journal papers, and two monographs *Theory of Creep Effect on Concrete Structures* (2015) and *Design Principles of Prestressed Concrete Structures* (second edition, 2019).

List of figures

List of tables

CHAPTER 1

Basic concepts of prestressed concrete

Contents

1.1 Basic concepts

Concrete, one of the most widely used materials in civil engineering structures, is strong in compression while it is weak in tension. As reinforcements are embedded in the tension zone to resist tension, reinforced concrete is formed which can be used more extensively than plain concrete. Even so, a reinforced concrete member subjected to tension, eccentric compression, bending, or torsion, inevitably exhibits cracking due to the low tensile strength of concrete. For this reason, reinforced concrete cannot be applied in those structures in which high crack resistance is required, such as oil storage tanks, nuclear power vessels, and structures in a severely corrosive environment. As for those structures in which cracking is allowed, on the other hand, the crack width must be controlled to a limited value to guarantee structural durability in the serviceability limit state. Corresponding to the typical maximum permissible values of crack width specified in the design codes, 0.3–0.4 mm, the tensile stress in reinforcement at a crack only reaches 250–400 MPa, showing that it is uneconomical to use high-strength materials in reinforced concrete. In addition, if reinforced concrete is employed to build a beam with

Analysis and Design of Prestressed Concrete
ISBN 978-0-12-824425-8, https://doi.org/10.1016/B978-0-12-824425-8.00001-4

a long span, the sectional dimensions and the reinforcement quantity have to be increased tremendously to control the crack width, resulting in a sharp increase in the proportion of the self-weight to the total design load, which eventually becomes an uneconomical scheme. Therefore, reinforced concrete cannot be used to build flexural structures with a long span.

Consider a reinforced concrete member subjected to axial tension as shown in Fig. 1.1A. The tensile stress in concrete quickly exceeds the tensile strength as the tension increases, with cracking becoming inevitable. If the reinforcements are replaced by several high-strength steel bars with outside plastic pipes which are embedded inside the concrete before casting, high compression in the concrete is generated as the bars are anchored at two member ends after they are stretched in high tension when the concrete attains sufficient strength, as shown in Fig. 1.1B. In this state, the normal stress of concrete in a prestressed section subjected to axial tension is derived as

$$\sigma_c = \sigma_{c,p} - \sigma_{c,T} = \frac{N_p}{A_c} - \frac{T}{A_c} \tag{1.1}$$

where σ_c = normal stress of concrete.

$\sigma_{c,p}$ = concrete stress due to compression generated by the stretched steel bars.

$\sigma_{c,T}$ = tensile stress of concrete due to external axial tension.

N_p = compression in concrete transferred from the tensioned steel bars.

T = external axial tension.

A_c = cross-sectional area of concrete.

In Eq. (1.1), if the introduced compression in concrete is greater than the external tension, tensile stress will not occur and cracking will be prevented in the tensile member.

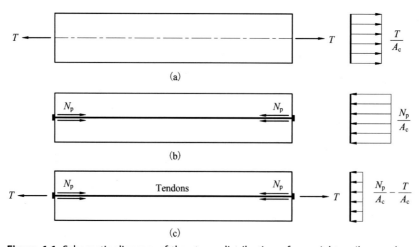

Figure 1.1 Schematic diagram of the stress distribution of an axial tensile member.

Consider a simply supported beam subjected to a uniformly distributed load. The beam is arranged with stretched parabolic steel wires which are anchored at the beam's ends, as shown in Fig. 1.2. The normal stress in concrete in the bottom face fiber at the midspan is given by

$$\sigma_c = \sigma_{c,p} - \sigma_{c,M} = \frac{N_p}{A_c} + \frac{N_p e_p y_1}{I_c} - \frac{M y_1}{I_c} \qquad (1.2)$$

where σ_c = stress in concrete in the bottom face fiber at the midspan section.

$\sigma_{c,p}$ = compressive stress in concrete in the bottom face fiber at the midspan section caused by the stretched wires.

$\sigma_{c,M}$ = tensile stress in concrete in the bottom face fiber at the midspan section caused by the uniformly distributed load.

M = bending moment at the midspan caused by the uniformly distributed load.

A_c = cross-sectional area of concrete.

I_c = moment of inertia of the concrete section.

e_p = eccentricity of steel wires with respect to the center of gravity of the concrete section (c.g.c.) at the midspan section.

y_1 = distance from the center of gravity of the concrete section to the bottom edge.

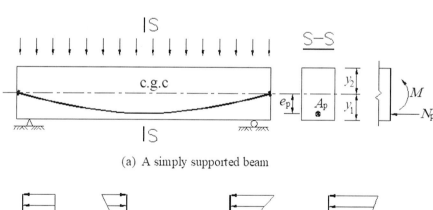

(a) A simply supported beam

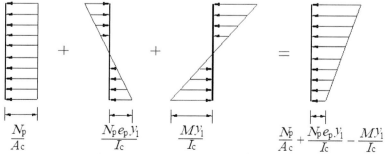

(b) Stresses in midspan section

Figure 1.2 Stress distribution at the midspan section of a simply supported beam.

Similarly, if the introduced compressive stress in concrete by the stretched wires is greater than the tensile stress caused by the uniformly distributed load, there will be no tensile stress at the tension zone.

The prestretched steel bars and wires are called prestressing steels or prestressing tendons, and the tensile force in the tendons is called the prestressing force. The stress achieved in concrete before it is subjected to a load is called the prestress. In a broad sense, prestressed concrete refers to concrete in which prestress is achieved deliberately before service. If the prestress is generated by the stretched tendons, the corresponding prestressed concrete can be treated as a particular form of reinforced concrete.

Obviously, the tensile member (Fig. 1.1C) and flexural member (Fig. 1.2A) arranged with prestressing tendons are prestressed concrete structures. The compression in concrete transferred from the prestressing force not only changes the internal forces of the concrete structure in service but also alters the stresses in sections. To fully or partially eliminate the tensile stress in concrete at service loads, the stress in the tendons generated by stretching should be high enough after all stress losses have taken place. Therefore, high tensile strength materials must be used for tendons, meanwhile, the high strength of tendons can be utilized fully.

1.2 The functions of prestress or prestressing force

Although the original purpose of introducing prestress in concrete was to fully eliminate the tensile stresses, there are several other functions. Take a reinforced concrete simply supported beam with a rectangular section only subjected to the self-weight as an example, as shown in Fig. 1.3. Assume that the load intensity due to self-weight of the beam along the entire span is constant, the tensile stress in concrete in the bottom face fiber at the midspan section is given by

$$\sigma_{c,M} = \frac{M \cdot \frac{h}{2}}{I} = \frac{\frac{1}{8} bh\gamma l^2 \cdot \frac{h}{2}}{\frac{1}{12} bh^3} = \frac{3\gamma l^2}{4h} \tag{1.3}$$

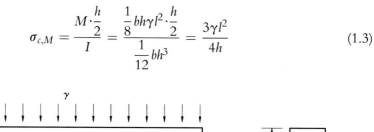
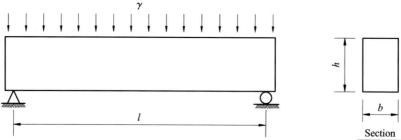

Figure 1.3 Schematic diagram of a simply supported beam.

where $\sigma_{c,M}$ = tensile stress in concrete in the bottom face fiber at the midspan section due to the self-weight of the beam.

γ = load intensity due to the self-weight of the beam.

h = depth of the rectangular section.

b = width of the rectangular section.

I = moment of inertia of the reinforced concrete section, $I = \frac{1}{12}bh^3$ is taken approximately.

l = effective span.

If cracking is not allowed, the maximum span of the simply supported beam can be derived from Eq. (1.3), as

$$l_{\max} = \sqrt{\frac{4}{3\gamma}} \cdot \sqrt{f_{ct,u}} \cdot \sqrt{h} \tag{1.4}$$

where $f_{ct,u}$ = tensile strength of concrete.

$f_{ct,u}$ could be the characteristic tensile strength of concrete, f_{ctk}, in the *fip* Model Code and the design codes in China, or modulus of rupture of concrete f_r, in the ACI code.

Eq. (1.4) indicates that the section depth will increase dramatically if a longer span beam is built without cracking, since the tensile strength of concrete is low and it fluctuates in a small range. As the external loads are exerted on the beam, the section depth has to be increased more dramatically. The huge depth not only generates heavy self-weight resulting in an uneconomical scheme, but also occupies excessive building space, indicating that reinforced concrete cannot be used to build long-span beams. If the compressive stress in concrete is introduced in the tension zone, Eq. (1.4) turns into

$$l_{\max} = \sqrt{\frac{4}{3\gamma}} \cdot \sqrt{f_{ct,u} + \sigma_{c,p}} \cdot \sqrt{h} \tag{1.5}$$

where $\sigma_{c,p}$ = introduced compressive stress in concrete in the bottom fiber at the midspan section.

Eq. (1.5) demonstrates that the span can increase effectively with the increase of introduced compressive stress in case of keeping the depth unchanged, namely, a higher span-to-depth ratio is achieved in the prestressed concrete beams. Meanwhile, the proportion of self-weight in total design load does not increase as sharply with the increase of span as that in the reinforced concrete beam. Therefore, the prestressed concrete can be used to build large-span flexural members with light self-weight.

The bending moment produced by the eccentric prestressing force, as shown in Fig. 1.2A, will generate upward camber. If the eccentricity forms a parabola (Fig. 1.4), it can be expressed as

$$e(x) = -\frac{4e_p}{l^2}x^2 + \frac{4e_p}{l}x \tag{1.6}$$

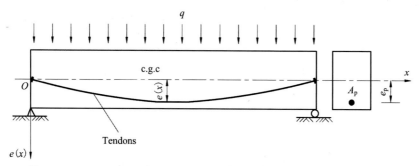

Figure 1.4 A prestressed concrete beam arranged with parabolic tendons.

where e_p = eccentricity of the tendons with respect to the center of gravity of the concrete section.

l = effective span.

x = distance from the left support to the section under consideration.

The total internal forces in a section of the prestressed concrete beam subjected to a uniformly distributed load are given by

$$M(x) = \left(\frac{qlx}{2} - \frac{qx^2}{2}\right) - \left(-\frac{4e_p}{l^2}x^2 + \frac{4e_p}{l}\right)N_p \tag{1.7}$$

$$V(x) = \left(\frac{ql}{2} - qx\right) - N_p \sin\theta(x) \tag{1.8}$$

$$N(x) = N_p \cos\theta(x) \tag{1.9}$$

where $\theta(x)$ = bending angle of the tendons with respect to the beam's axis.

q = uniform load due to the self-weight of the beam and external loads.

N_p = prestressing force.

$M(x)$ = bending moment in a section.

$V(x)$ = shear force in a section.

$N(x)$ = axial force in a section.

Assume that the modulus of elasticity of concrete is constant, the deflection at midspan can be derived by

$$f_{mid} = \int_0^l \frac{M(x)\overline{M}(x)}{EI}dx = \frac{5l^2}{48EI}\left(M_{mid} - N_p e_p\right) \tag{1.10}$$

where EI = sectional flexural rigidity.

M_{mid} = bending moment at midspan due to uniformly distributed load.

Eq. (1.10) demonstrates that the eccentric prestressing force can effectively adjust the deflection of the beam by altering the prestressing force N_p and its eccentricity e_p, respectively.

In general, the introduced prestress or prestressing force bears the following basic functions:

(1) To improve the working performance of concrete structures by actively counterbalancing fully or partially the tensile stress in concrete caused by the service loads.

(2) To enable concrete to be used to build long-span structures with light self-weight.

(3) To actively adjust the deflection of the flexural members by altering the prestressing force and its eccentricity, respectively.

(4) To enable high-strength materials to be employed efficiently and economically in concrete structures.

1.3 Prestress level

The prestress level, known as the degree of prestressing, refers to the ratio of prestressing-caused compressive stress in concrete in the extreme tension fiber to load-caused tensile stress, expressed as

$$\lambda = \frac{\sigma_{c,p}}{\sigma_{c,l}} \tag{1.11}$$

where $\sigma_{c,p}$ = compressive stress in concrete in the extreme fiber in the tension zone generated by the prestressing force.

$\sigma_{c,l}$ = tensile stress in concrete in the extreme fiber due to the service loads.

For flexural members, the prestress level can be expressed as

$$\lambda = \frac{M_0}{M} \tag{1.12}$$

where M_0 = decompression moment, the moment that eliminates the prestressing-caused compressive stress in concrete in the extreme tension fiber to zero.

M = bending moment caused by the service loads.

For axial tension members, the prestress level can be expressed as

$$\lambda = \frac{N_0}{N} \tag{1.13}$$

where N_0 = decompression axial force, the force that eliminates the prestressing-caused compressive stress in concrete to zero.

N = axial force caused by the service loads.

The prestress level is an important index in design, which can be used to control the concrete stress and cracking. There will be no tensile stress in the tension zone at service loads when $\lambda \geqslant 1$, while tensile stress occurs when $\lambda < 1$, and the cracking even develops when the tensile stress exceeds the tensile strength of concrete. Therefore, the prestress level makes the reinforced concrete a continuous spectrum which can be grouped into the following three categories:

Class I - fully prestressed concrete (FPC), $\lambda \geqslant 1$.

Class II - partially prestressed concrete (PPC), $0 < \lambda < 1$.

Class III - reinforced concrete (RC), $\lambda = 0$.

1.4 Classification of prestressed concrete

The prestressed concrete can be classified by stressing methods, prestress level, crack control, bonding condition, and the position of the prestressing tendons in the concrete section.

1.4.1 Classification by stressing methods

According to the process of casting concrete or stretching tendons first during construction, the prestressed concrete structures are classified into pretensioned and post-tensioned structures. The former refers to the structures that the tendons are stretched before casting concrete, while the tendons are stretched and anchored after the concrete is cast and hardens enough in the latter.

1.4.2 Classification by prestress level

As the prestress level is expressed by Eqs. (1.11)−(1.13), the prestressed concrete can be classified into fully prestressed concrete when $\lambda \geqslant 1$ and partially prestressed concrete when $0 < \lambda < 1$.

Fully prestressed concrete is broadly used in highway bridges, railway bridges, and civil and industrial buildings. For long-span bridges and those structures in which cracking is not allowed, such as oil tanks and nuclear vessels, fully prestressed concrete is one of the most preferable building materials. Since cracking is avoided in the service period, fully prestressed concrete bears high fatigue resistance, making it widely used in railway and highway beams, crane beams in factories, and other structures that come under repeated loading.

High-strength steel wires and strands without obvious yield points are widely used in prestressed concrete. To improve the structural ductility in earthquake-prone areas, a great quantity of reinforcing steels (nonprestressing steels) are commonly arranged in the concrete to replace part of the prestressing tendons, in this condition, tensile stress in concrete or cracking in the tension zone is allowed at service loads. These partially prestressed concrete structures can be classified into two types, Class A and Class B in the *Code for Design of Highway Reinforced Concrete and Prestressed Concrete Bridges and Culverts* (JTG 3362-2018). For Class A, the maximum tensile stress in concrete in the tension zone cannot exceed the allowable values prescribed in the code. For Class B, cracking is allowed while the crack width should be controlled to a limited value prescribed in the code.

1.4.3 Classification by the level of crack control

According to the level of crack control in serviceability limit state, the *Concrete structure design code* (GB 50010-2010) and *Prestressed concrete structure design code* (JGJ 369—2016) stipulate that the prestressed concrete can be classified into the following three categories:

Class I—no tensile stress in concrete in the tension zone is allowed in the service period.

Class II—limited tensile stress in concrete in the tension zone is allowed in the service period.

Class III—cracking is allowed while the maximum crack width cannot exceed the values specified in the code.

Three classes of the behavior of prestressed flexural members are defined in the ACI code. Class U members are assumed to behave as uncracked members, Class C members are assumed to behave as cracked members, and the behavior of Class T members is assumed to be in transition between uncracked and cracked.

1.4.4 Classification by bonding condition between the prestressing tendons and the concrete

The bonded prestressed concrete refers to those in which the prestressing tendons are arranged in the concrete and bond together, while the prestressing tendons do not contact the concrete directly in the unbonded prestressed concrete. If the unbonded prestressing tendons are arranged outside of the concrete in a structure, the corresponding structure is called the externally prestressed concrete structure.

1.5 Prestressed versus reinforced concrete

Compared to reinforced concrete, prestressed concrete bears the following advantages.

1.5.1 High crack resistance

The compressive stress in concrete generated by the longitudinal prestressing force can completely or partially eliminate the normal tensile stress caused by the service loads. In addition, the principal tensile stress in concrete due to the service loads can be effectively reduced or deleted by the inclined compression generated by the bent up longitudinal prestressing tendons and vertical prestressing tendons in the webs. Therefore, there is high crack resistance in the tension zone and flexural-shear zone in prestressed concrete structures.

1.5.2 High shear resistance

The vertical force generated by the bent up longitudinal prestressing tendons and vertical prestressing tendons counterbalances completely or partially the shear force caused by the design load, resulting in high resistance for shear force in service. On the other hand, the high shear bearing capacity helps to reduce the web thickness in the prestressed concrete structures, leading to lighter self-weight.

1.5.3 High durability

In a fully prestressed concrete structure, cracking is not allowed. In a partially prestressed concrete structure, although the cracking may occur under the most unfavorable live load, the cracks will close most of the time. Hence, the low chance of directly being in contact with moisture and air prevents the steel from corroding, thereby high durability is achieved in prestressed concrete.

1.5.4 High fatigue resistance

The fatigue resistance of the concrete structures depends mainly on the stress amplitude in tensile reinforcement. In the prestressed concrete structures, the variation of stress caused by the live load is relatively low compared to the existing stress in tendons in the service period, this low fluctuation gives prestressed concrete strong fatigue resistance.

1.5.5 Ability to control the deflection actively

According to Eq. (1.10), the deflection of a flexural member can be adjusted effectively by altering the prestressing force and its eccentricity, respectively. This advantage not only makes it possible to flexibly set the prestressing tendons according to the structure characteristics and cross-section forms in the design of new structures but also can adjust the deflection of existing structures by adding external prestressing tendons, which broadens the application of the prestressing technology.

1.5.6 Ability to build long-span structures with lighter self-weight

As shown in Eq. (1.5), a higher span-to-depth ratio is easily achieved in the prestressed concrete beams. In addition, the cracking and deflection of the beam can be controlled effectively by adjusting the longitudinal prestressing force and its eccentricity and the vertical prestressing force rather than by increasing the section depth, making prestressed concrete suitable for building long-span structures. At present, prestressed concrete has strong competitiveness in building bridges with spans from 30 to 300 m (Fig. 1.5).

1.5.7 Efficient utilization of high-strength materials

In a prestressed concrete member, the tendons must be stretched to a very high stress so that the permanent stress in tendons in the service period can be kept at a desired high

Figure 1.5 Picture of the Beipanjiang super bridge from Jiping Guo. *(This bridge, locating in Guizhou, China, is a prestressed concrete rigid frame with an effective span of 290 m. Design by* CCCC NO.2 Highway Survey Design & Research Institute Co. Ltd., *constructed by* Guizhou Road & Bridge Group Co. Ltd.*)*

value after many stress losses have occurred, so the high tensile strength of tendons is a must. Correspondingly, the concrete surrounding the prestressing tendons bears high compression at transfer, which requires high compressive strength in concrete. At flexural failure, the compressive strength of concrete should be high enough to match the high tensile strength of tendons in the ultimate limit state. Therefore, the high strength of materials is not only necessary but also can be fully utilized in prestressed concrete.

1.5.8 Good economy

Although special equipment and extra construction procedures are required during construction, and the adopted high-strength materials in the prestressed concrete structures are more expensive compared to reinforced concrete counterparts, the total cost for materials and construction in the former is still an economic program, especially for the highway and railway beams with medium or long spans under the same design live load. From another point of view, the longer beams used in a bridge result in a lower number of piers and less impact on the land under the bridge, which will also bring significant economic benefits. In China, viaducts with prestressed concrete box girders and pile foundations are broadly constructed in a high-speed railway with a ballastless track system to control the vertical deformation of tracks, where the length of the girder span seriously affects the cost of the substructure and the occupied land.

1.6 Concise history of prestressed concrete

It is pointed out in Chapter 36 of Tao Te Ching, a crystallization of ancient Chinese civilization written in approximately 500 BCE, "What to take, you have to give it first," which applies to the principle of prestressed concrete. For concrete structures at service loads, if high tensile capacity in the tension zone is pursued, the high compression in the tension zone should be introduced first, before service. It can be said that the application of the concept of prestress was employed in daily life in ancient times. For examples, vines, rattans, hemp ropes, or iron wires are stressed to tighten the wooden barrels to resist the tensile stress in circumferential wood plates caused by the radial expanding pressure due to water, and a flexible thin hacksaw blade is pretensioned to form a rigid blade to overcome the pressure generated by the wood when pulling the saw.

For the concept of prestressed concrete, it is generally believed that it was initiated by using prestressing steel tie rods to construct slabs and roofs with individual blocks and arches in an 1886 patent of P.H. Jackson, an engineer from California. Henceforth, some ideas and technologies on establishing and keeping the prestress in concrete were proposed and applied, while the prestress still diminished quickly, until time-dependent prestress loss due to shrinkage and creep of concrete was recognized and it could be overcome to a great extent by using high-strength steels, contributed by Eugene Freyssinet of France and other researchers, the practical application of prestressed concrete was achieved in the 1930s. Since a large number of damaged bridges and other structures needed to be rebuilt with an on-going steel shortage in Europe after World War II, prestressed concrete was widely applied, and accordingly the evolution of prestress technology and design theory was promoted rapidly. Thereafter, prestress technology was popularized and applied in the United States and other countries. Since the 1950s, engineers have successively developed the cantilever casting method, segment assembly method, and incremental launching method for prestressed concrete bridges with long spans, which have comprehensively improved the level of construction. Nowadays, prestressed concrete is one of the most widely used structural materials.

Suggested readings

[1]   Lin TY, Burns NH. Design of prestressed concrete structures. 3rd ed. New York: John Wiley and Sons; 1981.
[2]   Hu Di. Basic principles of prestressed concrete structure design. 2nd ed. Beijing: China Railway Publishing House; 2019.
[3]   Collins Michael P, Mitchell D. Prestressed concrete structures. Upper Saddle River: Prentice-Hall Inc; 1991.
[4]   Naaman Antoine E. Prestressed concrete structure analysis and design. 2nd ed. New York: McGraw Hill; 2004.
[5]   Nawy Edward G. Prestressed concrete: a fundamental approach. 5th ed. Upper Saddle River: Prentice-Hall Inc; 2006.

[6] ACI 318-19. Building code requirements for reinforced concrete. Farmington Hills, MI: American Concrete Institute; 2019.

[7] GB 50010: 2015. Code for design of concrete structures. Beijing: China Construction Industry Press; 2015.

[8] Q/CR 9300:2018. Code for design on railway bridge and culvert. Beijing: China Railway Publishing House; 2018.

[9] JTG 3362: 2018. Specifications for design of highway reinforced concrete and prestressed concrete bridges and culverts. Beijing: People's Communications Press; 2018.

CHAPTER 2

Prestressing materials

Contents

2.1 Concrete

2.1.1 Basic requirements for concrete

A typical prestressed concrete member consists of concrete, prestressing tendons, and reinforcing steels (nonprestressing steels). For post-tensioned members, anchorage systems and ducts for tendons are required.

Concrete is a mixture of cement, water, sand, and aggregate. Cement and water interact chemically to form a gel to bind sand and aggregate into a hardened stone-like material. To speed up the hardening of concrete during construction, admixtures such as water-reducing agent and early strength agents are commonly added. The proportion

Analysis and Design of Prestressed Concrete
ISBN 978-0-12-824425-8, https://doi.org/10.1016/B978-0-12-824425-8.00002-6

of mixtures not only determines the strength of the concrete, but also affects the time-dependent properties such as shrinkage and creep, and structural durability. For concrete used in prestressed structures, in addition to the general performance requirements of structural concrete, high compressive strength, low values of shrinkage and creep, good durability, and low content of chloride ions are the basic requirements that must be met.

From the construction stage to the service period, the prestressing tendons are always in a state of high tensile stress, which is necessary from the point of view of establishing the desired prestress in concrete or saving materials. Accordingly, high compressive strength is required for concrete in the regions surrounding the prestressing tendons. At the ultimate strength limit state, the high compressive strength of concrete in the compression zone is required to match the high tensile strength of prestressing tendons in the tension zone to provide high bearing capacity in a balanced flexural section. For post-tensioned members, the concrete under the anchorage bears high local compression, which also requires high compressive strength in concrete.

The structural concrete should have the required durability, such as chloride ion diffusion coefficient, crack resistance, reinforcement protection, corrosion resistance, frost resistance, wear resistance, and alkali—aggregate reaction resistance. For concrete in prestressed structures, the desired durability can be determined by the design service life and environmental categories along with their attack levels. Basically, the total content of chloride ions in concrete should not exceed 0.06% of the total amount of cementitious materials, and the minimum cement dosage should be 30 kN/m^3. When the fly ash is added, its dosage should not exceed 30%. Low-alkali Portland cement or low-alkali ordinary Portland cement mixed with high-quality fly ash and an appropriate amount of admixture should be used as grouting material for ducts, while aluminum powder or admixtures containing chloride, nitrate, and other harmful components should not be added, and the content of chloride ions brought into concrete by various raw materials should be strictly controlled at within 0.06% of the total amount of cementitious materials.

2.1.2 Concrete strength

The characteristic strength is defined as that the stress below which 5% of all the strength measurements for the standard concrete specimens may be expected to fall. Classification of concrete strength grade is based on the uniaxial compressive strength measurements taken from 150 mm concrete cubes specified in the design codes in China. Nevertheless, the characteristic axial compressive strength, f_{ck}, is derived from the test results of $150 \times 150 \times 300$ mm specimens. Unless specified otherwise, the compressive strength of concrete as well as its tensile strength, refers to the strength value obtained at a concrete age of 28 days under standard curing.

The relationship between cubic compressive strength and prism compressive strength specified in the *Codes for design of concrete structures* (GB50010-2010) is

$$f_{ck} = 0.88\alpha_{c1}\alpha_{c2}f_{cu,k} \tag{2.1}$$

where $f_{cu,k}$ = cubic compressive strength of concrete, tested from 150 mm concrete cubes.

f_{ck} = characteristic compressive strength of concrete, also called prism compressive strength, tested from prism specimens with $150 \times 150 \times 300$ mm.

α_{c1} = ratio of f_{ck} to $f_{cu,k}$, 0.76 is taken when the concrete grade is not greater than C50, $\alpha_{c1} = 0.78$ for C55 and $\alpha_{c1} = 0.82$ for C80, the values for intermediate strength between C55 and C80 can be obtained by the linear interpolation method.

α_{c2} = reduction coefficient of fracture when the concrete grade is greater than C40, $\alpha_{c2} = 1.0$ for C40 and $\alpha_{c2} = 0.87$ for C80, the values for intermediate strength between C40 and C80 can be obtained by the linear interpolation method.

The prism tensile strength can be obtained from the cubic compressive strength:

$$f_{ctk} = 0.88 \times 0.395\alpha_{c2}f_{cu,k}^{0.55}(1 - 1.645\delta)^{0.45} \tag{2.2}$$

where f_{ctk} = characteristic tensile strength of concrete, also called the prism tensile strength.

δ = coefficient of variation of concrete strength.

The design values can be obtained from the characteristic strengths:

$$f_{cd} = \frac{f_{ck}}{\gamma_c} \tag{2.3}$$

$$f_{td} = \frac{f_{ctk}}{\gamma_c} \tag{2.4}$$

where f_{cd} = design strength of concrete in compression.

f_{td} = design tensile strength of concrete.

γ_c = partial safety factor for concrete.

The characteristic strengths and design strengths of normal weight concrete, stipulated in the GB50010-2010 code and *Code for design on railway bridge and culvert* (Q/CR 9300-2018), are compiled in Tables 2.1 and 2.2, respectively. The characteristic strengths of concrete stipulated in the *Specifications for design of highway reinforced concrete and prestressed concrete bridges and culverts* (JTG 3362-2018) are the same as those in the GB 50010-2010 code. Tables 2.1 and 2.2 show that there are some differences in concrete strength values in different design codes.

Generally, the concrete grade in prestressed concrete should not be less than C40. In China, most of the railway and highway prestressed concrete girders and railway sleepers are made of C50 − C60 concrete at present. When the concrete grade exceeds C80, most of the current codes (such as GB 50010-2010, Q/CR 9300-2018, and JTG 3362-2018) specify that the part strength exceeding 80 MPa is not considered in the design, and it is treated as a strength reserve.

Table 2.1 Characteristic strengths of concrete (MPa).

Strength grade / Codes	C40	C45	C50	C55	C60	C65	C70	C75	C80
Characteristic axial compressive strength									
GB50010–2010	26.8	29.6	32.4	35.5	38.5	41.5	44.5	47.4	50.2
Q/CR 9300–2018	27.0	30.0	33.5	37.0	40.0	–	–	–	–
Characteristic axial tensile strength									
GB50010–2010	2.39	2.51	2.64	2.74	2.85	2.93	2.99	3.05	3.11
Q/CR 9300–2018	2.70	2.90	3.10	3.30	3.50	–	–	–	–

Table 2.2 Design values of concrete strength (MPa).

Strength grade / Codes	C40	C45	C50	C55	C60	C65	C70	C75	C80
Design values of axial compressive strength									
GB50010–2010	19.1	21.1	23.1	25.3	27.5	29.7	31.8	33.8	35.9
JTG 3362–2018	18.4	20.5	22.4	24.4	26.5	28.5	30.5	32.4	34.6
Q/CR 9300–2018	18.6	20.7	23.1	25.5	27.6	—	—	—	—
Design values of axial tensile strength									
GB50010–2010	1.71	1.80	1.89	1.96	2.04	2.09	2.14	2.18	2.22
JTG 3362–2018	1.65	1.74	1.83	1.89	1.96	2.02	2.07	2.10	2.14
Q/CR 9300–2018	1.80	1.93	2.07	2.20	2.33	—	—	—	—

2.1.3 Modulus of elasticity

The tangent modulus at the origin of the stress—strain curve of concrete under uniaxial compression is taken as the modulus of elasticity of concrete in Chinese codes. The modulus of elasticity of concrete can be obtained from the uniaxial compression tests performed on prism specimens with dimensions of 150 × 150 × 300 mm, or derived from the compressive strength based on their relationships achieved from experiments.

In the Q/CR 9300-2018 code, the modulus of elasticity of concrete can be obtained by

$$E_c = 10^4 (f_{ck})^{\frac{1}{3}} \tag{2.5}$$

where f_{ck} = characteristic compressive strength of concrete, MPa.

In the GB50010-2010 and JTG 3362-2018 codes, the modulus of elasticity of concrete can be obtained by

$$E_c = \frac{10^5}{2.2 + \dfrac{34.7}{f_{cu,k}}} \tag{2.6}$$

where $f_{cu,k}$ = cubic compressive strength of concrete cube, MPa.

The modulus of elasticity of concrete obtained from Eqs. (2.5) and (2.6) are listed in Table 2.3.

Actually, the modulus of elasticity of concrete is time-dependent, and grows fast in the early stage ($t < 28$ days) but relatively slowly in the later stage ($t > 28$ days). In some analyses, such as in calculations of the time-dependent deflections of a prestressed concrete flexural member considering the effect of concrete creep and shrinkage, the exact value of the modulus of elasticity at the time of loading should be provided. In the CEB FIP (Model Code 2010), the modulus of elasticity of concrete at age t (days) may be estimated by

$$E_c(t) = E_c \sqrt{e^{s\left(1 - \sqrt{\frac{28}{t}}\right)}} \tag{2.7}$$

where t = concrete age, days.

E_c = modulus of elasticity at an age of 28 days.

Table 2.3 Modulus of elasticity of concrete (10^4 MPa).

Codes	Concrete strength grade								
	C40	C45	C50	C55	C60	C65	C70	C75	C80
GB50010-2010, JTG 3362-2018	3.25	3.35	3.45	3.55	3.60	3.65	3.70	3.75	3.80
Q/CR 9300-2018	3.40	3.45	3.55	3.60	3.65	—	—	—	—

s = coefficient which depends on the strength grade of cement, 0.25 is taken for ordinary cement and quick hard cement, and 0.20 is taken for fast hard high-strength cement.

Experiments show that the concrete shear modulus G_c and Poisson's ratio ν_c are related to the stress, which can be approximated as a constant value if the concrete stress is less than $0.5f_{ck}$. The value of ν_c usually varies between 0.17 and 0.23, and 1/6 or 0.2 can be taken in most cases. Since G_c is difficult to obtain from the tests directly, it is calculated from $G_c = E_c/2(1 +\nu_c)$. Generally, $G_c = 0.4 - 0.43E_c$, $\nu_c = 0.2$ for uncracked concrete and $\nu_c = 0$ for cracked concrete.

2.1.4 Fatigue strength and fatigue modulus of concrete

The microcracks present in concrete before loading which will later develop under repeated loading can result in fatigue failure in concrete in the case that the concrete stress is less than the ultimate strength. Experimental studies show that concrete does not have an "endurance limit" corresponding to an infinite number of load repetitions. The fatigue strength of concrete is mainly related to the maximum stress and the characteristics of the stress cycle.

The GB50010-2010 code stipulates that the fatigue strength of concrete should be checked for prestressed concrete members subjected to repeated loading. The design values of axial compressive fatigue strength f_c^f, and tensile fatigue strength, f_t^f, shall be determined by the design strengths by multiplying the corresponding correction factor γ_ρ for fatigue strength. The correction factor γ_ρ shall be calculated based on the ratio of fatigue stresses compiled in Tables 2.4 and 2.5. When the concrete is subjected to tensile-compression fatigue stress, the fatigue strength correction factor γ_ρ is taken to be equal to 0.6.

Table 2.4 Correction coefficient γ_ρ for fatigue compressive strength.

ρ_c^f	$0 \leqslant \rho_c^f < 0.1$	$0.1 \leqslant \rho_c^f < 0.2$	$0.2 \leqslant \rho_c^f < 0.3$	$0.3 \leqslant \rho_c^f < 0.4$	$0.4 \leqslant \rho_c^f < 0.5$	$\rho_c^f \geqslant 0.5$
γ_ρ	0.68	0.74	0.80	0.86	0.93	1.0

Table 2.5 Correction coefficient γ_ρ for fatigue tensile strength.

ρ_c^f	$0 < \rho_c^f < 0.1$	$0.1 \leqslant \rho_c^f < 0.2$	$0.2 \leqslant \rho_c^f < 0.3$	$0.3 \leqslant \rho_c^f < 0.4$	$0.4 \leqslant \rho_c^f < 0.5$
γ_ρ	0.63	0.66	0.69	0.72	0.74
ρ_c^f	$0.5 \leqslant \rho_c^f < 0.6$	$0.6 \leqslant \rho_c^f < 0.7$	$0.7 \leqslant \rho_c^f < 0.8$	$\rho_c^f \geqslant 0.8$	
γ_ρ	0.76	0.80	0.90	1.00	

Table 2.6 Fatigue modulus of concrete (10^4MPa).

Concrete grade	C40	C45	C50	C55	C60	C65	C70	C75	C80
E_c^f	1.50	1.55	1.60	1.65	1.70	1.75	1.80	1.85	1.90

Table 2.7 Design values of fatigue compressive strength of concrete.

$\rho_{min} = \dfrac{\sigma_{min}}{f_{ck}}$	C40	C45	C50	C55	C60
0	13.2	15.1	16. 9	18.8	20.9
0.1	14.4	16.4	18.3	20.4	22.6
0.2	15.6	17.8	19.8	22.0	24.3
0.3	16.5	18.7	20.9	23.2	25.6
0.4	17.8	20.1	22.3	24.8	27.4
0.5	19.0	21.5	23.9	26.4	29.1

The ratio of concrete fatigue stresses is given by

$$\rho_c^f = \frac{\sigma_{c,min}^f}{\sigma_{c,max}^f} \tag{2.8}$$

where $\sigma_{c,min}^f$ = minimum concrete stress in the same fiber when fatigue is checked.

$\sigma_{c,max}^f$ = maximum concrete stress in the same fiber when fatigue is checked.

The fatigue moduli E_c^f of concrete stipulated in the GB50010-2010 code are compiled in Table 2.6.

The design values of concrete axial compressive fatigue strength can be obtained from Tables 2.2, 2.4, and 2.5, as compiled in Table 2.7, where σ_{min} is the minimum stress caused by the structural self-weight and additional dead load, and f_{ck} is the characteristic compressive strength of concrete. When ρ_{min} is an intermediate value in the table, it can be obtained by the linear interpolation method.

2.1.5 Creep

Under sustained loads with a fixed value, the deformation of concrete will continue to increase as the time increases; this phenomenon is called creep. Creep is a natural characteristic of concrete, which will not only cause prestress loss in prestressing tendons and reduce the crack resistance, but lead to additional long-term deflection or camber of the flexural structures. The change to concrete strain caused by creep is called the creep strain, and the change to deflection caused by creep is called the creep deflection.

The cement gel is made up of calcium silicate hydrates with pores containing absorbed water. Although the mechanism for the occurrence of creep is very complicated and is not yet fully understood, a consistent view that has been formed is that the creep is mainly related to the migration of water from the concrete to the environment and the plastic deformation of cement gel under loading. Hence, the influencing factors on concrete creep are mainly the concrete age at loading, duration of sustained loading, ambient relative humidity and notional size (affecting the migration rate of water from the concrete to the environment), water—cement ratio, and the properties and contents of aggregate and cement.

Experimental researches indicate that creep may be considered approximately proportional to stress (L'Hermite et al. 1958; Keeton 1965), provided that the applied stress is less than 40% of the concrete compressive strength. For a prestressed concrete structure at service loads, $\sigma_c \leqslant 0.4 f_{ck}$ is always satisfied, so the creep strains and creep deflections caused by loads applied at different ages are considered independent and can be added directly based on the superposition principle. Experiments also show that the creep develops rapidly in the earlier stage after loading while at a lower rate, later, the change rate of creep with time is very small. Generally, about 25% of the final creep develops in the first 2—3 weeks, 50% at 3 months, 75% at 1 year, and 90% after 2—3 years. Therefore, in time-dependent structural analysis for prestressed concrete structures, 5 years are commonly adopted as the final time. For some important prestressed structures or those that are sensitive to time-dependent concrete stress and deflection, the final time could be taken as 10 years or longer.

The concrete strain varies with time under loading, as shown in Fig. 2.1. The elastic strain $\varepsilon_e(t_0)$ in concrete is generated immediately at applying stress $\sigma_c(t_0)$ at a time t_0 (concrete age). At time $t(t > t_0)$, the total strain of concrete includes the elastic strain and the creep strain. Unloading at a time $t_1(t_1 > t_0)$, the elastic strain is immediately recovered and its value is close to the initial elastic strain $\varepsilon_e(t_0)$. It is observed that partial strain recovers with time and the other is irreversible. The reversible portion associated with nonaging behavior is called delayed elastic strain, $\varepsilon_{ce}(t_2, t_1)(t_2 \gg t_1)$, while the irreversible portion associated with aging behavior is called delayed plastic strain, $\varepsilon_{cp}(t_2, t_1)$.

When the concrete is subjected to a constant stress $\sigma(t_0)$ at an age of t_0, the total concrete strain at a time $t(t > t_0)$ (Fig. 2.1) is obtained as

$$\varepsilon_c(t, t_0) = \varepsilon_e(t_0) + \varepsilon_{cr}(t, t_0) = \frac{\sigma_c(t_0)}{E_c(t_0)} + \varepsilon_{cr}(t, t_0) \tag{2.9}$$

The total concrete strain at a time $t(t > t_0)$ also can be expressed as

$$\varepsilon_c(t, t_0) \approx \varepsilon_e(t_0) + \varepsilon_{ce}(t_2 \to \infty, t_1) + \varepsilon_{cp}(t_2 \to \infty, t_1) \tag{2.10}$$

where $\sigma_c(t_0)$ = constant concrete stress exerted at concrete age t_0.
$E_c(t_0)$ = modulus of elasticity of concrete at concrete age t_0.

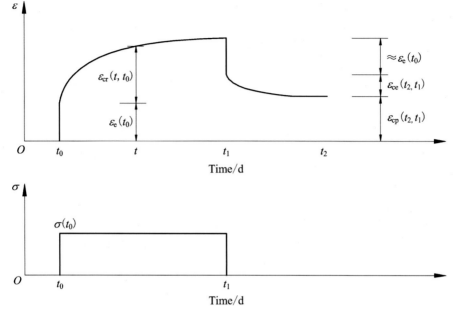

Figure 2.1 Schematic diagram of the time-dependent strain versus stress of concrete.

$\varepsilon_e(t_0)$ = elastic strain of concrete caused by applying stress at concrete age t_0.

$\varepsilon_{cr}(t, t_0)$ = creep strain at concrete age t due to creep under constant stress applied at concrete age t_0.

$\varepsilon_{ce}(t_2 \rightarrow \infty, t_1)$ = strain recovery after unloading.

$\varepsilon_{cp}(t_2 \rightarrow \infty, t_1)$ = unrecoverable strain after unloading.

From another point of view, $\varepsilon_{ce}(t_2 \rightarrow \infty, t_1)$ and $\varepsilon_{cp}(t_2 \rightarrow \infty, t_1)$ are generated at time period $(t - t_0)$ under stress $\sigma_c(t_0)$, hence, Eq. (2.10) can be rewritten as

$$\varepsilon_c(t, t_0) \approx \varepsilon_e(t_0) + \varepsilon_{ce}(t, t_0) + \varepsilon_{cp}(t, t_0) \tag{2.11}$$

The characteristic of creep is usually expressed by creep coefficient and creep function or creep compliance, defined as

$$\varphi(t, t_0) = \frac{\varepsilon_{cr}(t, t_0)}{\varepsilon_e(t_0)} \tag{2.12}$$

$$J(t, t_0) = \frac{1}{E_c(t_0)} + \frac{\varphi(t, t_0)}{E_c(t_0)} \tag{2.13}$$

where $\varphi(t, t_0)$ = creep coefficient, the ratio of the creep strain at a time t to the initial elastic strain.

$J(t, t_0) =$ creep function or creep compliance, total stress-dependent strain per unit stress.

Substituting Eqs. (2.12) and (2.13) into Eq. (2.9) yields:

$$\varepsilon_c(t, t_0) = \varepsilon_e(t_0)[1 + \varphi(t, t_0)] \tag{2.14}$$

$$\varepsilon_c(t, t_0) = \sigma(t_0) J(t, t_0) \tag{2.15}$$

Experimental studies show that the creep coefficient of concrete varies from 1 to 3 in an ordinary ambient environment, while it may be more than 4 in extremely dry environment (Brooks, 2005).

Concrete creep is accompanied by shrinkage, so they are almost studied and considered in structural analysis simultaneously. The proposed prediction models for creep and shrinkage popularly used include CEB-FIP MC78, CEB-FIP MC90-99 (Muller and Hillsdorf 1990; CEB 1991, 1993,1999), ACI 209R-92 (ACI Committee 209, 1992), B3—B4 (Bažant and Murphy, 1995; Wendner, Hubler and Bažant, 2013), and GL2000 (Gardner and Lockman 2001) models. In the CEB MC78 model, creep consists of delayed elastic creep and irreversible plastic creep, as shown in Eq. (2.11), which is still adopted in the Q/CR 9300-2018 code. The ACI 209R-92 and CEB-FIP MC90-99 models use a nominal value (ultimate value) corrected according to mixture proportioning and environment conditions along with a hyperbolic change with time for creep. In B3—B4 models, creep is distinguished into basic creep and drying creep, the former develops in conditions of no moisture exchange with the ambient environment, whereas the latter depends on moisture exchange.

The creep model in the JTG 3362-2018 code uses the CEB-FIP 1990 model. The creep coefficient is calculated by

$$\varphi(t, t_0) = \varphi_0 \cdot \beta_c(t - t_0) \tag{2.16}$$

$$\varphi_0 = \phi_{RH} \cdot \beta(f_{cm}) \cdot \beta(t_0) \tag{2.17}$$

$$\phi_{RH} = 1 + \frac{1 - \dfrac{RH}{RH_0}}{0.46 \left(\dfrac{h}{h_0}\right)^{\frac{1}{3}}} \tag{2.18}$$

$$\beta(f_{cm}) = \frac{5.3}{\left(\dfrac{f_{cm}}{f_{cm0}}\right)^{0.5}} \tag{2.19}$$

$$\beta(t_0) = \frac{1}{0.1 + \left(\dfrac{t_0}{t_1}\right)^{0.2}} \tag{2.20}$$

$$\beta_c(t - t_0) = \left[\dfrac{\dfrac{t - t_0}{t_1}}{\beta_H + \dfrac{t - t_0}{t_1}} \right]^{0.3} \tag{2.21}$$

$$\beta_H = 150 \left[1 + \left(1.2 \dfrac{RH}{RH_0} \right)^{18} \right] \left(\dfrac{h}{h_0} \right) + 250 \leq 1500 \tag{2.22}$$

where t_0 = concrete age at loading, days.

t = concrete age under consideration, days.

$\varphi(t, t_0)$ = creep coefficient.

φ_0 = notional creep coefficient.

$\beta_c(t - t_0)$ = coefficient to describe the development of creep with time after loading.

f_{cm} = mean cylinder compressive strength at the age of 28 days, $f_{cm} = 0.8 f_{cu,k} + 8$, MPa.

$f_{cu,k}$ = characteristic concrete compressive cubic strength at an age of 28 days, MPa.

β_{RH} = correction term for the effect of relative humidity on nominal creep coefficient.

RH = relative humidity of the ambient environment in a year, %.

β_{sc} = correction coefficient that depends on the type of cement, 5 is taken for common Portland cement or quick hardening cement, 8 is taken for fast-hardening high-strength cement.

H = notional size, mm, $H = 2A/u$, where A is the cross-sectional area and u is the peripheral length of the section in contact with the atmosphere.

$RH_0 = 100\%$.

$h_0 = 100$ mm.

$t_1 = 1$ day.

$f_{cmo} = 10$ MPa.

The notional creep coefficients for C20–C50 are compiled in Table 2.8.

Table 2.8 Notional creep coefficient.

	40% < RH < 70% Notional size/mm				70% < RH < 99% Notional size/mm			
Concrete age at loading/day	100	200	300	≥ 600	100	200	300	≥ 600
3	3.90	3.50	3.31	3.03	2.83	2.65	2.56	2.44
7	3.33	3.00	2.82	2.59	2.41	2.26	2.19	2.08
14	2.92	2.62	2.48	2.27	2.12	1.99	1.92	1.83
28	2.56	2.30	2.17	1.99	1.86	1.74	1.69	1.60
60	2.21	1.99	1.88	1.72	1.61	1.51	1.46	1.39
90	2.05	1.84	1.74	1.59	1.49	1.39	1.35	1.28

Note: When the actual notional thickness and the loading age are the intermediate values in the table, the nominal creep coefficient can be obtained by linear interpolation.

The creep model in the Q/CR 9300–2018 code is based on the CEB-FIP MC78 model, with the creep coefficient being given by

$$\varphi(t, t_0) = \beta_a(t_0) + 0.4\beta_d(t - t_0) + \varphi_f\left[\beta_f(t) - \beta_f(t_0)\right] \tag{2.23}$$

$$\beta_a(t_0) = 0.8\left[1 - \frac{f_{t_0}}{f_\infty}\right] \tag{2.24}$$

$$\varphi_f = \varphi_1 \cdot \varphi_2 \tag{2.25}$$

where $\varphi(t, t_0)$ = creep coefficient.

$\beta_a(t_0)$ = rapidly developed initial increase of the creep at loading.

$\beta_d(t - t_0)$ = developing in time of the nonaging part of creep (delayed elastic strain), which can be found in Fig. 2.2.

$\beta_f(t) - \beta_f(t_0)$ = developing in time of the aging part of creep (delayed plastic strain), which can be found in Fig. 2.3.

φ_f = flow plastic coefficient.

φ_{f1} = coefficient depends on ambient humidity, which can be checked in Table 2.9.

φ_{f2} = coefficient depends on notional size, which can be found in Fig. 2.4.

$\frac{f_{t_0}}{f_\infty}$ = ratio of concrete strength, which can be found in Fig. 2.5.

The notional size of the member can be obtained by

$$h = \lambda\frac{2A}{u} \tag{2.26}$$

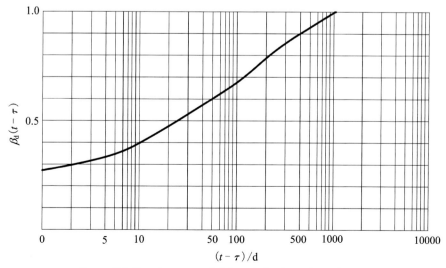

Figure 2.2 Development of delayed elastic strain with time.

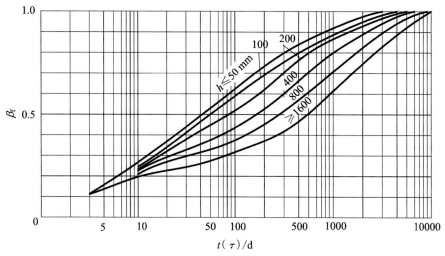

Figure 2.3 Development of delayed plastic strain with time.

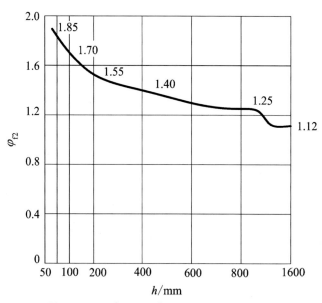

Figure 2.4 Influence of notional size on creep.

Table 2.9 φ_{f1} and λ.

Humidity	φ_{f1}	λ
—	0.8	30
90%	1.0	5
70%	2.0	1.5
40%	3.0	1.0

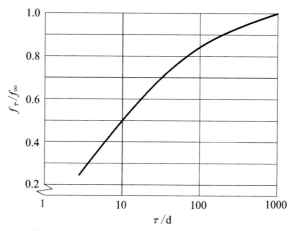

Figure 2.5 Development of strength with time.

Table 2.10 Notional creep coefficient and shrinkage strain.

	Notional creep coefficient φ_∞				Notional shrinkage strain $\varepsilon_\infty \times 10^6$			
	Notional size $\frac{2A}{u}$/mm				Notional size $\frac{2A}{u}$/mm			
Concrete age at loading/day	100	200	300	$\geqslant$ 600	100	200	300	$\geqslant$ 600
3	3.00	2.50	2.30	2.00	250	200	170	110
7	2.60	2.20	2.00	1.80	230	190	160	110
10	2.40	2.10	1.90	1.70	217	186	160	110
14	2.20	1.90	1.70	1.50	200	180	160	110
28	1.80	1.50	1.40	1.20	170	160	150	110
$\geqslant$ 60	1.40	1.20	1.10	1.00	140	140	130	110

Note: when the notional size and age of loading are intermediate, final values of creep coefficient can be obtained by the linear interpolation method.

where λ = coefficient for ambient humidity, which can be found in Table 2.10.

A = cross-sectional area of the member, mm^2.

u = perimeter exposed to the environment, mm.

The Q/CR 9300-2018 code provides the notional creep coefficient $\varphi(\infty, t_0)$ and shrinkage strain for commonly used members, as compiled in Table 2.10.

2.1.6 Shrinkage

Concrete shrinkage is the reduction in volume of a load-free and unrestrained specimen. The shrinkage is mainly related to the types of cement and aggregate, water-to-cement ratio, ambient humidity, and curing method. Shrinkage lasts for decades, generally, the long-term shrinkage reaches (300–800) $\mu\varepsilon$, and more than 1100 $\mu\varepsilon$ in an extreme dry ambient environment. In ordinary circumstances, the strain due to shrinkage is three to six times that of the ultimate axial tensile strain of concrete. Therefore, shrinkage is an important factor that causes cracking in a concrete structure. Concrete shrinkage and creep always occur simultaneously, resulting in prestress loss in prestressing tendons, time-dependent stresses in concrete, and reinforcements and deflections of the flexural structures.

Shrinkage can be classified into plastic shrinkage, autogenous shrinkage, drying shrinkage, thermal shrinkage, and carbonization shrinkage. Plastic shrinkage occurs in wet concrete before setting, while autogenous, thermal, drying, and carbonization shrinkage occur in hardened concrete after setting. Thermal shrinkage, also known as cold shrinkage, occurs when the temperature drops. Carbonization shrinkage is the result of a chemical reaction between cement paste and CO_2 in air (in the presence of water, the real medium is H_2CO_3), originating from that the Ca $(OH)_2$ crystal in the cement paste is carbonized into $CaCO_3$ precipitation. Carbonation shrinkage develops relatively late and is generally limited to the concrete surface. Autogenous shrinkage, also called chemical shrinkage or basic shrinkage, refers to the volume deformation of concrete due to the cement hydration reaction in a sealed specimen with no moisture exchange. Drying shrinkage is the reduction in volume caused principally by the loss of water during the drying process, which increases with time at a gradually decreasing rate and takes place in the months and years after casting. Among these shrinkages, drying shrinkage is the major part that is required to be considered in structural analysis. For high-strength concrete, the water-to-cement ratio is low and the amount of cement is high, so the autogenous shrinkage occurs earlier and faster, resulting in higher occupation in all shrinkage comparing to drying shrinkage, especially in the early stage after casting. In this case, autogenous shrinkage must be taken into account in structural analysis.

Concrete shrinkage can be described by notional shrinkage by multiplying the correction term for the effect of time on shrinkage, such as in the CEB-FIP MC90-99 and ACI 209R-92 models, or expressed by drying shrinkage and autogenous shrinkage, respectively, such as in the B3–B4 and *fib* Model Code 2010 models. The JTG 3362-2018 code stipulates that the shrinkage strain can be derived by

$$\varepsilon_{sh}(t, t_s) = \varepsilon_{sho}\beta_s(t - t_s) \tag{2.27}$$

$$\varepsilon_{sho} = \varepsilon_{sh}(f_{cm})\beta_{RH} \tag{2.28}$$

$$\varepsilon_{sh}(f_{cm}) = \left[160 + 10\beta_{sc}\left(9 - \frac{f_{cm}}{f_{cmo}}\right)\right] \times 10^{-6} \tag{2.29}$$

$$\beta_{RH} = 1.55\left[1 - \left(\frac{RH}{RH_0}\right)^3\right] \tag{2.30}$$

$$\beta_s(t - t_s) = \left[\frac{\dfrac{t - t_s}{t_1}}{350 \left(\dfrac{h}{h_0}\right)^2 + \dfrac{t - t_s}{t_1}} \right]^{0.5} \tag{2.31}$$

where t_s = concrete age at beginning of drying, 3—7 days is assumed.

t = concrete age under consideration, days.

$t - t_s$ = duration of drying.

$\varepsilon_{sh}(t, t_s)$ = shrinkage strain at concrete age t since the start of drying at an age t_s.

ε_{sho} = notional shrinkage coefficient, which can be found in Table 2.11.

$\beta_s(t - t_s)$ = coefficient describing the development of shrinkage with time.

β_{sc} = coefficient depending on the type of cement, 5 is taken for common Portland cement or quick hard cement, 8 is taken for fast-hardening high-strength cement.

β_{RH} = coefficient associated with the annual average relative humidity.

The shrinkage strain at an age of t, measured from an age of t_0, can be calculated by

$$\varepsilon_{sh}(t, t_0) = \varepsilon_{sho}[\beta_s(t - t_s) - \beta_s(t_0 - t_s)] \tag{2.32}$$

The GB50010-2010 code provides the notional shrinkage, as compiled in Table 2.11.

The notional shrinkage strain of concrete provided in the Q/CR 9300-2018 code is compiled in Table 2.12.

Table 2.11 Notional shrinkage $\varepsilon_{sho}(\times 10^{-6})$.

40% ⩽ RH < 70%	70% ⩽ RH < 90%
529	310

Notes: (1) This table applies to concrete made of general silicate cement or quick hard cement.
(2) This table applies to the average temperature of seasonal changes of $-20°C - +40°C$.
(3) The values in this table are calculated according to C40 concrete, for concrete C50 and above, the value should be multiplied by $\sqrt{32.4/f_{ck}}$.
(4) If the ambient relative humidity is in 40% ⩽ RH < 70%, RH = 55% is taken, and RH = 80% if 70% ⩽ RH < 99%.

Table 2.12 Notional shrinkage strain $\varepsilon_u(\times 10^{-4})$.

Concrete age at applying prestressing force/day	Notional size $\frac{2A}{u}$/mm			
	100	200	300	⩾ 600
3	2.50	2.00	1.70	1.10
7	2.30	1.90	1.60	1.10
10	2.17	1.86	1.60	1.10
14	2.00	1.80	1.60	1.10
28	1.70	1.60	1.50	1.10
⩾ 60	1.40	1.40	1.30	1.00

Notes: (1) The concrete age at applying prestressing force takes 3—7 days for pretensioned members and 7—28 days for post-tensioned members.
(2) When the notional size and concrete age are the intermediate values in the table, it can be obtained by the linear interpolation.

2.1.7 Temperature effects

Temperature affects the hydration rate of concrete, and thereby affects the development of concrete strength and modulus of elasticity, and also affects the time-dependent creep and shrinkage. On the other hand, as a material, thermal expansion occurs when ambient temperature increases, which is related to the type of aggregate, mix proportion, and the moisture state of concrete. The coefficient of thermal expansion of concrete varies from $0.6 \times 10^{-5}/°C$ to $1.3 \times 10^{-5}/°C$ in most cases. Generally, $1.0 \times 10^{-5}/°C$ can be taken for normal weight concrete and 0.8×10^{-5} for lightweight concrete.

The thermal expansion of concrete may be calculated by

$$\varepsilon_{c,T} = \alpha \Delta T \tag{2.33}$$

where $\varepsilon_{c,T}$ = thermal strain.

α = coefficient of thermal expansion of concrete.

ΔT = temperature change in concrete.

The values of concrete strength and modulus of elasticity, the formulas for creep and shrinkage, discussed above, are valid for a mean temperature between approximate $-20°C$ and $+40°C$. Therefore, the influence of temperature deviation must be considered in those structural analyses where the results should have high accuracy. In the *fip* Model Code 2010, the effect of deviation from a mean concrete temperature of $20°C$ for the range of approximately $-20°C$ to $+80°C$ is dealt with.

2.2 Prestressing tendons

2.2.1 Basic requirements for prestressing tendons

Numerous factors cause the reduction of tensile stress in prestressing tendons during construction and in the service period, and the value of stress loss may reach 20%–40% of the initial stress in tendons at stretching. Therefore, the initial stress should be high enough that the remaining tensile stress after all stress losses can be kept at a required high level, therefore high-strength materials must be used for prestressing tendons.

For a fixed-length prestressing tendon under high tension, the phenomenon that the stress in the tendon reduces with time is called stress relaxation. The relaxation lessens the prestressing force, resulting in a decrease in the crack resistance in the precompressed concrete zones and an increase in the time-dependent deflection of the flexural members. This may seriously affect the long-term structural performance. As the time-dependent stress loss due to concrete creep and shrinkage is taken into account further, it is harder to predict exactly the long-term behavior of a prestressed concrete structure. From this point of view, the prestressing tendons must be made of low-relaxation materials.

Steel strands and wires, in which the diameter of a single wire varies from 4 to 7 mm, broadly used as prestressing tendons, are sensitive to corrosion. Once a steel wire under high tension is slightly rusted, stress concentration will occur in the vicinity of the corrosion pit, which is very likely to cause a sudden fracture. Therefore, the prestressing materials must bear good corrosion resistance.

For convenience in construction and obtaining agreeable mechanical properties in the service period, other properties such as flexibility of arrangement, good fatigue resistance, and sufficient ductility before failure are also required for prestressing tendons.

2.2.2 Classification of prestressing tendons

The prestressing tendons, also called the prestressing reinforcements, are fabricated by steels or nonsteels. The prestressing steels refer to high-strength wires, strands, and threaded bars. One strand consists of several wires. The commonly used nonsteel prestressing tendons are made of fiber-reinforced polymer (FRP), such as carbon fiber and aramid fiber, known as FRP tendons. FRP tendons mainly include CFRP (carbon fiber-reinforced polymer) tendons, AFRP (aramid fiber-reinforced polymer) tendons, and GFRP (glass fiber-reinforced polymer) tendons.

According to the bond state between the tendons and the concrete, the prestressing tendons can be classified into bonded and unbonded tendons. Prestressing steels can be employed as both bonded and unbonded tendons, while special anticorrosion protections must be made for unbonded prestressing steels. FRP tendons can also be used both for bonded and unbonded tendons.

2.2.3 Bonded prestressing steels

2.2.3.1 High-strength steel wires

In terms of the surface shape, steel wires are classified into plain round wires, helical rib wires (Fig. 2.6A), and indented wires (Fig. 2.6B). Along the direction of length, there are

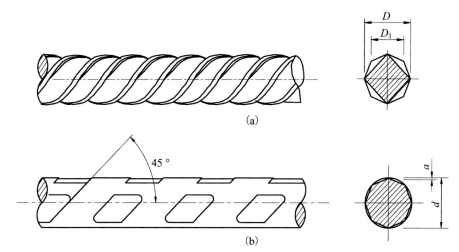

Figure 2.6 Helical rib wire and indented wire.

continuous and regularly helical ribs on the surface of helical rib wires and indentations on the surface of indented wires, both have good bond properties with concrete and are widely used in pretensioned members.

The high-strength wires are grouped into cold-drawn wires and stress-relieved wires according to the processing methods. A cold-drawn wire (WCD) is produced by drawing a hot-rolled steel rod through dies at normal temperature, while a stress-relieved wire is obtained by drawing along with short-time heat treatment. The stress-relieved wires are classified further into normal-relaxation wires (WNR) and low-relaxation wires (WLR). A WNR is obtained by short-time heat treatment at an appropriate temperature after the wire is straightened. A WLR is obtained by short-time heat treatment when the steel is in the plastic state to eliminate the residual stress generated during the drawing process. The relaxation in a WLR is only one-fourth to one-third of that in a WNR. The steel wires are manufactured to comply with the requirements of the relevant codes or specifications, such as *Steel wire for prestressing of concrete* (GB/T5223-2014) in China and ASTM A421 in the USA.

Wires with a diameter from 4 to 12 mm and characteristic strength from 1470 to 1860 MPa are commonly used. Wires with a diameter of 5 mm, 7 mm, and 9 mm are widely adopted in prestressed concrete structures. In the GB/T5223-2014 code, the strength grades of steel wires are distinguished into 1470 MPa, 1570 MPa, 1670 MPa, 1770MPa, and 1860 MPa.

As a product, a prestressing steel wire should be marked with the following information: nominal diameter, strength grade, processing method, surface shape (*P* for plain round wire, *H* for ribbed wire, *I* for indented wire), and the specification edition. For example, a cold-drawn wire with a diameter of 5 mm and characteristic tensile strength of 1860 MPa is labeled as 5.00-1860-WCD-P-GB/T5223-2014.

2.2.3.2 Steel strands

One steel strand is fabricated by twisting a plurality of prestressing steel wires in the same direction. One strand could be composed of two wires (1 × 2), three wires (1 × 3), seven wires (1 × 7), or 19 wires (1 × 19), as shown in Fig. 2.7. If the diameter of the single steel wire in one strand is indicated by ϕ (mm), one strand can be described by the number and the diameter of the wire together. For instance, $7\phi5$ refers to one strand composed of seven wires with a diameter of 5 mm for each wire, as shown in Fig. 2.7C and D.

The commonly used steel wires in the strands have diameters varying from 2.5 to 6.0 mm, of which $\phi4$ and $\phi5$ are the most commonly used. The nominal tensile strengths of strands are classified into the following grades: 1470 MPa, 1570 MPa, 1670 MPa, 1720 MPa, 1770 MPa, 1820 MPa, 1860 MPa, and 1960 MPa.

The $7\phi5$ strand with characteristic tensile strength of 1860 MPa is most extensively used in railway and highway prestressed concrete structures. One $7\phi5$ strand is composed

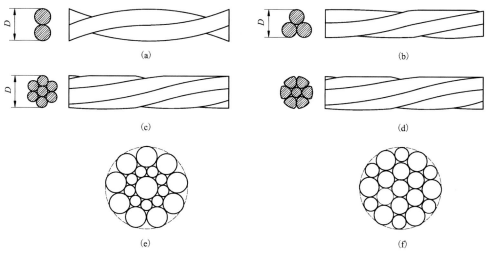

Figure 2.7 Cross-sections of prestressing steel strands.

of seven wires, of which six wires with a diameter of 5 mm are wound around a central seventh that has a slightly larger diameter (Fig. 2.7C). One standard strand is composed of plain cold-drawn wires while one indented strand is composed of indented wires. When one standard strand is drawn through dies, it turns into one compact strand (Fig. 2.7D). During drawing through dies, the wires in a strand squeeze each other, then the internal gaps decrease, resulting in a smaller outer diameter which requires less spacing in concrete, making it convenient to be arranged in a section.

In terms of the structure characteristics, the prestressing steel strands can be classified into eight categories, as compiled in Table 2.13.

Table 2.13 Category of prestressing steel strand.

Structural codes for strands	Components in a strand	Schematic sections
1×2	Two wires	Fig. 2.7(a)
1×3	Three plain round wires	Fig. 2.7(b)
$1 \times 3I$	Three indented wires	Fig. 2.7(b)
1×7	Seven plain round wires (standard strand)	Fig. 2.7(c)
$1 \times 7I$	Six indented wires + one plain round wire	Fig. 2.7(c)
$(1 \times 7)C$	Seven plain round wires (compact strand)	Fig. 2.7(d)
$1 \times 19S$	$(1 + 9 + 9)$ wires	Fig. 2.7(e)
$1 \times 19W$	$(1 + 6 + 6/6)$ wires	Fig. 2.7(f)

As a product, one prestressing steel strand should be marked with the following information: prestressing steel strand (PSS), nominal diameter, strength grade, structural codes, and the specification edition. For instance, one standard strand with a nominal diameter of 15.2 mm and characteristic tensile strength of 1860 MPa can be labeled as PSS 1 × 7-15.2-1860-GB/T5224-2014.

2.2.3.3 High-strength threaded steel bars

The screw-thread bar is made with alloy steel through hot rolling, and is widely used in prestressed concrete structures. The discontinuous external threads on the whole surface of the bar (Fig. 2.8) make it easy to be anchored by a nut at any point along the length or be linked to another threaded bar through a coupler. The mechanical properties of the hot-rolled threaded bar are similar to reinforcing steels in that there are obvious yield points and yielding plateau in the stress—strain curve. The threaded steel bars are manufactured to conform to the requirements of relevant codes or specifications, such as *Screw-thread steel bars the prestressing of concrete* (GB/T 20065-2016).

According to the yield strength, the prestressed steel bars (PSBs) are classified into five grades: PSB785, PSB830, PSB930, PSB1080, and PSB1200 (the number after PSB indicates the yield strength). The range of the nominal diameter of PSBs varies from 15 to 75 mm, of which 25 and 32 mm are most commonly used in prestressed concrete structures.

2.2.4 Unbonded prestressing steels

When special measures for anticorrosion protection are employed for high-strength steel wires, threaded bars, and strands, they can be used as unbonded prestressing steels, in which the unbonded strands are extensively used in unbonded prestressed concrete structures. The unbonded prestressing steels can be arranged inside or outside the concrete.

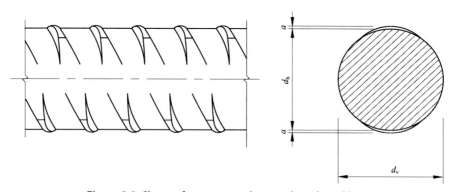

Figure 2.8 Shape of a prestressed screw-thread steel bar.

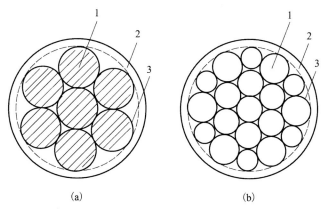

Figure 2.9 Cross-section of unbonded prestressed strands: (1) strands; (2) sheathings; (3) coating.

2.2.4.1 Unbonded prestressing steels arranged inside the concrete

Generally, coatings or sheathing on prestressing steels made during manufacture are adopted as a measure to protect steel from corrosion. The coating material, which can be a metallic coating or organic coating, should have good chemical stability, no erosion to the surrounding materials, impermeable, nonhygroscopic, strong corrosion resistance, high lubricity, good brittle resistance at low temperature, and certain flexibility. The sheathings should have sufficient flexibility, abrasion and impact resistance, no erosion to the surrounding materials, good chemical stability at high temperature, and good brittle resistance at low temperature. High-density polyethylene (HDPE) and polypropylene (PE) are commonly used to fabricate the sheathings.

The unbonded prestressing steels broadly used inside the concrete are composed of seven or 19 strands, as shown in Fig. 2.9.

2.2.4.2 External prestressing steels arranged outside the concrete

The prestressing steels arranged outside the concrete are called the external prestressing steels. Considering the external prestressing steels exposed to the natural environment during their service life, extra anticorrosion protection is required. Generally, there are three anticorrosion protection measures, namely, coatings on prestressing steels, exterior sheathings, and fillers insider the sheathings, as shown in Fig. 2.10. The external prestressing steels and the corresponding anticorrosion protections form an external cable.

The unbonded strands can adopt smooth strands, hot-dip galvanized strands, and epoxy-coated strands. The zinc coating and epoxy coating on strands can be treated as "internal protection," the outer sheathing using HDPE pipe and galvanized steel pipe can be treated as "external protection," and the grease or cement mortar filled in the pipe can be treated as "intermediate protection."

The external cables are classified into the finished cables, composing of unbonded prestressing steels and HDPE sheath, as shown in Fig. 2.10A, which are fabricated in

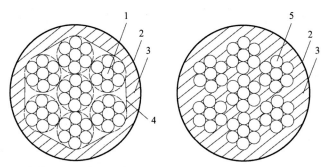

Figure 2.10 Structural diagram of typical external cables: (1) unbonded strand; (2) HDPE sheath; (3) filling material; (4) outer sheath; (5) bonded strand.

the factory by a hot extruding process, and the nonfinished cables fabricated in the construction location, as shown in Fig. 2.10B. The steel strands (characteristic tensile strength should be greater than 1720 MPa) are generally adopted to make nonfinished cables, and the HDPE or steel pipe is used for the outer side sheath. The finished cables are fabricated to conform to *Hot-extruded PE protection paralleled high strength wire cable for cable-stayed bridge* (GB/T18365-2018).

2.2.5 Mechanical properties of prestressing steels

2.2.5.1 Stress—strain curve

The prestressing threaded bars are made by a hot-rolling process, so their stress—strain curve is similar to that of the reinforcing steels under uniaxial tension, as shown in Fig. 2.11, where there is an obvious yield point B (f_y) and yielding plateau. The stresses corresponding to points A and D are the proportional limit, f_p, and the ultimate strength, f_u. f_y is very close to f_p, hence, the strength grade of the prestressing threaded bars can be classified by yield strength, f_y.

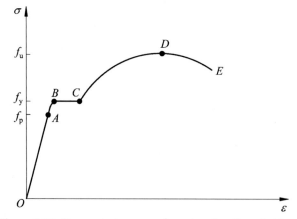

Figure 2.11 Stress—strain curve of prestressing threaded bars.

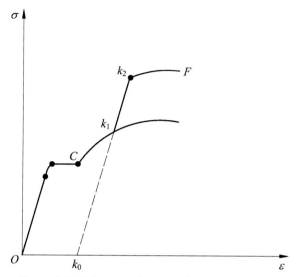

Figure 2.12 Schematic diagram of cold-drawn aging.

If the steel bar is tensioned to point k_1 locating between points C and D at room temperature, as shown in Fig. 2.12, plastic strain ε_{k_0} is generated, and this operation is described as cold drawn. As the tension is released, the elongated point retracts back to the point k_0 rather than its origin. If the steel bar is tensioned again just after releasing, the point starts from k_0 to k_1, then to point D finally if the tension increases continuously. Under such circumstances, the new yield point at second tension increases to k_1 while the ultimate strength is the same as that in the first tension. In the case of the second tension operated after some time since the first tension, the point starts from k_0 to k_1, then to k_2 and F as the tension increases continuously, showing that both the yield strength and ultimate strength increase. This phenomenon is called the cold-drawing aging. High-strength wires are fabricated by drawing at room temperature, so their stress–strain relations are demonstrated by $k_0-k_1-k_2-F$, as shown in Fig. 2.12.

Actually, the yield point of wires is not obvious as point k_2 in Fig. 2.12. Fig. 2.13 shows the typical stress–strain curve of prestressing steel wires under uniaxial tension. Point a corresponds to the proportional limit, while point b corresponds to the stress that the permanent residual plastic strain in the wire is 0.2% when the tension is released, noted as $f_{0.2}$. In Fig. 2.13, the stress–strain curve can be divided into three portions, a linear portion oa, a nonlinear portion with a gradually decreasing slope up to point b, and a final portion up to failure.

$f_{0.2}$ is called the 0.2% proof stress of prestressing steel wires. Considering the yield point in Fig. 2.13 is not obvious, $f_{0.2}$ is treated as the "notional yield stress" in China's specifications. $f_{0.2}$ equals approximately 85% of the ultimate tensile strength, which is used to define the design strength of prestressing steel wires. In some other codes, 0.1% proof stress is used to define the "yield stress."

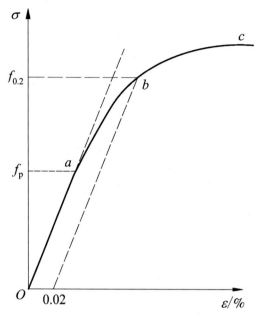

Figure 2.13 Stress—strain curve of prestressing steel wires.

2.2.5.2 Mechanical properties

The mechanical properties of the stress-relieved plain round steel wires and ribbed wires commonly used in concrete structures shall comply with the requirements at Table 2.14 specified in the GB/T5223-2014 code. For stress-relieved steel wires, the bending times shall be greater than three times.

The mechanical properties of prestressing steel strands with seven wires and 19 wires commonly used in concrete structures shall comply with the requirements in Tables 2.15 and 2.16 specified in the GB/T5224-2014 code, respectively.

The mechanical properties of prestressing threaded steel bars shall comply with the requirements in Table 2.17 specified in the GB/T20065-2016 code.

2.2.5.3 Modulus and strength of prestressing steels

The moduli of elasticity of steel wires and strands can be taken as the tangential modulus of line *oa* in Fig. 2.13. The moduli of elasticity of prestressing steels and nonprestressing steels are compiled in Table 2.18.

The characteristic tensile strengths of prestressing steels are compiled in Table 2.19.

The design strengths of prestressing steels specified in the GB50010-2010 and JTG 3362-2018 codes are compiled in Table 2.20.

The characteristic strengths and design strengths of prestressed steels in the Q/CR 9300-2018 code are compiled in Tables 2.21 and 2.22, respectively.

Table 2.14 Mechanical properties of stress-relieved plain round steel wires and ribbed wires.

Nominal diameter d_m/mm	Nominal tensile strength R_m/MPa	0.2% yield strength $F_{P0.2}$/KN ≥	Elongation under maximum tension (L_0 = 200 mm) A_{gt}/% ≥	Performance under repeated bending		Relaxation performance	
				Number of bending (time/180 degrees) ≥	Bending radius R/mm	Ratio of initial tension to maximum tension/%	Relaxation rate in 1000 h/% ≤
4.00	1470	16.22	3.5	3	10	70	2.5
5.00		25.32		4	15		
6.00		36.47		4	15		
7.00		49.64		4	20	80	4.5
8.00		64.84		4	20		
9.00		82.07		4	25		
10.00		101.32		4	25		
4.00	1570	17.37		3	10		
5.00		27.12		4	15		
6.00		39.06		4	15		
7.00		53.16		4	20		
8.00		69.44		4	20		
9.00		87.89		4	25		
10.00		108.51		4	25		
4.00	1670	18.47		3	10		
5.00		28.85		4	15		
6.00		41.54		4	15		
7.00		56.55		4	20		
8.00		73.86		4	20		
9.00		93.50		4	25		
4.00	1770	19.58		3	10		
5.00		30.58		4	15		
6.00		44.03		4	15		
7.00		59.94		4	20		
7.50		68.81		4	20		
4.00	1860	20.57		3	10		
5.00		32.13		4	15		
6.00		46.27		4	15		
7.00		62.98		4	20		

Table 2.15 Mechanical properties of 1 × 7 steel strands.

Steel strands	Nominal diameter D_m/mm	Nominal tensile strength R_m/MPa	0.2% yield strength $F_{P0.2}$/KN ≥	Elongation under maximum tension (L_0 ≥ 500 mm) A_{gt}/% ≥	Relaxation performance	
					Ratio of initial tension to maximum tension/%	Relaxation rate in 1000 h/% ≤
1 × 7	15.20	1470	181	3.5	70	2.5
	4.80	1570	194			
	5.00	1670	206		80	4.5
	9.50	1720	83.0			
	11.10		113			
	12.70		150			
	15.20		212			
	17.80		288			
	15.70	1770	234			
	21.60		444			
	9.50		89.8			
	11.10		121			
	12.70	1860	162			
	15.20		229			
	15.70		246			
	17.80		311			
	18.90		360			
	21.60		466			
	9.50	1960	94.2			
	11.10		128			
	12.70		170			
	15.20		241			
1 × 7I	12.70	1860	162			
	15.20		229			
(1 ×7)C	12.70	1860	183			
	15.20	1820	264			
	18.00	1720	338			

Table 2.16 Mechanical properties of 1 × 19 steel strands.

Steel strands	Nominal diameter D_m/mm	Nominal tensile strength R_m/MPa	0.2% yield strength $F_{P0.2}$/KN ≥	Elongation under maximum tension (L_0 ≥ 500 mm) A_{gt}/% ≥	Relaxation performance	
					Ratio of initial tension to maximum tension/%	Relaxation rate in 1000 h/% ≤
1 × 19S (1 + 9 + 9)	28.60	1720	805	3.5	70	2.5
	17.8	1770	334			
	19.3		379		80	4.5
	20.3		422			
	21.8		488			
	28.6		829			
	20.3	1810	432			
	21.8		499			
	17.8	1860	341			
	19.3		400			
	20.3		444			
	21.8		513			
1 × 19W (1 + 6 + 6)	28.6	1720	805			
		1770	829			
		1860	854			

Table 2.17 Mechanical properties of prestressing threaded steel bars.

Grade	Yield strength R_{cL}/MPa $\geqslant$	Tensile strength R_m/MPa $\geqslant$	Elongation after fracture A/% $\geqslant$	Elongation under max. Tension A_{gt}/% $\geqslant$	Relaxation performance	
					Initial stress	Relaxation rate in 1000 h/%
PSB785	785	980	8	3.5	$0.7R_m$	$\leqslant 4.0\%$
PSB830	830	1030	7			
PSB930	930	1080	7			
PSB1080	1080	1230	6			
PSB1200	1200	1330	6			

Table 2.18 Modulus of elasticity of steels.

Type of steels	Modulus of elasticity/MPa
Prestressing stress–relieved wire	2.05×10^5
Prestressing steel strand	1.95×10^5
Prestressing threaded steel bar	2.0×10^5
HPB300 (nonprestressing steel)	2.1×10^5
HRB400, HRBF400, RRB400 (nonprestressing steel)	2.0×10^5

Table 2.19 Characteristic tensile strength of prestressing steels.

Types of prestressing steels		Diameter/mm	Symbol	f_{ptk}/MPa
Strand	1×3	8.6, 10.8, 12.9	ϕ^S	1470, 1570, 1720, 1860, 1960
	1×7	9.5, 12.7, 15.2, 17.8 21.6		1720, 1860, 1960 1860
Stress-relieved wire	Smooth and ribbed surface	5	ϕ^P	1570, 1770, 1860
		7	ϕ^H	1570
		9		1470, 1570
Threaded bar		18, 25, 32, 40, 50	ϕ^T	785, 930, 1080

2.2.6 Prestressing FRPs and their mechanical properties

Fiber-reinforced polymers (FRPs) made from fibers and binders were first developed in the 1960s and have been used as reinforcements in civil engineering structures since the 1980s. Nowadays, they are widely used in the concrete structures.

Table 2.20 Design tensile and compressive strengths of prestressing steels (MPa).

Types of prestressing steels		f_{pd}	f'_{pd}
Strand (1×7)	$f_{ptk} = 1720$	1170 (1220)	390
	$f_{ptk} = 1860$	1260 (1320)	
	$f_{ptk} = 1960$	1330 (1390)	
Stress-relieved wire	$f_{ptk} = 1470$	1000 (1040)	410
	$f_{ptk} = 1570$	1070 (1110)	
	$f_{ptk} = 1770$	1200	
	$f_{ptk} = 1860$	1260 (1320)	
Threaded bar	$f_{ptk} = 540$	450	400
	$f_{ptk} = 785$	650	
	$f_{ptk} = 930$	770	
	$f_{ptk} = 1080$	900	

Note: The data without parentheses in the table are recorded from JTG 3362 code. The data in parentheses are the different values in the GB50010 code.

Table 2.21 Characteristic strength of prestressing steels.

Prestressing steels		Nominal diameter/mm	Characteristic tensile strength/MPa
Types	Models		
Wire	—	4—5	1470, 1570, 1670, 1770, 1860
		6—7	1470, 1570, 1670, 1770
Strand	Standard	12.7	1770, 1860, 1960
	(1×7)	15.2	1470, 1570, 1670, 1720, 1860, 1960
		15.7	177, 1860
	Compacted	12.7	1860
	(1×7)C	15.2	1820
	Threaded bar	PSB830	—
		830	
	PSB980	—	980

Table 2.22 Design strength of prestressing steels (MPa).

Prestressing steels		Nominal diameter/mm	Design tensile strength	Design strength in compression
Types	Models			
Wire	$f_{ptk} = 1470$	4, 5, 6, 7	945	410
	$f_{ptk} = 1570$	4, 5, 6, 7	1010	
	$f_{ptk} = 1670$	4, 5, 6, 7	1075	
	$f_{ptk} = 1770$	4, 5, 6, 7	1140	
	$f_{ptk} = 1860$	4, 5	1200	
Strand	Standard	12.7	1140, 1200, 1260	390
	(1×7)	15.2	945, 1010, 1075, 1105, 1200, 1260	
		15.7	1140, 1200	
	Compacted	12.7	1200	
	(1×7)C	15.2	1170	
Threaded bar	PSB830	—	660	400
	PSB980	—	780	

FRP tendons have many desirable properties, such as light weight, high strength, high corrosion resistance, and fatigue resistance, which are preferred to be used in a severe corrosive environment. Meanwhile, there is no obvious yield point or yielding plateau in the stress—strain curve of FRPs, and the plastic deformation is small before a sudden failure. Thus, the reinforcing steels are always required to be arranged in a prestressed concrete structure with FRP tendons.

Table 2.23 Mechanical properties of fiber ribs.

Types of fiber	Characteristic tensile strength/MPa	Modulus of elasticity/MPa	Elongation rate at failure/%
Carbon fiber tendon	$\geqslant 1800$	$\geqslant 1.4 \times 10^5$	$\geqslant 1.50$
Aramid fiber tendon	$\geqslant 1300$	$\geqslant 0.65 \times 10^5$	$\geqslant 2.00$

FRP tendons are fabricated to conform to *PAN-based carbon fiber* (GB/T 26752-2020) and *Fiber reinforced composite bars for civil engineering* (GB/T 26743-2011). The main mechanical properties of commonly used unbonded FRP tendons are compiled in Table 2.23.

The cross-sectional area of unbonded prestressing FRP is calculated by the nominal diameter. The design tensile strength can be derived using

$$f_{fpd} = \frac{f_{fpk}}{\gamma_f \gamma_e} \tag{2.34}$$

where f_{fpd} = design tensile strength of prestressing FRP.

f_{fpk} = characteristic tensile strength of prestressing FRP.

γ_f = partial material coefficient for unbonded prestressing FRP, $\gamma_f = 1.4$

γ_e = ambient impact coefficient on prestressing FRP, the values can be taken from Table 2.24.

2.2.7 Relaxation

The tendon relaxation is related to the initial tensile stress, tension duration, type of tendon, and ambient temperature. The relaxation accelerates with the increase in the initial tensile stress and the ambient temperature.

Table 2.24 Ambient impact coefficient on prestressing FRP.

Ambient environment type	Fiber ribs	
	Carbon fiber tendon	Aramid fiber tendon
Class IV	1.0	1.2
Classes IIa, IIb	1.1	1.3
Classes IIIa, IIIb, IV, V	1.2	1.5

Experiments have shown that the stress relaxation is relatively small, and can be omitted in calculations when the initial tensile stress in the tendon is less than $0.5f_{ptk}$ (f_{ptk} is the characteristic tensile strength). As the initial tensile stress gets closer to f_{ptk}, the relaxation develops fast and the ultimate value increases sharply. From this point of view, the initial tensile stress in tendons at stretching during construction should be controlled within a reasonable range. Generally, for prestressing steels in the prestressed concrete structures, the total stress loss due to relaxation may reach 2.0%–8.0% of the ultimate tensile strength.

The ultimate value and the development of relaxation in tendons can be obtained by experiments at a nominal temperature of 20°C for a period of 1000 h from an initial force of 70% of ultimate tensile strength. The ultimate relaxation of the steel strands and wires can be estimated by

$$\sigma_{REL} = \xi_1 \xi_2 \left(\lambda_1 \frac{\sigma_p}{f_{ptk}} - \lambda_2 \right) \sigma_p \tag{2.35}$$

where σ_{REL} = ultimate relaxation of the prestressing steels.

ξ_1 = coefficient related to the tensioning method, 1.0 is taken for tensioning once, 0.9 is taken for over-tension.

ξ_2 = coefficient related to the type of prestressing steels, the values in different specifications are different, as compiled in Table 2.25.

λ_1 = coefficient related to the type of prestressing steels.

λ_2 = coefficient related to the type of prestressing steels.

f_{ptk} = characteristic tensile strength of prestressing steels.

σ_p = initial tensile stress in prestressing steels.

The ultimate relaxation of the prestressing steels determined from Eq. (2.35) in the Q/CR 9300-2018, JTG 3362-2018, and GB50010-2010 codes is compiled in Table 2.25.

In simplified analysis for general prestressed concrete structures, the JTG 3362-2018 code specifies that the ultimate relaxation can be taken as the value occurring at 1000 h after stretching, and the intermediate can be derived using the ratios in Table 2.26.

For prestressing threaded steel bars, the ultimate relaxation can be taken as $0.05\sigma_{con}$ when the tendons are stretched once and $0.035\sigma_{con}$ when the tendons are overstretched.

For high-precision time-dependent effect analysis, the time-dependent expression for tendon relaxation is required. Statistical regression analysis of tested results based on exceeding 9-year experiments (Magura and Sozen) shows the development of the relaxation of prestressed steels (Fig. 2.14) can be expressed as

$$\sigma_{REL}(t) = \frac{\log(t)}{K} \left(\frac{\sigma_p}{f_{py}} - 0.55 \right) \sigma_p \tag{2.36}$$

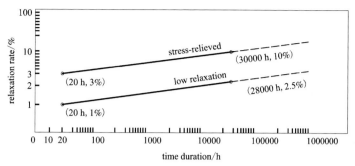

Figure 2.14 Stress relaxation versus loading duration.

where $\sigma_{REL}(t)$ = stress relaxation of the prestressing steels.

f_{py} = yield stress of the prestressing steels.

σ_p = initial stress in prestressing steels.

K = coefficient related to the type of prestressing steels, 10 is taken for general stress-relieved steels, 40−50 is taken for low-relaxation types.

The *fip* Model Code 2010 code recommends that the relaxation loss after infinite time may be assumed to be two to three times the value at 1000 h. Also, the ultimate value and development rate of the tendon relaxation vary with the temperature, and the temperature effect on the final value of the relaxation values is demonstrated by Fig. 2.15.

For prestressing FRP tendons, the ultimate relaxation can be estimated by

$$\sigma_r(t) = r(t) \cdot \sigma_{con} \tag{2.37}$$

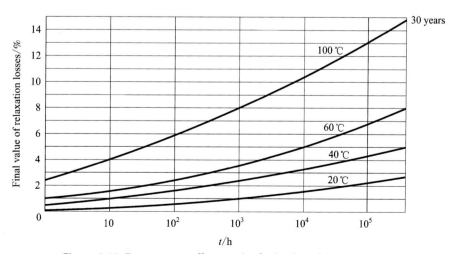

Figure 2.15 Temperature effect on the final value of the relaxation.

Table 2.25 Ultimate relaxation of steel strands and wires (MPa).

Grade of prestressing steels	JTG 3362-2018		Q/CR 9300-2018, GB50010-2010
Class I	$\psi\zeta\left(0.52\dfrac{\sigma_{pe}}{f_{ptk}} -0.26\right)\sigma_{pe}$	$\zeta = 1.0$	$0.4\psi\left(\dfrac{\sigma_{con}}{f_{ptk}} -0.5\right)\sigma_{con}$
Class II		$\zeta = 0.3$	$\lambda_1\left(\dfrac{\sigma_{con}}{f_{ptk}} -\lambda_2\right)\sigma_{con}$ $\quad\begin{array}{l}\sigma_{con}\leq 0.7f_{ptk}: \lambda_1 = 0.125,\\ \lambda_2 = 0.5\\ \sigma_{con} > 0.7f_{ptk}:\\ \lambda_1 = 0.2, \lambda_2 = 0.575\end{array}$
Notes	$\psi = 1.0$ for tension once, $\psi = 0.9$ for super tension		

Table 2.26 Ratio of intermediate value to ultimate relaxation.

Time/day	2	10	20	30	40
Ratio	0.50	0.61	0.74	0.87	1.00

Table 2.27 Relaxation rate for prestressing FRP.

Type of prestressing FRP	Relaxation rate r
CFRP (carbon fiber-reinforced polymer) tendon	2.2%
AFRP (aramid fiber-reinforced polymer) tendon	16%

where $r(t) =$ relaxation rate.

Generally, the time-dependent stress loss due to relaxation can be derived from

$$\sigma_{l5}(t) = \frac{a + b \log t}{100} \sigma_{con} \tag{2.38}$$

where $a =$ coefficient, 0.231 is taken for prestressing CFRP, and 3.38 is taken for prestressing AFRP.

$b =$ coefficient, 0.345 is taken for prestressing CFRP, and 2.88 is taken for prestressing AFRP.

For prestressing FRP with a service life of 50 years, the relaxation can be taken from Table 2.27. For prestressing AFRP, over 1 hour shall be carried out under jacking force before anchoring, otherwise, 20% should be taken for the relaxation rate in Table 2.27.

2.2.8 Fatigue strength of prestressing tendons

For prestressed concrete structures under repeated loading, such as bridges, the stress in prestressing tendons fluctuates frequently over the service period. The ability of prestressing tendons and their anchorages to resist such stress fluctuations is called fatigue resistance. This mainly depends on the stress fluctuation range, the characteristics of the stress cycle, the number of cycles, and the initial defects in the tendons.

Fatigue failure originates from microcrack propagation at the region of damage (surface or interior). The indented steel wires are prone to cause micro-damage during processing, so their fatigue strength is lower than that of the round steel wires. The ends of the prestressing tendons in the post-tensioned members are fixed by anchorage devices, and the micro-damage possibly occurs during anchoring, therefore, the fatigue strength of the tendons in the anchorage region is lower than in other parts.

The fatigue stress ratio of the prestressing tendons is defined as

Table 2.28 Limits of stress amplitude of the prestressing tendons (MPa).

Type of prestressing steels	Limit of stress amplitude			
	GB50010-2010 ($f_{ptk} = 1570$ MPa)			
	$\rho_p^f = 0.7$	$\rho_p^f = 0.8$	$\rho_p^f = 0.9$	Q/CR 9300-2018
Stress–relieved wires	240	168	88	150
Strands	144	118	70	140
Threaded bar	—			80

$$\rho_p^f = \frac{\sigma_{p,\min}^f}{\sigma_{p,\max}^f} \tag{2.39}$$

where $\sigma_{p,\max}^f$ = maximum stress of the layer of prestressing tendons in fatigue checking.

$\sigma_{p,\min}^f$ = minimum stress of the layer of prestressing tendons in fatigue checking.

The amplitude of fatigue stress is defined as

$$\Delta\sigma_p^f = \sigma_{p,\max}^f - \sigma_{p,\min}^f \tag{2.40}$$

Corresponding to a certain number of cycles (usually two million times) and the typical minimum stress in the structure, the fatigue failure of the prestressing tendons can be prevented by controlling the stress amplitude. Table 2.28 lists the limits of the stress amplitude of the prestressing tendons in fatigue check stipulated in the Q/CR 9300-2018 and GB50010-2010 codes, respectively.

Suggested readings

[1]   GB 50010: 2015. Code for design of concrete structures. Beijing: China Construction Industry Press; 2015.
[2]   Q/CR 9300:2018. Code for design on railway bridge and culvert. Beijing: China Railway Publishing House; 2018.
[3]   TB 10092: 2017. Code for design of concrete structures of railway bridge and culvert. Beijing: China Railway Publishing House; 2017.
[4]   JTG 3362: 2018. Specifications for design of highway reinforced concrete and prestressed concrete bridges and culverts. Beijing: People's Communications Press; 2018.
[5]   JGJ 369: 2016. Prestressed concrete structure design code. Beijing: China Construction Industry Press; 2016.
[6]   GB/T5223: 2014. Steel wire for prestressing of concrete. Beijing: China Construction Industry Press; 2014.
[7]   GB/T 20065: 2016. Screw-thread steel bars for the prestressing of concrete. Beijing: China Construction Industry Press; 2016.
[8]   GB/T 18365: 2018. Hot-extruded PE protection paralleled high strength wire cable for cable-stayed bridge. Beijing: China Construction Industry Press; 2018.
[9]   GB/T 26752: 2020. PAN-based carbon fiber. Beijing: China Construction Industry Press; 2020.
[10]   GB/T 26743:2011. Fiber reinforced composite bars for civil engineering. Beijing: China Construction Industry Press; 2011.

[11] CEB–FIP model code for concrete structures 1978. Paris: Comité Euro - International du Béton/Fédération International de la Préconstrainte; 1978.

[12] CEB–FIP model code for concrete structures 1990. Paris: Comité Euro-International du Béton/Fédération International de la Préconstrainte; 1990.

[13] Fip model code 2010 — final draft, volume 1. Fib Bulletin N.65, Ernst & Sohn; 2012. Fib International federation for structural concrete: fib model code for concrete structures 2010. Berlin: Ernst & Sohn; 2013.

[14] Eurocode 2. Design of concrete structures. European standard, prEN 1992-1-1. European Committee for Standardization; 2002.

[15] ACI 209.2R:08. Guide for modeling and calculating shrinkage and creep in hardened concrete. America Concrete Institute. ACI Committee 209. Farmington Hills, MI; 2008.

[16] Bažant ZP. Mathematical modeling of shrinkage and creep of concrete. New York: Jhon Willey & Sons Ltd; 1988.

[17] Gardner NJ, Lockman MJ. Design provisions for drying shrinkage and creep of normal-strength concrete. ACI Mater J 2001;98:159—67.

[18] Wendner R, Hubler MH, Bažant ZP. The B4 model for multi-decade creep and shrinkage prediction. In: Mechanics and physics of creep, shrinkage, and durability of concrete; 2013. p. 429—36.

[19] Magura DD, Sozen MA, Siess CP. A study of stress relaxation in prestressing reinforcement. PCI J 1964;9(2):13—57.

[20] Brooks J. 30-year creep and shrinkage of concrete. Mag Concr Res 2005;57(9):545—56.

[21] Hu Di. Basic principles of prestressed concrete structure design. 2nd ed. Beijing: China Railway Publishing House; 2019.

CHAPTER 3

Prestressing methods and anchorage systems

Contents

3.1 Prestressing methods

The approaches to introducing prestress in concrete can be divided into two categories, one is to stretch the tendons by a jacking system and anchor them in the concrete structures, and the other refers to all other methods. According to the sequence of stretching tendons and pouring concrete during construction, the former is classified into two groups: the pretensioning method in which the tendons are stretched before pouring concrete, and the post-tensioning method in which the tendons are stretched after the concrete attains sufficient strength. The latter includes electrothermal, self-stressing, and external loading approaches, etc.

The electrothermal approach, based on the principle of expansion due to an increase in temperature, and contraction due to a decrease in temperature of materials, may be used when the construction space for operating tension equipment such as jacks is limited. First, a high current is passed to the prestressing tendons arranged inside or

Analysis and Design of Prestressed Concrete
ISBN 978-0-12-824425-8, https://doi.org/10.1016/B978-0-12-824425-8.00003-8

outside the hardened concrete section to increase the temperature, thereby elongating the tendons. The tendons are then anchored followed by cutting off the power supply, and the prestress in concrete is achieved by the contraction when the temperature of the tendons decreases. If the tendons are embedded in the expansive concrete before curing, the volume expansion of concrete during curing will drive the tendons to extend together. In this condition, the tendons are tensioned while the concrete is compressed in a self-balancing system, thereby the prestress in concrete is generated to form a self-stressing concrete. If a steel girder with a precamber is forced by external loads to a straight member, and the concrete is cast to encase the flange or both the flange and web of this girder, partial camber of the steel girder recovers once the external loads are removed when the bond between the concrete and the steel girder attains sufficient strength, resulting in compression in concrete on the concave side or whole section. Thus, prestress in concrete is imparted and a prestressed concrete composite beam is fabricated. The above approaches to imparting prestress in concrete are also often adopted in engineering practice.

In this chapter, the most widely used pretensioning and post-tensioning methods, along with the corresponding anchorage systems, are discussed.

3.2 Pretensioning method

The typical construction sequence for a pretensioned concrete member is shown in Fig. 3.1, including the following key steps:

(1) Arranging the tendons according to the design drawings (profile and amount of the tendons) in a stressing bed system.

(2) Fixing temporarily one end of the tendons and stretching the tendons at the other end by a jacking system.

(3) Anchoring the stretching side of the prestressing tendons temporarily by a gripping device when the jacking force reaches the design value.

(4) Casting concrete and curing.

(5) Releasing the temporary anchorage at two ends when the concrete attains sufficient strength, generally exceeding 80% of the design strength in compression, and cutting off the excess prestressing tendons outside the member.

In pretensioned concrete, the transfer of the prestressing is achieved by the cohesive force between the tendons and the concrete, so high-strength indented wires, helical rib wires, strands, and threaded steel bars are commonly used in pretensioned concrete. According to the characteristics of the construction process of the pretensioning method, the tendons are arranged in a straight or broken line with two segments in most cases. For short members with straight tendons by standardized design, multiple members

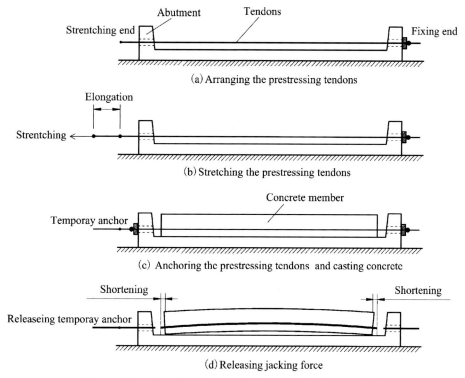

Figure 3.1 Schematic diagram of the pretensioning method.

may be fabricated simultaneously in a long precast bed. To reduce the construction cycle period, heat treatment, such as steaming, is commonly adopted during curing for concrete.

For short pretensioned concrete members, such as railway sleepers, it is required to achieve a high cohesive force in a short length at the ends to transfer the prestressing quickly and safely; in this case, a low level of jacking stress may be taken so that a larger cross-section area of the prestressing tendons can be provided in the case of equivalent prestressing, or more steel wires with small diameters are replaced to increase the cohesive force. However, for long members with eccentric straight tendons, such as simply supported beams with a long span, it is necessary to reduce the effect of the eccentric prestressing on the concrete in the vicinity of the beam end to avoid cracking on the top edge. In this case, an effective approach may be taken, that is, a certain length of the tendons near the end is covered with a plastic sheath or coated with grease (called "insulation") to decrease or eliminate the local eccentric prestressing.

The pretensioning method is suitable for making short- and medium-span concrete members arranged with straight-line or polyline prestressing tendons.

3.3 Post-tensioning method

The typical construction sequence for a post-tensioned concrete member is shown in Fig. 3.2, including the following key steps:

(1) Fixing the ducts for prestressing tendons in the right position.

(2) Casting concrete and curing.

(3) Stretching the tendons when the concrete attains sufficient strength, generally exceeding 80% of the design strength in compression.

(4) Anchoring the prestressing tendons.

(5) Grouting the ducts for the bonded tendons with cement mortar to make the tendons and concrete bond together.

(6) Casting concrete to cover the anchorages embedded in the concrete to prevent from corroding, or making special protection for anchorages of unbonded prestressing tendons or external prestressing tendons.

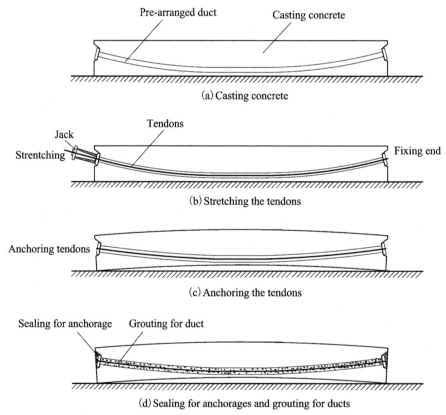

(a) Casting concrete

(b) Stretching the tendons

(c) Anchoring the tendons

(d) Sealing for anchorages and grouting for ducts

Figure 3.2 Schematic diagram of post-tensioning method.

It should be mentioned that the tendons are stretched by the ram of the jack which reacts against the anchorage embedded in the concrete, so the flexural member arranged with eccentric tendons, as shown in Fig. 3.2, will generate upward camber during stretching tendons if all the tendons are stretched at the same time, since the eccentric prestressing force is designed to counterbalance the self-weight of the member, the weight applied subsequently (such as a deck system in railway and highway bridges) and live loads. In this case, the data shown in the oil gauge for the jack are equal to the value of tension at the jacking point which considers the effect of the self-weight of the member.

Prestressing ducts usually consist of a spirally wound sheet, called corrugated metal or plastic ducts. The prestressing tendons may be placed into the ducts before casting concrete. In this case, the tendons must be dragged or rotated to prevent the tendons from being consolidated by leaking cement mortar after the concrete is cast. If the prestressing tendons are placed into the ducts after casting concrete, the walls should be dredged immediately that the concrete is cast to ensure that the walls are unblocked. In addition, the ducts for tendons with a short or medium length can be formed by an extractable duct maker. The rubber rods are fixed in the positions for ducts before casting concrete first, then the rods are pulled at one end by a piece of equipment when the concrete hardens sufficiently. Under high tension, the plastic rod contracts laterally due to the Poisson effect, thereby the rod can easily be pulled out and the tendon duct is formed.

The tendons may be stretched simultaneously at both ends, or stretched at one side and the other end is fixed by anchorage first, depending on many factors such as the length of the tendons, the magnitude of the frictional resistance by the duct wall against the tensioning tendons, and the space limit for the operation of stretching equipment, etc.

Theoretically, any profiles of the tendons can be used in post-tensioned concrete, since the tendon ducts can be easily arranged before casting concrete, and the hardened concrete can bear the radial pressure and concentrated force caused by the curved tendons at stretching. Therefore, post-tensioned concrete is broadly used in precast concrete structures and cast-in-situ concrete structures (Figs. 3.3 and 3.4).

The anchorages are embedded inside the concrete or connected with the concrete structure permanently, and no additional external reaction system is required when stretching and anchoring the tendons, therefore post-tensioning technology is also widely used to reinforce existing concrete structures.

3.4 Anchorage system

The successful development of anchorage systems for prestressing tendons is the basis for the application of prestressed concrete in engineering. Since the first practical anchorage system was devised about a century ago, many anchorage systems, such as Freyssinet, Dywidag, VSL, BBRV, and OVM, have been developed, and new systems are being developed.

Figure 3.3 Workers arranging the ducts for tendons in a box girder in a prefabrication plant.

Figure 3.4 Workers arranging the ducts for tendons in the web of a continuous box girder.

The function of the anchorage system is to transmit the prestressing from the pre-stressing tendons to the concrete reliably and safely. The anchorage system can be divided into two groups, permanent and temporary systems, the former is embedded or connected with the post-tensioned members permanently and the latter is used to temporarily anchor the tendons during the construction of pretensioned members. In the post-tensioned members, the anchorage used in the stretching side is called the tensioning-end anchorage, while the anchorage used on the fixed side is called the fixing-end anchorage in which this end of the tendons is fixed firmly before the other end is stretched. The gripping device, also known as the tool anchorage, is used in the construction of pretensioned members to temporarily hold the prestressing tendons in the anchorage fixed with the abutment. The coupler is a device to connect the segmental prestressing tendons into a long continuous tendon, which are commonly used in the continuous beams with long spans constructed by the segmental cantilever method.

In a typical tensioning-end anchorage system, a steel anchor plate or anchor ring and corresponding clips, wedges, nuts, or plugs are included at least, and a steel anchor backing plate (also called a bearing plate) with a through-hole inside is also commonly included. In a fixing-end anchorage system, except by looping the wires around the concrete to achieve the anchorage, a steel anchor plate or bearing plate is required at least. For the developed anchorage systems, the anchoring principle can be roughly divided into four types:

(1) Wedge action type—the prestressing steel and fiber-reinforced polymer (FRP) wires and strands are anchored by frictional grip by the clips, wedges, or tapered plugs in the steel anchor plate or anchor ring.

(2) End bearing type—the prestressing steel wires and threaded bars are anchored in the steel bearing plate by button heads or nuts at the end of the tendons.

(3) Wrapping type—one end of the prestressing steel strands with embossed flowers is anchored in the concrete.

(4) Adhesive type—the prestressing FRP wires, strands, and plates are anchored by the adhesive force between the tendons and the filling adhesive material.

3.4.1 Performance of anchorage system

An anchorage system, consisting of an anchorage and grips, sometimes including couplers, is required to bear reliable anchoring performance, such as sufficient bearing capacity and good applicability. The performance of an anchorage system under static loading should be illustrated by the anchorage efficiency coefficient measured from the static loading test of the prestressing tendon—anchorage assembly, and the total elongation of the prestressing tendons corresponding to the ultimate tension resistance.

3.4.1.1 Efficiency coefficient and elongation of the prestressing tendon—anchorage assembly

The efficiency coefficient of the prestressing tendon—anchorage assembly is defined as

$$\eta_a = \frac{F_{Tu}}{n \times F_{pm}} \tag{3.1}$$

where η_a = efficiency coefficient of the tendon—anchorage assembly.

n = number of prestressing tendons in the tendon—anchorage assembly.

F_{Tu} = measured ultimate tensile forces in the tendon—anchorage assembly.

F_{pm} = measured average ultimate tensile force of single tendon.

f_{pm} = measured average ultimate tensile strength of tendons.

A_p = nominal cross-sectional area of single tendon.

In the *Anchorage, grip, and coupler for prestressing tendons* (GBT 14370-2015), the efficiency coefficient and total elongation of the tendon—anchorage assembly should satisfy [1]

$$\eta_a \geqslant 0.95 \tag{3.2}$$

$$\varepsilon_{Tu} \geqslant 2.0\% \tag{3.3}$$

where ε_{Tu} = total elongation of the prestressing tendons corresponding to the ultimate tension resistance.

The prestressing tendon—anchorage assembly needs to pass the performance test by fatigue loading, where the maximum stress in prestressing tendons should be 65% of the characteristic tensile strength and the fatigue stress amplitude should not be less than 80 MPa. After two million repeated loadings, the anchorage should not be subject to fatigue failure and the area of the prestressing tendons that shows fatigue failure due to gripping should not be less than 5% of the total area of the prestressing tendons in a tendon—anchorage assembly. For post-tensioned structures that antiseismic properties are required, the tendon—anchorage assembly should also pass the cyclic loading test with a cycle number of 50 times. After 50 loading cycles, the prestressing tendons in the gripping region do not break.

The efficiency coefficient of the gripping device is defined as

$$\eta_g = \frac{F_{Tu}}{F_{ptk}} \tag{3.4}$$

where F_{Tu} = measured ultimate tensile force in the prestressing tendon—grip assembly.

F_{ptk} = nominal ultimate tensile force of the prestressing tendons.

The efficiency coefficient of the gripping device should satisfy

$$\eta_g \geqslant 0.95 \tag{3.5}$$

3.4.1.2 Efficiency coefficient and elongation of the prestressing FRP—anchorage assembly

The efficiency coefficient of the prestressing FRP—anchorage assembly is defined as

$$\eta_a = \frac{F_{Tu}}{F_{ptk}} \tag{3.6}$$

where η_a = efficiency coefficient of the prestressing FRP—anchorage assembly.

F_{Tu} = measured ultimate tensile force in the prestressing FRP—anchorage assembly.

F_{ptk} = measured average ultimate tensile force of single prestressing FRP, $F_{ptk} = f_{ptk}A_{pk}$.

f_{ptk} = characteristic tensile strength of prestressing FRP.

A_{pk} = nominal cross-sectional area of the prestressing FRP in a prestressing tendon—anchorage assembly.

In the GBT 14370-2015 code, the efficiency coefficient and total elongation of the prestressing FRP—anchorage assembly should satisfy

$$\eta_a \geq 0.90 \tag{3.7}$$

$$\varepsilon_{Tu} \geq 2.0\% \tag{3.8}$$

3.4.2 Anchorage system for prestressing steels

3.4.2.1 Wedge-type anchorage

The wedge-type anchorages for prestressing steels used both as tensioning-end and fixing-end anchorages can be subdivided into clip-type and conical plug-type anchorages.

A typical clip-type anchorage system for prestressing steel strands is composed of a steel anchor plate or anchor ring, clips, and a steel anchor backing plate with a through-hole inside, as shown in Fig. 3.5. The corrugated duct is connected to the end of the backing plate, and a continuous spiral reinforcement with a certain length around the backing plate and corrugated duct is arranged to restrain the radial

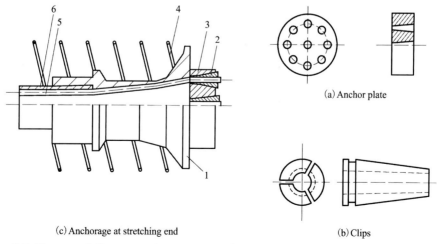

Figure 3.5 Diagram of clip-type anchorage system for strands: (1) anchor backing plate; (2) clip; (3) anchorage plate; (4) spiral reinforcement; (5) strands; (6) corrugated duct.

deformation of concrete under high local compression caused by the anchor plate subjected to the concentrated prestressing force.

When the jacking force is released, the strands retract and bring the clips to draw in the anchor plate or anchor ring, then the clips are wedged with the strands into the holes tightly to achieve anchoring of the strands. A clip consists of two or three pieces, as shown in Fig. 3.6, in which many tooth dents are made inside to improve the gripping force for strands.

A clip-type anchorage can be used to anchor a single strand or a bundle of strands at the tensioning end, as shown in Fig. 3.7. In a thin concrete structure, such as a flange of girder and thin plate, a flat anchorage with the strands arranged in a row is commonly used, as shown in Fig. 3.7C. When a single strand is anchored, a steel anchor ring rather than an anchor plate is used, as shown in Fig. 3.7A.

The conical anchorage consists of an anchor ring and a tapered plug, as shown in Figs. 3.8 and 3.9, which can be used to anchor $\phi5$ and $\phi7$ steel wires and strands. This type was devised first by Freyssinet in 1939, so it is also known as an F-anchorage.

(a) A clip with two pieces   (b) A clip with three pieces

Figure 3.6 Structure of a clip.

(a) Anchorage for a single strand   (b) Anchorage for a group of strands   (c) Flat anchorage

Figure 3.7 Clip-type anchorage system.

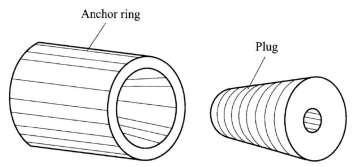

Figure 3.8 Diagram of the conical anchor structure.

Figure 3.9 Conical anchor.

The wires or strands retract to drive the plug back into the anchor ring once the jacking force is released, then the tendons are anchored tightly. To ensure that the driving force due to friction between the tendons and the plugs is greater than the friction resistance between the tendons and the interior walls of the hole in the anchor ring, many tooth dents on the outer surface of the cone plug are made (Fig. 3.9).

Anchor rings and plugs are usually made of high-quality carbon steel No. 45. Some manufacturers use 20Cr or other steels with a higher strength to make plugs. The anchor plates are commonly made of Q235 steel.

The cone plug anchor is simple in structure, low price, and convenient to construct, but its disadvantage is that the tendons have a large amount of retraction, resulting in large prestress loss.

3.4.2.2 End bearing anchorage

As a threaded steel bar passes through a steel anchor plate, it can be fixed on the plate by a nut at the end, as shown in Fig. 3.10. If one end of each wire is upset to form a rivet head at natural ambient temperature, the wires in a group passing through the anchor plate can also be fixed on the plate by the end rivet heads, as shown in Figs. 3.11 and 3.12. The end nut or cold-formed rivet head and the anchor plate make up an end-bearing anchorage system which is used at tensioning and fixing ends in post-tensioned concrete. The nut-bearing anchorage, known as the Dywidag system, has been used worldwide since 1965. The rivet-head-bearing anchorage, known as the BBRV system, has also been used widely since 1949.

The screw-threads can be rolled out on the surface of a coarse steel bar over the entire length, in which the bar can be cut off and fixed by a nut at any point, or connected by a coupler to form a long bar, as shown in Fig. 3.13A. Alternatively, the fine threads can be cut by the machine in the vicinity of the end of the coarse steel bar, by which the bar is fixed on the anchor plate with a nut, as shown in Fig. 3.13B. For the machine-threaded

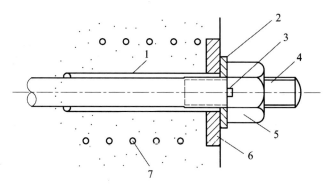

Figure 3.10 Diagram of the screw end rod anchor: (1) reserved hole; (2) washer; (3) row groove; (4) threaded steel bar; (5) anchor nut; (6) bearing plate; (7) spiral reinforcement.

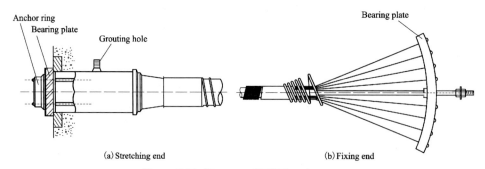

(a) Stretching end   (b) Fixing end

Figure 3.11 Diagram of BBRV anchorage.

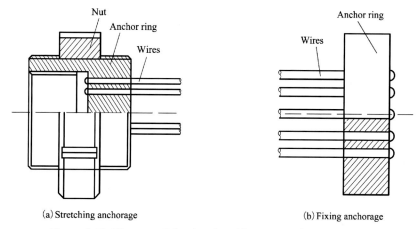

(a) Stretching anchorage   (b) Fixing anchorage

Figure 3.12 Diagram of the rivet-head-bearing anchorage system.

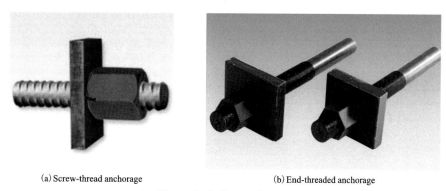

(a) Screw-thread anchorage   (b) End-threaded anchorage

Figure 3.13 Nut anchorage.

coarse bars, the length and fine threads at two ends should be cut accurately in the factory, which not only requires sufficient dimensional accuracy during the construction of the concrete structure to avoid inability when installing bolts but also causes inconvenience in the transportation of the long steel bars to the construction site from the processing site, so screw-thread coarse steel bars with high strength are preferred.

The structure of the anchorage system for anchoring a group of riveted wires at the tensioning end and fixing end is different, where an adjustable nut is needed in the tensioning end to fix the wires after jacking, as shown in Figs. 3.12A and 3.14. To make the stress distribution under the anchorage more uniform, the steel wire can be arranged separately, as shown in Fig. 3.11B.

(a) End-rivet anchorage system   (b) End-rivet anchorage at tensioning end

Figure 3.14 End-rivet anchorage system.

3.4.2.3 Extruding anchorage

Wedge-type anchorage for strands at the fixing end is broadly used, as shown in Figs. 3.5 and 3.7. In addition, extruding anchorage, also called H-type anchorage can be effectively used to anchor the strands at the fixing end, as shown in Fig. 3.15. To protect the steel sockets while casting concrete, a steel plate connected with the anchor plate is commonly used to cover the sockets.

After a strand wound with steel wire lining cloth outside is placed into a steel socket, the strand and steel socket are extruded together tightly when they are put into a squeezing machine, then an extruded anchor-head is fabricated, as shown in Figs. 3.15B and C. The strands can be easily anchored on the bearing plate by the extruded anchor-heads, so the extruding anchorage system can be treated as a kind of end-bearing anchorage system.

3.4.2.4 Wrapping anchorage

If the ends of strands can be distributed in a scattered manner in the concrete, the strands with cold-formed flower, bulb, or pear shapes in the ends can be directly anchored in the concrete, as shown in Figs. 3.16 and 3.17. The flower shapes can be produced by a portable squeezing machine on the construction site. The processed strands are fixed

(a) Anchor ring   (b) Flat extruding anchorage   (c) Circular extruding anchorage

Figure 3.15 Extruding anchorage system.

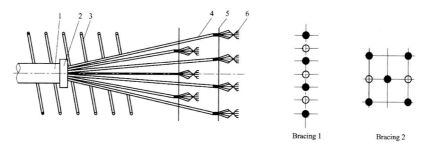

Figure 3.16 Diagram of flower anchorage: (1) duct; (2) restrain ring; (3) spiral reinforcement; (4) strands; (5) bracing; (6) bulb or flower portion.

Figure 3.17 Flower anchorage system.

in the correct position before casting concrete, then they are stretched on the other side when the concrete attains sufficient strength.

The strands are anchored by wrapping the flower-shaped strand end with surrounding concrete, and so it is called the wrapping anchor, or the flower anchor.

3.4.2.5 Coupler

Sometimes, several segments are required to connect into a long tendon using a coupler or numerous couplers. The typical couplers for strands are shown in Fig. 3.18, and can be divided into flat and circular types. A coupler plays a role both as a tensioning-end anchorage and a fixing-end anchorage, so a typical coupler should be composed of a steel anchor plate, numerous clips, and extruded anchor-heads in the subsequently connected strands. When a large bonded prestressed concrete continuous beam is constructed by the segmental method, the cement mortar must be grouted in the duct of previously anchored strands before stretching the subsequent strands, as shown in Fig. 3.18D.

3.4.3 Anchorage systems for prestressing FRP

The anchorage systems for prestressing FRP are also distinguished as tensioning-end and fixing-end anchorage systems. The commonly used prestressing FRP is made into wires, strands, rods, and narrow plates, which require different anchorage systems.

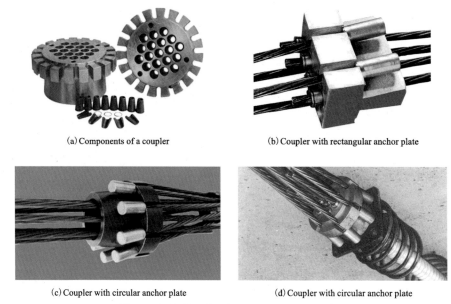

(a) Components of a coupler   (b) Coupler with rectangular anchor plate

(c) Coupler with circular anchor plate   (d) Coupler with circular anchor plate

Figure 3.18 Coupler for prestressing strands.

For the developed anchorage systems for prestressing FRP, the anchor can be achieved by mechanical wedge action, adhesive force, or a combination of the two. In a wedge anchorage system, FRP rods, wires, and narrow plates are anchored by wedges in the anchor hole as for prestressing steels. A wedge anchorage system consists of an anchor plate or ring with a through-hole, several pieces of clips, and a bearing plate that can be used to anchor FRP wires and narrow plates, as shown in Figs. 3.19 and 3.20. To avoid fracturing due to the low shear strength of prestressing FRP tendons under high tension and inclined compression, the slope in the inside cone should be gentle.

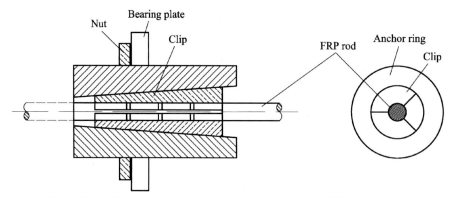

Figure 3.19 Mechanical clamping anchorage for a single FRP rod or wire.

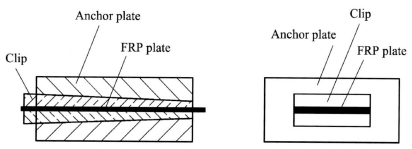

Figure 3.20 Mechanical clamping anchorage for a single FRP plate.

The clips in the mechanical anchorage system can be replaced by an adhesive material to form the adhesive anchorage system, as shown in Fig. 3.21.

Although Fig. 3.20 shows the anchorage system for prestressing FRP narrow plates, the FRP plate cannot be stretched directly, so a corresponding stretching system should be employed. Fig. 3.22 demonstrates a typical anchorage system for FRP narrow plates, both at the fixing end and stretching end. Before stretching, the FRP plate has been fixed at both ends. As the pump system starts work, the fixed FRP plate is elongated along two threaded steel rods by the movement of the jack ram, and two anchor nuts are tightened when the jacking force reaches the design value.

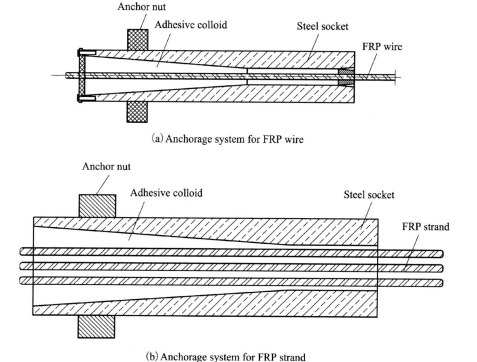

(a) Anchorage system for FRP wire

(b) Anchorage system for FRP strand

Figure 3.21 Adhesive anchorage system.

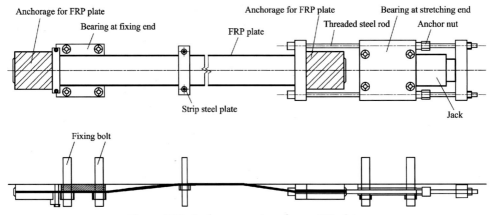

Figure 3.22 Anchorage system for an FRP plate.

3.5 Jacks for stretching tendons

The hydraulic jacks along with the associated oil pump system are used to stretch the pre-stressing tendons. The double-acting jacks with a through core are most widely used for a bundle of wires and strands (Fig. 3.23). Before the stressing operation, the wires and strands are temporarily anchored in a steel plate over the hydraulic ram, and the jack head is held against the bearing plate in concrete. As the pump works, the ram drives the temporary anchor plate forward (Fig. 3.24). When the data shown in the oil gauge reach the design value of jacking force and the elongation of the tendons achieves the predicted value, the pump stops charging oil. Once the oil pressure is released, the ram draws in and the stressing tendons retract, and the clips are driven into the anchor hole tightly, thus the anchor for the strands is achieved.

Figure 3.23 Double-acting jack with through core.

Figure 3.24 Temporary anchor plate moves with ram.

An intelligent tensioning system has been developed and is being used increasingly widely. The system is mainly composed of a jack, data control system, oil–pump control system, touch screen, pressure sensor, and displacement sensor. The internal oil pressure of the jack measured by the pressure sensor and the elongation of the prestressing tendons detected by the displacement sensor are taken as the feedback parameters to automatically adjust the tension in the stretching process by the control systems. Fig. 3.25 shows stretching of the tendons by the intelligent tensioning system.

If a single strand or a bundle of wires needs to be stretched, the portable jack that the temporary anchorage locates at the front (Fig. 3.26) can be used. This head–anchor jack is widely employed to stretch a single strand in the group anchorages one by one, checking whether a specific strand is tensioned enough, conducting supplementary tensioning for a specific strand or releasing some anchored strands.

Figure 3.25 Workers operating the intelligent tensioning system to stretch the tendons.

Figure 3.26 Head-anchor jack.

Figure 3.27 Pull rod jack.

A piston rod is used as a tension element in the pull rod jack (Fig. 3.27), which is used for stretching threaded steel bars. As the rod end inside the jack is connected with the steel bar and the jack head is placed against the concrete, the bar is elongated by the piston rod when the jack works. The pull rod jack is lightweight, portable, and easy to be operated, therefore it is widely used in construction sites.

3.6 Duct grouting and anchorage sealing in post-tensioned concrete

When the prestressing tendons are anchored in post-tensioned concrete, the grouting for ducts should be completed as quickly as possible. The grouting bears several important functions, such as:

(1) Filling the space in the duct to avoid accumulation of water freezing.
(2) Protecting the prestressing steels against corrosion.
(3) Making the concrete cross-section into a whole integrity and increasing the section stiffness.
(4) Providing an effective bond between the prestressing steels and the concrete.

The most commonly used material for grouting is cement mortar, in which the water—cement ratio of the mortar is less than that of concrete, and this ratio should not exceed 0.33 in railway bridges, as prescribed in the *Technical specification of cable grouts on post-prestressed concrete railway girder* (QCR 409-2017) [2]. In order to protect prestressing steel against corrosion, the mortar must be alkaline and have low permeability and low shrinkage, good workability, and sufficient compressive strength and bond strength.

The mouth for grouting is located at the steel anchor plate, as shown in Fig. 3.28. An intelligent grouting system, including a mortar production system (Fig. 3.29), has been developed and is widely used.

Generally, an appropriate amount of admixtures, such as water-reducing agents, retarders, air-entraining agents, corrosion inhibitors, and microexpansion agents, should be blended into cement mortar. The water-reducing agent and retarder help to improve the fluidity of the mortar under a given water—cement ratio, reduce the bleeding water, prevent segregation under high pressure during grouting, and delay the consolidation of

Figure 3.28 Mouth for grouting.

(a) Cement mortar is grouting

(b) Intelligent operation system and mortar production system

Figure 3.29 Cement mortar grouting using an intelligent grouting system.

Table 3.1 Required performance of cement mortar.

Serial number	Items for testing		Requirements
1	Setting time/h	Initial setting	$\geqslant 4$
2		Final setting	$\leqslant 24$
3	Fluidity	Fluidity from machine	18 ± 4
4		Fluidity in 30 min	$\leqslant 28$
5	Bleeding rate	Free bleeding rate	0
6		Bleeding rate in 3 h	$\leqslant 0.1\%$
7	Bleeding rate under pressure	0.22 MPa (the vertical height of duct $\leqslant 1.8$ m)	$\leqslant 3.5\%$
8		0.36 MPa (the vertical height of duct >1.8 m)	
9	Filling degree		Pass
10	Strength at 7 days/MPa	Antiflexural strength	$\geqslant 6.5$
11		Compressive strength	$\geqslant 35$
12	Strength at 28 days/MPa	Antiflexural strength	$\geqslant 10$
13		Compressive strength	$\geqslant 50$
14	Free expansion rate (24 h)		0%–3%
15	Air content		2%–4%
16	Chloride ion content		0.6%

mortar. The air-entraining agent can improve the antifreezing capacity during consolidation. The expansion agent causes the cement mortar to expand slightly to compensate for the shrinkage and to ensure that the ducts are dense.

The performance of cement mortar for grouting should conform to the stipulations in QCR 409-2017, which are compiled in Table 3.1. The compressive strength and flexural strength of the mortar are determined by the test of standard specimens under standard curing.

The *Specifications for design of highway reinforced concrete and prestressed concrete bridges and culverts* (JTG 3362-2018) prescribes that the measured compressive strength of cement mortar for grouting should not be less than 50 MPa for a specimen with a dimension of 40 mm × 40 mm × 160 mm under 28 days standard curing [3].

Before grouting, the compressed air is input to check whether the duct is blocked or leaking, if necessary the water can be used to wash the duct in hot weather, in this case, the effect of water accumulated in the duct on the cement mortar should be taken into account. The grouting must be completed continuously for a duct, and the vacuum-assisted grouting process may be used for a long duct. As for grouting velocity, the lower value should be adopted for vertical ducts and the higher value can be taken for long or large ducts, and sometimes the higher value may be required in hot climates.

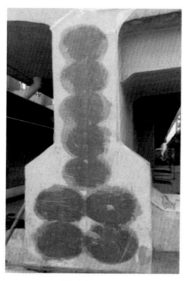

(a) Before sealing   (b) After sealing

Figure 3.30 Before and after sealing for anchorages.

For post-tensioned concrete structures, the anchorages embedded in the concrete must be enclosed by concrete, and the structural reinforcement is required to be arranged around the anchorages to connect with the reinforcement in the structure after the prestressing tendons are anchored, this process is called anchorage sealing (Fig. 3.30). The strength of concrete for sealing should be greater than 80% of the concrete strength of the structure.

References

[1] GBT 14370: 2015. Anchorage, grip, and coupler for prestressing tendons. Beijing: China Standards Press; 2015.
[2] QCR 409: 2017. Technical specification of cable grouts on post-prestressed concrete railway girder. Beijing: China Railway Publishing House; 2017.
[3] JTG 3362: 2018. Specifications for design of highway reinforced concrete and prestressed concrete bridges and culverts. Beijing: People's Communications Press; 2018.

CHAPTER 4

The strategy of analysis and design

Contents

4.1 Knowledge system of prestressed concrete

To carry out a complete analysis for a prestressed concrete structure, the following information should be known first: material properties, anchorage systems, stressing technology, concrete curing methods, structural systems, ambient environment, service loads (actions and combinations), time of loading, etc. When designing, the complete analysis for the preliminary design scheme is required to be fulfilled first, then the specifications adopted to verify the derived design parameters to obtain a safe and functional prestressed concrete structure, all of which compose the knowledge system of prestressed concrete, as illustrated briefly in Fig. 4.1.

4.2 Analysis of prestressed concrete structures

4.2.1 Contents of analysis

One of the main purposes of analysis is to guarantee a prestressed concrete structure that is in the required states of safety, functionality, and durability. Generally, safety can be ensured by sufficient strength, rigidity, and stability, which can be achieved by verifying the required indexes at the ultimate limit state. A structure is built to meet specific functions. Take a prestressed concrete railway beam as an example, the flexural rigidity should

Analysis and Design of Prestressed Concrete
ISBN 978-0-12-824425-8, https://doi.org/10.1016/B978-0-12-824425-8.00004-X

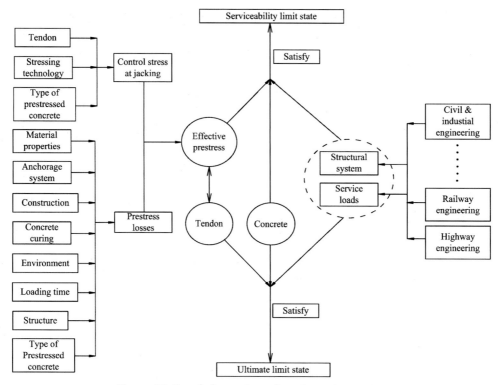

Figure 4.1 Knowledge system of prestressed concrete.

be sufficient to control the midspan deflection due to train loads to a limited range to keep the rails at a required smoothness, which not only affects the comfort of passengers but is also related to safe operation. The durability can be ensured by having sufficient crack resistance or limiting the crack width, and the high resistance ability of concrete and tendons to the attacks by the various ambient environments. Namely, the functionality and durability of a prestressed concrete structure can be accomplished by verifying the required indexes at the serviceability limit state. Therefore, the main contents of the analysis are to derive the design parameters required to verify the design indexes at the ultimate limit state and the serviceability limit state.

When analysis of a prestressed concrete structure is carried out, the following parameters should be obtained.

At the ultimate limit state:

(1) Flexural bearing capacity of the normal sections (flexural structures).
(2) Shear bearing capacity of the oblique sections (flexural structures).
(3) Tension or compression bearing capacity (tension or compression members).
(4) Torsion bearing capacity (torsion members).

(5) Overall stability (slender columns and flexural members at stretching tendons or transfer).

(6) Local compression bearing capacity (in the anchorage zone).

(7) Local tension bearing capacity (in the overall region of the anchorage zone).

At the serviceability limit state:

(1) Short-term deflection (flexural structures).

(2) Long-term deflection (flexural structures).

(3) Rotation angle of beam end (railway beams).

(4) Crack width (flexural, tension, and eccentric compression members).

(5) Natural frequency of beams (railway beams).

Crack resistance at the serviceability limit state:

(1) Normal stress in concrete.

(2) Shear stress in concrete.

(3) Principal stress in concrete.

(4) Stress change in outermost reinforcement in the tension zone from decompression state to critical cracking state (flexural structures).

Safety of prestressing tendons after stretching:

Tensile stress in tendons.

At the fatigue limit state:

(1) Fatigue stress of concrete (beams under repeated loading).

(2) Fatigue stress of prestressing tendons (beams under repeated loading).

(3) Fatigue stress of reinforcing steels (beams under repeated loading).

4.2.2 Stress analysis based on the knowledge system

The first and most important thing that needs to be addressed in the analysis is to calculate exactly the values of the existing prestressing force in the structure after the transfer, which is the foundation to derive the exact values of the design parameters, such as the stress in concrete for verifying the crack resistance and the midspan deflection under live loads for verifying the flexural rigidity. However, the prestressing force provided by the tendons depends on the tendon stresses which change from jacking to the entire service period. Now we know that almost all the factors related to the prestressed concrete structures, such as material properties (concrete creep and shrinkage, tendon relaxation), stressing methods, curing methods for concrete, reinforcement ratio in sections, time of loading (time of stretching the tendons and applying the deck systems), ambient humility, and prestressed concrete type (bonded and unbonded systems), affect the development and magnitude of the stress reduction in prestressing tendons. Therefore, the existing stress in prestressing tendons at any time under consideration, known as the effective stress or effective prestress, should be calculated from the moment of jacking to the service period step by step based on the knowledge system shown in Fig. 4.1.

As the effective stress in prestressing tendons is achieved, the prestressing force is obtained. By treating the prestressing force as a special external force exerted on the concrete structure, the stress analysis at service loads at different stages can be carried out using the superposition principle.

4.2.3 General strategy for analysis

4.2.3.1 The functions of the prestressing tendons

In prestressed concrete structures, the prestressing tendons play two important roles. First, the prestressing force is provided by the elongated tendons, which determines the short-term and long-term performance of the concrete structures at the serviceability limit state. Second, at the ultimate strength limit state, the prestressing tendons in the tension zone, same as the nonprestressing steels, and the concrete in compression provide the flexural bearing capacity. In the oblique sections, the bent-up prestressing tendons, transverse prestressing tendons, and concrete provide shear resistance. Therefore, the optimized quantity and layout of prestressing tendons should be determined on the premise of providing sufficient prestressing force and contributing enough bearing capacity in the case of the lowest cost.

In short, the prestressing force provided by the prestressing tendons can be treated as a kind of external force exerted on the concrete structure at the serviceability limit state, while the prestressing tendons participate to resist the design load at the ultimate limit state.

4.2.3.2 Characteristics in stress analysis in prestressed concrete structures

There are four obvious characteristics in stress analysis in prestressed concrete structures compared with reinforced concrete structures.

First, the tensile stress in prestressing tendons changes with time since the tendons are stretched. The creep and shrinkage of concrete, and relaxation of tendons inevitably lead to tensile stress reduction in tendons. Accordingly, the prestressing force exerted on the concrete structures changes with time. It remains difficult to obtain the exact time-dependent stresses in tendons at any time under consideration, however, fortunately, this is not necessary for general structures in most cases. The approach of calculating stress in tendons provided in the present design codes, such as *Code for design on railway bridge and culvert* (Q/CR 9300-2018) and *Specifications for design of highway reinforced concrete and prestressed concrete bridges and culverts* (JTG 3362-2018), is a simplified method.

Second, for post-tensioned structures, the stresses in prestressing tendons over their entire lengths are different due to friction between the tendons and the duct, and the stresses in different tendons with different profiles have a significant difference. When the time-dependent effects on the tendon stress due to concrete creep and shrinkage are taken into account, it is very complicated and tedious to derive the exact values of tendon stress and the corresponding prestressing force at any section at any time by hand, especially for long-span highway and railway beams. Generally, the design codes present simplified approaches.

Third, the position of maximum tensile stresses of concrete during construction and in the service period may be different. Considering a post-tensioned simply supported highway beam, it generates upward camber (see Fig. 3.2) at transfer. In this case, the maximum tensile stress of concrete occurs in the top edge fiber at the midspan section caused by the prestressing force and the self-weight of the beam, while the maximum tensile stress of concrete occurs in the bottom edge fiber when all design loads are applied in the service period. Therefore, the tensile concrete stresses are required to be calculated and verified in different positions at different critical sections at different stages.

Finally, the sectional properties of post-tensioned and pretensioned structures in the calculations are different. At transfer, the converted section to concrete should be used in the calculations in the pretensioned structures since the tendons and the concrete have been bonded together to bear the prestressing force. In the post-tensioned structures, the net section in which the duct area is subtracted from the whole cross-section should be taken since the tendons and the concrete do not bond together, and the net concrete section bears the prestressing force only. This difference requires different formulas to be established for post-tensioned and pretensioned structures in the analysis, respectively.

4.2.3.3 Characteristics in the calculations of deflection of prestressed concrete members

There are three obvious characteristics in the calculations of deflection of prestressed concrete flexural members.

First, the member exhibits an upward camber during stretching tendons or at transfer. Because the eccentric prestressing tendons are designed to resist the design load (including the self-weight of the member, the subsequently applied second weight such as deck structure, the live load, and other actions), the member will inevitably generate a camber when all tendons are stretched or at transfer. This camber should be calculated exactly since it significantly affects the accuracy in predicting the long-term camber or deflection due to concrete creep and shrinkage and tendon relaxation.

Second, for post-tensioned and pretensioned members at transfer, the sectional properties and the effective stress in prestressing tendons used to calculate the cambers are different. As discussed above, the net sectional properties should be used for post-tensioned members, while converted sectional properties are needed for pretensioned members. Considering the characteristics of stress distribution over the entire length of the tendons, even for a simply supported prestressed concrete beam in the railway or highway bridge, the deflection calculation is also very complex.

Third, when all the tendons are stretched and anchored, for the flexural member, whether in a statically determinate system or a statically indeterminate system, there is a tendency for an upward camber. After several years of service, it is found that the camber in a statically determinate system continues to develop due to concrete creep and shrinkage. In contrast, the camber in a statically indeterminate system with long spans possibly turns into downward deflection gradually, and the deflection continues to develop later. One of the important reasons for this phenomenon is the proportion of the self-weight in the total design load. Thus, the simplified analysis or formulas for long-term deflection in a statically determinate or indeterminate system should be distinguished.

4.2.3.4 Some professional terms

In a prestressed concrete structure, both the design load and the prestressing force provided by the prestressing tendons cause section stresses and deflections. In most cases, these generated stresses and deflections have opposite directions, such as in the tension and flexural members; however, they may have the same directions, such as in the small eccentric compression members, which makes the description lengthy or repetitive, and sometimes confusing. Therefore, it is necessary to define some commonly used terms.

Tension zone and compression zone

The tension or compression zone is that portion of a prestressed concrete member where flexural tension or compression would occur under service loads if the prestressing force was not present. The strict distinction between tension and compression zones depends on the load and sectional properties, while these parameters cannot be accurately determined before analysis. Therefore, in most cases, the tension or compression zone generally refers to the region on the side of tension or compression due to the service loads.

Camber and deflection

The camber refers to the upward vertical deformation of a flexural member in the horizontal state while the deflection can be used to describe the upward or downward vertical deformation by adding positive or negative sign. In this book, when the vertical deformation is determined to be upward, it is described by the camber rather than by the negative deflection or upward deformation. Otherwise, it is described as deflection whether the direction of the vertical deformation is downward or indeterminate.

Net section and converted section

For prestressed concrete structures, whether constructed by the pretensioning or post-tensioning method, the reinforcing steels (nonprestressing steels) and the concrete are bonded together at transfer. Thus, the net section in the post-tensioned members or structures at transfer is composed of the concrete and the reinforcing steels converted to concrete, in which the cross-sectional area of the ducts for prestressing tendons are subtracted. As for the converted section, it is composed of the concrete and all reinforcements (prestressing tendons and reinforcing steels) converted to concrete.

Centroid and center of gravity of the section

A concrete section or a converted section can be treated as a homogeneous section with an identical density, so the center of gravity of the section coincides with the centroid of the section. For easy description in the drawings, the abbreviation "c.g.c," representing the center of gravity of the cross-section of the concrete or converted concrete, is used in this book.

4.2.3.5 Sign conventions

In the calculations of the bearing capacity at the ultimate limit state, both the tensile strength and compressive strength of prestressing tendons, reinforcing steels, and concretes are positive.

In the calculations at the serviceability limit state, the following sign conventions are commonly adopted:

Stress in prestressing tendons: positive (+) for tension, negative (−) for compression.

Stress in reinforcing steels: positive (+) for tension, negative (−) for compression.

Stress in concrete: positive (+) for compression, negative (−) for tension.

Bending moment in section: positive (+) for bending moment resulting in downward deflection, negative (−) for bending moment resulting in camber. However, in stress analysis, the bending moment generated by the eccentric prestressing tendons is always taken positive (+) when it produces compression in the tension zone.

Deflection: positive (+) for downward deformation for horizontal flexural members, negative (−) for upward deformation.

Eccentricity of the prestressing tendons: positive (+) when measured downward from the centroid of the concrete cross-section, negative (−) when measured upward from the centroid of the concrete cross-section.

4.2.3.6 Units

The International System of Units (SI) is adopted in this book, and the default length unit in the drawings is mm. When describing the prestressing tendons and the nonprestressing steels, the default unit for diameter and radius is mm, and the symbol ϕ stands for the diameter. For example, a prestressing wire with a diameter of 5 mm is abbreviated as "a $\phi5$ wire," and a prestressing strand composed of seven wires with a diameter of 5 mm for each wire is abbreviated as "a $7\phi5$ strand." A prestressing strand with a diameter of 15.2 mm, composed of seven wires with a diameter of 5 mm for each wire, also can be written as "a $\phi^S15.2$ strand," where the superscript "s" represents the strand. In this case, "a $7\phi5$ strand" and "a $\phi^S15.2$ strand" have the same diameter.

4.3 General issues for the design of prestressed concrete structures

4.3.1 Design objectives

In general, the design objectives for a new structure include the following three categories:

(1) Meet the basic functionality required by the investors.
(2) Meet the requirements of society.
(3) Meet the provisions in the relevant technical codes or specifications.

The basic functionality required by the investors at least includes the following aspects:
(1) The usages of the structure under certain environmental conditions.
(2) Economics.
(3) Special requirements.

According to the usages of the structure, the required relevant design codes or specifications in design, action types, environmental category, impact level on structure, etc., are determined. The special requirements may involve some particular demands, such as sustainability and maintainability, or ease in extending for future new usage.

Social requirements involve the architectural form and aesthetic appearance of the structure which are coordinated with the surrounding environment, decreasing the impact on the land and surrounding areas, and reducing the impact on nearby residents during service (such as the noise emitted by high-speed trains), etc.

Once the relevant design codes along with the estimated design actions are determined, an optimized preliminary design considering the economics and social requirements can be achieved, then the design parameters or indexes listed in Section 4.2.1 can be calculated and verified to achieve sufficient safety, functionality, and durability.

4.3.2 Actions and combinations

The types of actions and their combinations that may be applied to the structure should be determined before design according to the usage of the structure. When the limit state method is used for design, the typical combinations of actions are as follows:
(1) Fundamental combinations of actions (or characteristic combinations of actions)

$$S_d = \gamma_0 S \left(\sum_{i=1}^{n} \gamma_{Gi} G_{ik} + \gamma_{Q1} Q_{1k} + \sum_{j=2}^{m} \gamma_{Qj} \psi_{cj} Q_{jk} \right) \tag{4.1}$$

(2) Accidental combinations of actions

$$S_d = S \left(\sum_{i=1}^{n} G_{ik} + A_d + \left(\psi_{f1} \text{ or } \psi_{q1} \right) Q_{1k} + \sum_{j=2}^{m} \psi_{qj} Q_{jk} \right) \tag{4.2}$$

(3) Seismic combinations of actions

$$S_d = S\left(\sum_{i=1}^{n} G_{ik} + \gamma_I A_{Ek} + \sum_{j=1}^{m} \psi_{qj} Q_{jk}\right) \tag{4.3}$$

(4) Frequent combinations of actions

$$S_d = \sum_{i=1}^{n} G_{ik} + \psi_{f1} Q_{1k} + \sum_{j=2}^{m} \psi_{qj} Q_{jk} \tag{4.4}$$

(5) Quasi-permanent combinations of actions

$$S_d = \sum_{i=1}^{n} G_{ik} + \sum_{j=1}^{m} \psi_{qj} Q_{jk} \tag{4.5}$$

where S_d = design action-effect.

γ_0 = importance coefficient of the structure.

$S()$ = function of design action-effect, where $\sum$ and "+" represent a linear combination.

γ_{Gi} = ith partial factor of the permanent action.

G_{ik} = characteristic value of the ith permanent action.

γ_{Q1} = factor of dominant variable action.

Q_{1k} = characteristic value of dominant variable action.

γ_{Qj} = partial factor of the jth variable action.

ψ_{cj} = jth combination factor of the variable action, 1.0 is taken generally.

Q_{jk} = characteristic value of the jth variable action.

n = number of the permanent action participating in the combination.

m = number of the variable action participating in the combination.

A_d = design value of the accidental action.

ψ_{q1} = factor of dominant quasi-permanent action, 1.0 is taken generally.

ψ_{qj} = factor of the jth quasi-permanent action, 1.0 is taken generally.

γ_I = importance coefficient of earthquake action.

A_{Ek} = characteristic value of the earthquake action.

At the ultimate limit state, the fundamental combinations are applied for persistent and transient situations, the accidental combinations are applied for accidental situations and the seismic combinations are applied for seismic situations. At the serviceability limit state, characteristic combinations, frequent combinations, and quasi–permanent combinations are adopted. In the calculations of the combinations, all actions possibly applying simultaneously on the structure should be included in the combination to find the most disadvantageous effects.

The action types, partial factors, and rules of combinations are specified in the relevant design codes. For prestressed concrete structures in a railway, the action types and their partial factors stipulated in the *Code for design on railway bridge and culvert* (Q/CR 9300–2018) are compiled in Tables 4.1 and 4.2, respectively.

Table 4.1 Types of action for a railway bridge.

No.	Type of action		Name of action
1.	Permanent action		Self-weight of the structure
2.			Other weights (weights of equipment, sidewalk, etc.)
3.			Prestressing force
4.			Concrete shrinkage and creep
5.			Uneven settlement in adjacent piers
6.			Soil pressure
7.			Hydrostatic pressure and buoyancy
8.	Variable action	Basic variable action	Trainload
9.			Road (urban road) load
10.			Vertical impact force due to running trains
11.			Centrifugal force due to running trains
12.			Lateral swing force due to running trains
13.			Soil pressure caused by trains
14.			Bending force
15.			Pedestrian load
16.			Aerodynamic force
17.		Other variable action	Train braking or traction
18.			Wind load
19.			Temperature action
20.			Contractility
21.			Frictional resistance of bearing
22.			The pressure of water flow
23.			Wave force
24.			Ice pressure
25.			Frost heaving pressure
26.			Construction load
27.	Accidental action		Train derailment load
28.			Rail breaking force
29.			Ship or raft impact force
30.			Vehicle impact force
31.	Seismic action		Seismic action

Table 4.2 Partial factors of the actions for a railway bridge.

| No. | Name of action | | Ultimate limit states Fundamental combinations | | | | | |
			I	II	III	IV	V	VI
1.	Self-weight		1.1 (1.2)					
2.	Additional weight	Ballast deck	1.4					1.2
		Other decks	1.1					
3.	Prestressing force		1.0 (1.35)					
4.	Effect of concrete shrinkage and creep		1.1					
5.	Uneven settlement effect		1.0					0.5
6.	Soil pressure		1.2					1.1
7.	Hydrostatic pressure and buoyancy		1.1					
8.	Trainload and dynamic action		1.5	—	1.2	1.2	1.2	—
9.	The centrifugal force of the train		1.5	—	1.2	1.2	1.2	—
10.	Lateral swing force of the train		1.5	—	1.2	1.2	1.2	—
11.	Soil pressure caused by train		1.5	—	1.2	1.2	1.2	—
12.	Wind load		—	1.4	1.1	—	0.75	0.75
13.	Temperature action	Uniform temperature	—	—	—	1.3	1.0	1.0
		Gradient temperature	—	—	—	1.0	0.8	0.8
14.	The pressure of water flow		—	—	0.8	1.0	0.8	1.1
15.	Ice pressure		—	—	1.1	1.1	1.1	1.0
16.	Construction load		—	—	—	—	—	1.15

4.3.3 Major steps of design

The major steps in designing a prestressed concrete member or structure are outlined in Fig. 4.2.

A structure is designed for a special purpose in a given site under a certain environment, and so the relevant codes and specifications used to guide the design are determined, and the major information for the design is also known, such as the design service life, action types, environmental categories, geological data, and the design parameters or indexes are required to be calculated and verified at the ultimate limit state and serviceability limit state (as listed in Section 4.2.1).

To start a design, the designers must call on design experiences, refer to the designs of similar structures, and use the above data and design specifications to obtain a preliminary design scheme, including the structural system, sectional dimensions, materials, and the corresponding prestressing system. Then, the structural analysis based on the preliminary scheme is carried out to derive the internal forces.

In the analysis, a set of formulas for calculating the design parameters (such as the concrete stress) and the sectional strength are established. When the requirements for the design parameters stipulated in the codes or specifications are introduced, the expressions

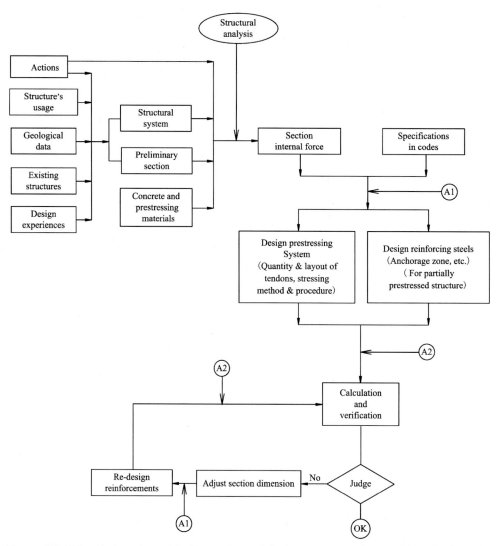

Figure 4.2 Major design steps. A1 - Expressions of design parameter by rewriting the formulas obtained in the analysis; A2 - Formulas in the analysis.

for estimating the required concrete cross-sectional dimensions or geometric properties, cross-sectional area and eccentricity of the prestressing tendons, and cross-sectional area of the reinforcing steels in the anchorage zone can be derived by rewriting the established formulas, which can be used to carry out the preliminary design. Then, the comprehensive analysis of the preliminary design scheme can be accomplished from construction to the service period, and the corresponding verification also can be fulfilled. If all the design parameters satisfy the requirements stipulated in the codes, the final design scheme is achieved. Otherwise, it is necessary to modify the cross-section of concrete or the quantity and layout of prestressing tendons or reinforcing steels and to carry out the calculations and verifications again, until the new prestressed concrete structure bears sufficient strength, rigidity, stability, durability, and predicted functionality to meet all the requirements stipulated in the codes.

Suggested readings

[1]   GB 50010: 2015. Code for design of concrete structures. Beijing: China Construction Industry Press; 2015.

[2]   Q/CR 9300: 2018. Code for design on railway bridge and culvert. Beijing: China Railway Publishing House; 2018.

[3]   TB10621: 2014. Code for design of high-speed railway bridge. Beijing: China Railway Publishing House; 2014.

[4]   JTG 3362: 2018. Specifications for design of highway reinforced concrete and prestressed concrete bridges and culverts. Beijing: People's Communications Press; 2018.

[5]   JGJ 369: 2016. Prestressed concrete structure design code. Beijing: China Construction Industry Press; 2016.

[6]   CCES 01: 2018. Durability design and construction of concrete structure. Beijing: China Construction Industry Press; 2018.

[7]   TB100005: 2010. Code for durability design on concrete structure of railway. Beijing: China Railway Publishing House; 2010.

[8]   Hu D. Basic principles of prestressed concrete structure design. 2nd ed. Beijing: China Railway Publishing House; 2019.

[9]   Naaman Antoine E. Prestressed concrete structure analysis and design. 2nd ed. New York: McGraw Hill; 2004.

[10]   Nawy Edward G. Prestressed concrete: a fundamental approach. 5th ed. Upper Saddle River: Prentice-Hall, Inc.; 2006.

CHAPTER 5

Calculation of effective stress in prestressing tendons

Contents

Analysis and Design of Prestressed Concrete
ISBN 978-0-12-824425-8, https://doi.org/10.1016/B978-0-12-824425-8.00005-1

5.1 Concept of effective stress in prestressing tendons

For the compressive load in prestressed concrete imposed by highly tensioned tendons reacting on concrete, any kind of reduction of elongation in prestressing tendons will cause prestress loss. Take a post-tensioned member as an example, the retraction of tendons at the seating of anchorage and the shortening due to shrinkage and creep of concrete as time passes lead to a reduction of the stress in tendons. These decreased tensile stresses in tendons are called the prestress losses, and the remaining stress in tendons at a given time after the losses have taken place is defined as the effective stress, expressed by

$$\sigma_{pe}(x, t) = \sigma_{con} - \sum \sigma_{li}(x, t) \tag{5.1}$$

where $\sigma_{pe}(x, t)$ = effective stress in prestressing tendons at a distance x from the end of anchorage device at the time t under consideration.

σ_{con} = tensile stress in prestressing tendons at the jacking point prior to transfer, also called the control stress in prestressing tendons at stretching.

$\sigma_{li}(x, t)$ = prestress losses, relating to stressing methods, time, and point under consideration.

x = distance from the end of the anchorage device to the point under consideration.

The control stress in prestressing tendons can be derived by the jacking force divided by the total area of the prestressing tendons stretched at a time, as

$$\sigma_{con} = \frac{T_{jack}}{A_p} \tag{5.2}$$

where T_{jack} = jacking force shown in the pressure gauge for the jack prior to transfer.

A_p = total cross-sectional area of the prestressing tendons stretched at a time.

5.1.1 Control stress in tendons at stretching

The tensile stress in tendons at stretching must be controlled in a reasonable range. It is uneconomical if the tensile stress is too low since the high strength of the tendons cannot be fully utilized, while it brings adverse consequences if the stress is too high. A high value of σ_{con} possibly causes the following issues.

(1) The larger the σ_{con} is, the more likely the prestressing tendons will snap. When a group of steel wires or several strands are stretched by a jack at one time, the tensile stress in prestressing tendons is an average value produced by jacking force, as given by Eq. (5.2), meaning that each wire or strand will have different tensile stress because there is a different amount of initial slack in the wires before stretching whether they are prestretched or not, and those that have low slack bear higher tensile stress. Furthermore, the tensile stress in each wire in a strand is also different in a similar way. Therefore, it commonly occurs that some of the wires are snapped when their stress exceeds the tensile strength even when σ_{con} is less than the tensile strength if σ_{con} is relatively high.

(2) The relaxation of prestressing tendons develops faster and the total amount of relaxation is greater under a high value of σ_{con}, especially when σ_{con} is close to the nominal ultimate tensile strength. These not only result in serious prestress loss but bring difficulty in predicting and controlling the long-term behavior of prestressed concrete structures.

(3) Wires have no obvious yield point, so a high value of σ_{con} lessens the ductility of prestressing tendons, resulting in low brittle fracture resistance in the prestressed concrete structures.

The control stress in tendons made from prestressing steels, both by pretensioning and post-tensioning, should satisfy the requirement specified in the codes, expressed as

$$\sigma_{con} \leqslant \lambda f_{ptk} \tag{5.3}$$

where f_{ptk} = characteristic tensile strength of prestressing tendon.

λ = coefficient of control stress in prestressing tendons at stretching.

λ has a different value in different codes. The values of λ compiled in Table 5.1 are prescribed in the *Code for design on railway bridge and culvert* (Q/CR 9300-2018), *Specifications for design of highway reinforced concrete and prestressed concrete bridges and culverts* (JTG 3362-2018), and *Codes for design of concrete structures* (GB 50010-2010).

Table 5.1 Coefficient of control stress in prestressing tendons at stretching.

Item	Steel wires and strands	Screw-thread steel bars
For civil and industrial structures (GB50010-2010)	0.75	0.85
For highway bridge (JTG 3362-2018)		
For railway bridges (Q/CR 9300-2018)	0.75	0.90

Note: $\lambda = 0.70$ for medium-strength steel wire (GB50010-2010).

When overtensioning is adopted, the maximum control stress in tendons at the jacking point should not exceed $(\lambda + 0.05)\sigma_{con}$.

On the other hand, from an economic point of view, the control stress at stretching should not be less than $0.40f_{ptk}$ in wires and $0.50f_{pk}$ in prestressing threaded steel bars.

5.1.2 Types of prestress loss

Generally, the prestress losses are classified in the following seven categories:

(1) Prestress loss due to friction between the prestressing tendons and the duct wall at stretching, $\sigma_{l1}(x)$.

(2) Prestress loss due to retraction of the tendon caused by a slip of clippers or wedges, compressive deformation of the anchorage device, and compaction of joints between assembled blocks at the seating of anchorage, $\sigma_{l2}(x)$.

(3) Prestress loss due to heat treatment curing if the prestressing bed is fixed with earth, $\sigma_{l3}(x)$.

(4) Prestress loss due to elastic shortening at transfer or by stretching subsequent tendons, $\sigma_{l4}(x)$.

(5) Prestress loss due to relaxation of the prestressing tendon, $\sigma_{l5}(x, t)$.

(6) Prestress loss due to creep of concrete, $\sigma_{l6,cr}(x, t)$.

(7) Prestress loss due to shrinkage of concrete, $\sigma_{l6,sh}(x, t)$.

Considering that the prestress losses due to shrinkage and creep are interdependent, they are generally denoted together as $\sigma_{l6}(x, t) = \sigma_{l6,cr}(x, t) + \sigma_{l6,sh}(x, t)$ in many design codes, such as GB50010-2010, Q/CR 9300-2018, and JTG 3362-2018.

For an annular member (such as a pole or a circular vessel) prestressed with spiral tendons, prestress loss will occur due to radial compaction of concrete when stretching tendons, denoted as σ_{l7} in the GB50010-2010 code. The value of σ_{l7} is inversely proportional to the diameter of the member, generally, $\sigma_{l7} = 30$ MPa is taken when the diameter is less than 3 m and $\sigma_{l7} = 0$ when the diameter is larger than 3 m.

The emergence and development of prestress loss with time in the post-tensioned and pretensioned members are drawn in Fig. 5.1A and B, respectively. In Fig. 5.1A, t_0 refers to the time (or concrete age) that the partial tendons of the post-tensioned members are stretched and anchored, and t_1 refers to the time that all the subsequent tendons are stretched and anchored. In Fig. 5.1B, t_0 refers to the time that the tendons of the pretensioned members are stretched and temporarily anchored in the prestressing bed, and t_1 refers to the time at transfer of the prestressing force to the concrete. t_s refers to the time in the service period that most of the prestress losses due to relaxation of the tendons and shrinkage and creep of concrete have taken place. For the pretensioned members under curing, the relaxation of the prestressing tendons occurs, so part of the relaxation ($\xi_1 \sigma_{l5}$) has taken place before transfer.

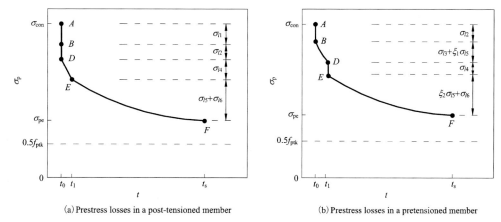

(a) Prestress losses in a post-tensioned member   (b) Prestress losses in a pretensioned member

Figure 5.1 Prestress losses in prestressed concrete members. (A) Prestress losses in a posttensioned member. (B) Prestress losses in a pretensioned member. (Part of the relaxation $\xi_1\sigma_{l5}$ takes place before transfer, the other $\xi_2\sigma_{l5}$ takes place after transfer).

In Fig. 5.1A and B, the prestress losses can be grouped into two categories, those that occur immediately during construction, called the instantaneous loss including $\sigma_{l1}(x)-\sigma_{l4}(x)$, and those that occur over an extended period, called the time-dependent or long-term loss including $\sigma_{l5}(x,t)$ and $\sigma_{l6}(x,t)$.

5.2 Effective stress in prestressing tendons immediately after transfer

5.2.1 Prestress loss due to friction between the tendons and the anchorage

When a large number of prestressing wires or strands are anchored by clips or cone plugs in an anchorage, the diameter of the outermost anchor holes in the anchor plate is larger than that of the duct ($D_a > D_d$), thereby two turns of the prestressing tendons are inevitable within the anchorage, as shown in Fig. 5.2. The loss of stress will occur due to friction at the turning points when stretching tendons, denoted as $\sigma_{l,01}$. The value of $\sigma_{l,01}$ should be determined by testing in situ, while $\sigma_{l,01} = 2\% - 3\%\sigma_{con}$ may be adopted approximately in most cases. Therefore, the initial stress in prestressing tendons at the end of the anchorage device before the transfer is

$$\sigma_{con,a} = \sigma_{con} - \sigma_{l,01} \qquad (5.4)$$

where $\sigma_{l,01}$ = prestress loss due to friction between the anchorage device and the tendons at stretching.

Thus, σ_{con} in Eq. (5.1) is replaced by $\sigma_{con,a}$ when the prestressing tendons are anchored by an anchor plate with a large diameter.

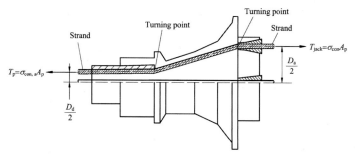

Figure 5.2 Two turns of tendons within the anchorage device.

5.2.2 Prestress loss due to friction between the tendons and the duct wall

The prestress loss $\sigma_{l1}(x)$ due to friction between the prestressing tendons and the duct wall occurs when stretching the tendons in the post-tensioned members.

The profile of prestressing tendons usually consists of several curves, including curved segments and straight lines. When the prestressing tendons are stretched, the tendons are elongated and slide through the curved duct wall where the frictional resistance is developed. In addition, numerous chords with small bending angles due to unintentional misalignment are inevitably generated in any segment of the tendons during construction, where additional friction (called wobble friction) will be produced during stretching. Therefore, the prestress loss due to friction includes two parts, one is caused by curvature friction and the other is caused by wobble friction.

The classical friction theory is commonly used to calculate $\sigma_{l1}(x)$.

5.2.2.1 Friction caused by the curvature

Take an infinitesimal length dx of a prestressing tendon at the location with a total angular change θ from the jacking end, as shown in Fig. 5.3B. Set the tensile force in

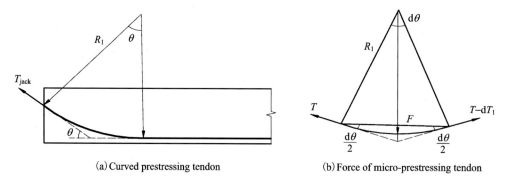

(a) Curved prestressing tendon   (b) Force of micro-prestressing tendon

Figure 5.3 Diagram of frictional loss calculation. (A) Curved prestressing tendon. (B) Force of micro-prestressing tendon.

the prestressing tendon T at tensioning side and $T - dT_1$ at another side, where dT_1 is the change of tensile force due to friction. If the influence of dT_1 on the radial pressure in concrete is disregarded, the radial pressure is given by

$$F = 2T \sin \frac{d\theta}{2} \approx 2T \frac{d\theta}{2} = Td\theta$$

The frictional resistance generated by radial pressure is obtained as

$$dT_1 = -\mu dF = -\mu Td\theta \tag{5.5}$$

where μ = coefficient of friction between the prestressing tendons and the duct wall.

5.2.2.2 Friction caused by unintentional displacement

Assume that the average random radius of unintentional angular displacement of the duct is R', and the corresponding change of angle is $d\theta'$ in an infinitesimal length, the frictional resistance may be taken as

$$dT_2 = -\mu Td\theta' = -\mu T \frac{dx}{R'} = -\frac{\mu}{R'} Tdx$$

Let $k = \frac{\mu}{R'}$ be the coefficient of wobble friction reflecting the impact of unintentional displacement, and the above formula becomes

$$dT_2 = -kTdx \tag{5.6}$$

5.2.2.3 Total prestress loss due to friction

The total frictional resistance due to curvature friction and wobble friction in an infinitesimal length can be derived from Eqs. (5.5) and (5.6), as

$$dT = dT_1 + dT_2 = -\mu Td\theta - kTdx = -T(\mu d\theta + kdx)$$

From which

$$\frac{dT}{T} = -\mu d\theta - kdx \tag{5.7}$$

Integrate Eq. (5.7) and consider the boundary conditions, $T = T_{jack} (= A_p\sigma_{con})$ when $x = 0$ and $\theta = 0$, resulting in

$$T(x) = T_{jack}e^{-(\mu\theta + kx)} \tag{5.8}$$

Then the tensile stress in prestressing tendons is

$$\sigma(x) = \frac{T(x)}{A_p} = \sigma_{con}e^{-(\mu\theta + kx)} \tag{5.9}$$

Hence, the prestress loss due to friction is attained as

$$\sigma_{l1}(x) = \sigma_{con} - \sigma(x) = \sigma_{con}\left[1 - e^{-(\mu\theta + kx)}\right] \tag{5.10}$$

where σ_{con} = control stress in prestressing tendons at jacking, MPa.

μ = coefficient of friction between the prestressing tendons and the duct wall.

κ = coefficient of wobble friction, rad/m.

x = tendon length between the end of the anchorage device and the point under consideration, for slender members the projected distance from the anchorage device end to the point under consideration on the member's axis can be adopted, m.

θ = sum of the angular change over a tendon length x, rad.

μ, κ should be determined by experimental data obtained in situ. If reliable data are not available, values in Table 5.2a or Table 5.2b can be adopted.

If $\mu\theta + kx \leq 0.3$, $\sigma_{l1}(x)$ can be calculated approximately by

$$\sigma_{l1}(x) = \sigma_{con}(\mu\theta + kx) \tag{5.11}$$

For spatial tendons, the sum of the angular change can be calculated by vectorially adding the total vertical angular change and horizontal angular change. When the projected plan and elevation of the tendons are parabolic or circular, the sum of the angular change in Eq. (5.10) can be obtained by

$$\theta = \sqrt{\alpha_v^2 + \alpha_h^2} \tag{5.12}$$

When the developed plan and elevation of the tendons are generalized curves, the tendon can be split into small chords, then Eq. (5.12) turns into

$$\theta = \sum \sqrt{\Delta\alpha_v^2 + \Delta\alpha_h^2} \tag{5.13}$$

where α_v = sum of the vertical angular change of the prestressing tendons.

α_h = sum of the horizontal angular change of the prestressing tendons.

$\Delta\alpha_v$ = vertical angular change at each chord in the generalized curve prestressing tendons.

$\Delta\alpha_h$ = horizontal angular change at each chord in the generalized curve prestressing tendons.

Table 5.2a The values of μ and κ.

Type of duct	μ	κ
Drawing rubber pipe to form duct	0.55	0.0015
Drawing steel pipe to form duct	0.55	0.0015
Smooth iron pipes	0.35	0.0030
Corrugated metal ducts	0.20–0.26	0.0020–0.0030

Note: The data in this table were recorded from the Q/CR 9300-2018 code.

Table 5.2b The values of μ and κ.

Type of duct		κ	μ Steel wires and strands	Threaded steel bars
Internal prestressing tendons	Corrugated metal ducts	0.0015	0.20–0.25	0.50
	Corrugated plastic ducts	0.0015	0.14–0.20	–
	Smooth iron tube	0.0030	0.35	0.40
	Smooth steel pipes	0.0010	0.25–0.30	–
	Core forming	0.0014–0.0015	0.55	0.60

Note: The data in this table were recorded from the JTG 3362-2018 and GB 50010-2010 codes.

In order to reduce the prestress loss due to friction, the following measures are often adopted during construction:

(1) Stretching the tendon at two ends simultaneously. For symmetrical tendons, the length and total angular change of the tendons at the middle are half of that of the single-end tensioning, which can effectively reduce the loss due to friction. Two-end tensioning is broadly used in the flexural structures with long spans.

(2) Segmentally tensioning and anchoring for long tendons. To reduce the excessive prestress loss due to friction when stretching long tendons in multispan prestressed concrete continuous beams, several short segments are stretched one by one and are linked by the couplers to form the required long tendon.

(3) Overstressing for compensating the frictional loss. Overstressing is an effective approach to compensate the prestress loss due to friction, which is commonly used in highway bridges if needed, while it is banned for high-speed railway bridges. The JTG 3362-2018 stipulates that the maximum stress in tendons at overstretching should not exceed $1.05\sigma_{con}$. When overstressing is used, the typical stretching process is as follows: $0 \rightarrow 0.1 - 0.15\sigma_{con} \rightarrow 1.05\sigma_{con}$ (duration of 2 min) $\rightarrow 0 \rightarrow \sigma_{con}$.

5.2.3 Prestress loss due to anchorage set

When the jacking force is released, the prestressing force is transferred from the jack to the concrete accompanied by retraction of the prestressing tendons due to the slip of wedges or clips and compressed deformation of anchorage. The retraction causes the loss of stress in tendons, denoted as $\sigma_{l2}(x)$, occurring both in pretensioned and post-tensioned structures.

Table 5.3 Values of tendon retraction due to anchorage set (mm).

Type of anchorage		Manifestation	Calculated value
Tapered anchorage for wires		Retraction of wedges and deformation of anchorage	6—8
Clip-on anchorage	Pre compression	Retraction of clips and deformation of anchorage	4—5
	No pre compression		6—8
Nut-type anchorage		Compaction of gap	1—3
A gap of each back pad		Compaction of gap	1—2
Heading anchor		Compaction of gap	1

Typical values of tendon retraction due to anchorage set are 3—8 mm, as compiled in Table 5.3, depending on the types of anchorage system and stressing methods.

For short prestressing tendons, such as the tendons in railway sleepers and the vertical tendons in the webs of a box girder, $\sigma_{l2}(x)$ is sensitive to the magnitude of retraction, since $\sigma_{l2}(x)$ may account for a large proportion of σ_{con}. For continuous girders with long spans constructed by the balanced cantilever method, as shown in Fig. 5.4, the curved tendons are stretched and anchored at the end of each segment, the retraction of the tendons is possibly counterbalanced by the reverse-friction resistance in the vicinity of the end resulting in an uneven distribution of $\sigma_{l2}(x)$ which obviously affects the local prestressing force.

Hence, the prestress due to retraction of the prestressing tendons at seating of anchorage should be calculated deliberately in some cases.

5.2.3.1 $\sigma_{l2}(x)$ in the pretensioned members
The prestressing tendons are arranged as a straight line or broken lines in the pretensioned members, so the assumption that the retraction is uniformly distributed over the entire

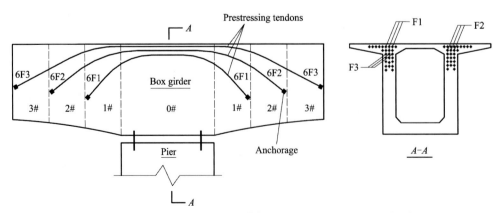

Figure 5.4 Schematic diagram of the balanced cantilever method.

length of the tendon is adopted, thereby the prestress loss due to anchorage set temporarily can be given by

$$\sigma_{l2}(x) = \frac{\Delta l}{l} E_p \qquad (5.14)$$

where Δl = retraction of the prestressing tendons due to slip of wedges or clips.

l = effective length of the prestressing tendons.

E_p = modulus of elasticity of prestressing tendon.

5.2.3.2 $\sigma_{l2}(x)$ in the post-tensioned members

In the post-tensioned members, Eq. (5.14) can be also used to calculate $\sigma_{l2}(x)$ for straight-line prestressing tendons. For curved tendons, the loss of prestress in tendons due to retraction is decreased away from the anchorage due to the friction opposite to the direction of elongation, so the reverse friction possibly occurs in the vicinity of the anchorage when the curvature of the tendons is large or the change in the bending angle is dramatic, as shown in Fig. 5.5. In Fig. 5.5, let the point A' be the initial tendon stress at the end of anchorage device given by Eqs. (5.2) or (5.4), $A'SH$ and $A''SH$ represent the distributions of tendon stress immediately prior to and after anchorage set, respectively. $A'SH$ is symmetrical with $A''SH$ along AS when the assumption that the coefficient of reverse friction is the same as that of friction at the stretching tendons is adopted. The

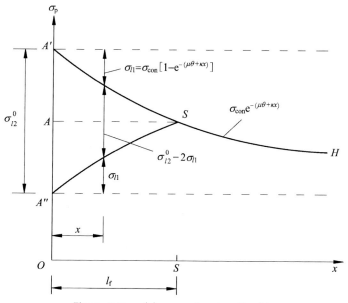

Figure 5.5 $\sigma_{l2}(x)$ occurs in a length of l_f.

distance between the end of the anchorage device and the fixed-point S is called the influence length of tendon retraction, denoted as l_f.

Considering that the retraction is counterbalanced by the change of strain in the tendons over the length of l_f, gives

$$\Delta l = \int_0^{l_f} \frac{\sigma_{l2}(x)}{E_p} dx \tag{5.15}$$

where Δl = tendon retraction due to slip of wedges or clips and compressed deformation of anchorage device.

l_f = influence length of tendon retraction.

l_f and $\sigma_{l2}(x)$ can be derived from Eq. (5.15) by the unified analytical method theoretically, while an approximate approach is commonly adopted for simplification, such as the methods provided in the Q/CR 9300-2018 and JTG 3362-2018 codes.

Unified analytical method

Assume that the typical prestressing tendons are composed of multicurves (x_i, R_i) $(R_i = \infty$ in the straight line segment), where x_i is the distance from the end of the anchorage device to the point under consideration and R_i is the radius of the ith segment. Based on the assumption that the coefficient of reverse friction is the same as that at stretching tendons, the reverse frictional resistance at retraction is equal to the frictional resistance at elongation under the same tension force in the tendons $(\sigma_{con} A_p$ or $\sigma_{con,a} A_p)$, that is $AA' = AA''$ shown in Fig. 5.5, gives

$$\sigma_{l2}(x) = \sigma_{l2}^0 - 2\sigma_{l1}(x) = \sigma_{l2}^0 - 2\sigma_{con}\left[1 - e^{-(\mu\theta + \kappa x)}\right] \tag{5.16}$$

where σ_{l2}^0 = prestress loss of the prestressing tendons at the end of the anchorage device due to retraction.

$\sigma_{l2}(x)$ = prestress loss of the prestressing tendons at x away from the end of the anchorage device due to retraction considering the influence of reverse friction.

Substituting Eq. (5.15) into Eq. (5.16) yields

$$\Delta l = \frac{\left(\sigma_{l2}^0 - 2\sigma_{con}\right) l_f}{E_p} + \frac{2\sigma_{con}}{E_p} \int_0^{l_f} e^{-(\mu\theta + \kappa x)} dx$$

Considering the boundary condition, $\sigma_{l2} = 0$ when $x = l_f$, resulting in

$$\sigma_{l2}^0 = 2\sigma_{con}\left[1 - e^{-(\mu\theta + \kappa l_f)}\right] \tag{5.17}$$

For prestressing tendons consisting of numerous straight lines and quadratic curves, θ is a linear function of x, giving

$$-[\mu\theta(x) + \kappa x] = \alpha x + \beta \tag{5.18}$$

where α, β = geometric coefficients at the point under consideration.

For high-order curved segments, the prestressing tendons may be split into numerous small intervals which can be expressed approximately as a straight line or a quadratic curve.

Let $c = \frac{\Delta l E_p}{2\sigma_{con}}$, substituting Eq. (5.18) into Eq. (5.17) yields

$$l_f = \frac{1}{\alpha} - \left(\frac{c}{e^\beta} + \frac{1}{\alpha}\right) e^{-\alpha l_f} \tag{5.19}$$

Substituting Eqs. (5.17) and (5.19) into Eq. (5.16) yields

$$\sigma_{l2}(x) = 2\sigma_{con}\left(e^{\alpha x + \beta} - e^{\alpha' l_f + \beta'}\right) \tag{5.20}$$

In the region of $x \leqslant l_f$, the effective stress in prestressing tendons immediately after anchorage set is

$$\sigma_{pe}(x) = \sigma_{con} - \sigma_{l1}(x) - \sigma_{l2}(x)$$

$$= \sigma_{con}\left(2e^{\alpha' l_f + \beta'} - e^{\alpha x + \beta}\right) \tag{5.21}$$

where α', β' = geometric coefficients at the fixed-point S, as shown in Fig. 5.5.

l_f can be attained by the iterative method from Eq. (5.19). When the term $e^{-\alpha l_f}$ is expanded by the Taylor series, an approximate formula for l_f is achieved when the first two items are kept, as

$$l_f = \frac{1}{2\alpha'}\left(-w - \sqrt{w^2 + 4w}\right) \tag{5.22}$$

Then Eq. (5.20) turns into

$$\sigma_{l2}(x) = 2\sigma_{con}\left(e^{\alpha x + \beta} - e^{\frac{-w - \sqrt{w^2 + 4w}}{2} + \beta'}\right) \tag{5.23}$$

where $w = -\dfrac{2}{\dfrac{e^{\beta'}}{c\alpha'} + 1}$.

For prestressing tendons with a typical profile as shown in Fig. 5.6, where AB, CD and EF are straight lines, while BC and DE are curves, α and β are calculated by

$$\alpha = -\kappa - \frac{\mu}{R_i} \tag{5.24}$$

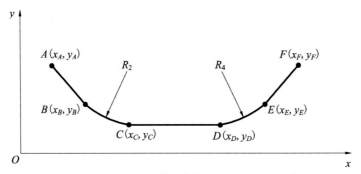

Figure 5.6 Typical profile of the prestressing tendon.

$$\beta = -\sum_{i=1}^{m} \frac{x_{i+1} - x_i}{R_i} \mu \qquad (5.25)$$

where m = total number of segments of the prestressing tendons.

R_i = radius of the ith segment of the prestressing tendons.

x_i = distance from the starting point at the ith segment of the prestressing tendons to the end of the anchorage device.

x_{i+1} = distance from the ending point at the ith segment of the prestressing tendons to the end of the anchorage device.

For straight tendons, curved tendons with a small change of bending angle, or short tendons, the reverse friction occurs over the entire length of the tendon l_p, in this case, $l_f = l_p$ for stretching at one end and $l_f = l_p/2$ for stretching at both ends.

Simplified linear method

Take the prestressing tendons with a typical profile shown in Fig. 5.6 stretched at both ends as the analysis object. Assume that the frictional prestress loss in each segment is given by Eq. (5.11), so the distribution of the tendon stress immediately prior to the anchorage set is a broken line. The different location of fixed-point S gives different formulas on prestress loss due to retraction.

(1) When $x_B < l_f < x_C$

The tendon stress prior to anchorage set is illustrated by $A'B'SC$ (Fig. 5.7) if the fixed-point S is located within the segment BC.

Eq. (5.15) and Fig. 5.7 give

$$\Delta l = \int_0^{l_f} \frac{\sigma_{l2}(x)}{E_p} dx = \Delta l_{AB} + \Delta l_{BS} \approx \overline{\varepsilon_{AB}} x_1 + \overline{\varepsilon_{BS}} (l_f - x_1)$$

$$= \left[2 \times \frac{1}{2} \left(\frac{\sigma_{con} - \sigma_S}{E_p} + \frac{\sigma_B - \sigma_S}{E_p} \right) \right] x_B + \left(2 \times \frac{1}{2} \frac{\sigma_B - \sigma_S}{E_p} \right) (l_f - x_B) \qquad (5.26)$$

where $\overline{\varepsilon_{AB}}$ = average strain in the segment AB due to tendon retraction.

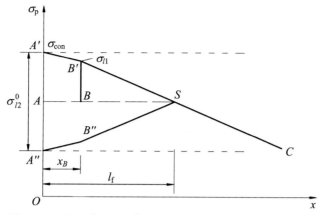

Figure 5.7 Distribution of tendon stress at the anchorage set.

$\overline{\varepsilon}_{BS}$ = average strain in the segment BS due to tendon retraction.
σ_B = tendon stress at the point B at seating of anchorage.
σ_S = tendon stress at the point S at seating of anchorage.
x_B = length of segment AB projected on the member's axis.

The tendon stresses prior to anchorage set at points B and S can be obtained from Eq. (5.11), as

$$\sigma_B = \sigma_{con}(1 - \kappa x_B) \tag{5.27}$$

$$\sigma_S = \sigma_{con}\left\{1 - \left[\kappa l_f + \frac{\mu}{R_2}\left(l_f - x_B\right)\right]\right\} \tag{5.28}$$

Substituting Eqs. (5.27) and (5.28) into Eq. (5.26) yields

$$l_f = \sqrt{\dfrac{\dfrac{\Delta l E_p}{\sigma_{con}} + \dfrac{\mu}{R_2}x_B^2}{k + \dfrac{\mu}{R_2}}} \tag{5.29}$$

Hence, the tendon stress at the end of the anchorage device at seating of anchorage is acquired as

$$\sigma_{p,0} = \sigma_{con}\left[1 - 2\kappa l_f - \frac{2\mu}{R_2}\left(l_f - x_B\right)\right] \tag{5.30}$$

And the tendon stress at the fixed-point S is

$$\sigma_{p,s} = \sigma_{con}\left[1 - \kappa l_f - \frac{\mu}{R_2}\left(l_f - x_B\right)\right] \tag{5.31}$$

(2) When $x_C < l_f < l_p/2$

Similarly, when $x_C < l_f < l_p/2$ (l_p is the length of the prestressing tendons), there are

$$l_f = \sqrt{\frac{1}{\kappa}\left[\frac{\Delta l E_p}{\sigma_{con}} - \frac{\mu}{R_2}\left(x_C^2 - x_B^2\right)\right]} \tag{5.32}$$

$$\sigma_{p,0} = \sigma_{con}\left[1 - 2\kappa l_f - \frac{2\mu}{R_2}(x_C - x_B)\right] \tag{5.33}$$

$$\sigma_{p,s} = \sigma_{con}\left\{1 - \left[\kappa l_f + \frac{\mu}{R_2}(x_C - x_B)\right]\right\} \tag{5.34}$$

$$\sigma_{l2}^0 = \sigma_{con}\left[2\kappa l_f + \frac{2\mu}{R_2}(x_C - x_B)\right] \tag{5.35}$$

(3) When $l_f > l_p/2$

$l_f > l_p/2$ means that the fixed-point S is located at the middle of the segment CD (Fig. 5.6) where $l_f = l_p/2$, hence

$$\sigma_{l2}^0 = \sigma_{con}\left[1 - \frac{\kappa l_p}{2} - 2\mu\frac{(x_C - x_B)}{R_2}\left(1 - \frac{x_C}{l_p} - \frac{x_B}{l_p}\right)\right] - \frac{2\Delta l E_p}{l_p} \tag{5.36}$$

If the further assumption that the prestress loss due to tendon retraction is distributed linearly over the entire length between the end of the anchorage device and the fixed-point, the influence length of the single-end stretching tendon can be obtained by referring to Eq. (5.15), as

$$l_f = \sqrt{\frac{\Delta l E_p}{\Delta \sigma_d}} \tag{5.37}$$

where $\Delta\sigma_d$ = prestress loss due to friction per unit length, given by

$$\Delta\sigma_d = \frac{\sigma_{l1,m}}{\dfrac{l_p}{2}} \tag{5.38}$$

where $\sigma_{l1,m}$ = prestress loss due to friction at the middle of the prestressing tendons at stretching.

When $l_f < l_p/2$, the prestress loss due to anchorage set considering the effect of reverse friction is

$$\sigma_{l2}^0 = \Delta\sigma_d l_f \tag{5.39}$$

$$\sigma_{12}(x) = \sigma_{12}^0 \frac{l_f - x}{l_f} \tag{5.40}$$

If $l_f > l_p/2$, the reverse friction occurs in the half-length of the prestressing tendon, giving

$$\Delta\sigma_x' = \sigma_{12}' - 2x'\Delta\sigma_d \tag{5.41}$$

where σ_{12}' = prestress loss under the anchorage considering the reverse friction in the case of $l_f < l_p$.

If the entire length of the prestressing tendons is an arc with a radius of R_p, $\theta = x/R_p$, Eqs. (5.37)–(5.40) give

$$l_f = \sqrt{\frac{\Delta l E_p}{\sigma_{con}\left(\frac{\mu}{R_p} + \kappa\right)}} \tag{5.42}$$

$$\sigma_{12}(x) = 2\sigma_{con}l_f\left(\frac{\mu_f}{R_p} + \kappa\right)\left(1 - \frac{x}{l_f}\right) \tag{5.43}$$

where R_p = radius of the arc prestressing tendons.

For the parabolic prestressing tendons with corresponding central angle $\theta \le 45$, Eqs. (5.42) and (5.43) can still be used.

Eqs. (5.29)–(5.36) are provided in the Q/CR 9300-2018 code and Eqs. (5.37)–(5.43) are provided in the JTG 3362-2018 code.

5.2.4 Prestress loss due to heat treatment curing

Heat treatment is commonly used during concrete curing, which may produce prestress loss in the pretensioned members.

Before casting concrete, the tendons are stretched and temporarily anchored on the abutments of the prestressing bed at ambient temperature. During heat treatment curing, the prestressing tendons expand with the increase in temperature. As the temperature drops back after heat treatment curing has been completed, the tendons and the concrete which are bonded together shrink simultaneously. If the abutments are fixed with the earth, the generated expansion in tendons due to increased temperature cannot be recovered, resulting in the loss of stress in tendons. Moreover, relaxation of the tendons will be accelerated by high temperature during heat treatment curing giving rise to additional loss of stress. Experimental data show that the prestress loss due to relaxation during the heat treatment may reach up to 75% of the total value of relaxation loss.

The tendon deformation due to direct thermal origin can be calculated by

$$\Delta l = \alpha(T_2 - T_1)l = \alpha \cdot \Delta T \cdot l \tag{5.44}$$

where $\alpha =$ coefficient of thermal expansion of the prestressing tendons, generally $\alpha = 1.0 \times 10^{-5}/°C$.

$T_1 =$ ambient temperature before heating.

$T_2 =$ maximum temperature of the concrete during heat treatment curing.

$\Delta T =$ change of temperature in concrete.

$l =$ effective length of the prestressing tendons (distance between two temporary anchorages).

Hence, the prestress loss due to the direct thermal effect during heat treatment curing is obtained as

$$\sigma_{l3}(x) = \frac{\Delta l}{l}E_p = \alpha E_p(T_2 - T_1) \tag{5.45}$$

where $E_p =$ modulus of elasticity of prestressing tendon, $E_p = 2.0 \times 10^5$ MPa.

Substituting $\alpha E_p = 2.0$ MPa/°C into Eq. (5.45) yields

$$\sigma_{l3}(x) = 2(T_2 - T_1)(\text{MPa}) \tag{5.46}$$

The *fip* Model Code 2010 provides the approach to calculate the additional relaxation loss due to heat treatment curing by adding to the time duration defined by

$$t_{ep} = 1.04t_{p1}(T_2 - 20) \tag{5.47}$$

$$t_{p1} = (T_2 - 20)^{-1} \int_0^{t_a} [T(t) - 20]dt \tag{5.48}$$

where $t_{ep} =$ equivalent duration for calculating relaxation.

$t_{p1} =$ mean duration of the heating cycle.

$t_a =$ age of the concrete when its temperature returns to ambient temperature.

$T(t) =$ temperature in concrete at time t, °C

When t_{ep} given in Eq. (5.47) is substituted into Eq. (2.36), the total relaxation occurring during heat treatment curing is obtained.

In the case that the tendons and the concrete are bonded completely, the change in temperature does not lead to prestress loss in tendons. Therefore, raising the temperature by stages during curing can effectively reduce the losses due to the direct thermal origin. In the first stage, the temperature of the concrete is controlled at a low level (about 20°C) after casting, the small change in temperature results in a small amount of prestress loss in tendons. As the concrete hardens enough and the adhesion between the concrete and the tendons is sufficient to prevent the tendon sliding, prestress loss due to the direct thermal origin will not occur when the temperature is further increased in the second stage, since the tendon deformation caused by temperature change is the same as in concrete.

5.2.5 Prestress loss due to elastic shortening

The prestress loss due to elastic shortening $\sigma_{l4}(x)$, occurs both in the pretensioned and post-tensioned members, although its mechanism is different.

5.2.5.1 $\sigma_{l4}(x)$ in the pretensioned members

As the concrete attains sufficient strength, the prestressing force is transferred from temporary anchorages to the concrete when the jacking force is released, then a pretensioned concrete member is fabricated. Since the concrete and the tendons have been bonded together, they bear the prestressing force together at transfer and shorten simultaneously, so the strain change due to elastic shortening can be calculated in this newly balanced condition, as

$$\varepsilon_c(x) = \frac{N_{p0} - \varepsilon_c(x)E_pA_p}{E_cA_0} + \frac{N_{p0}e_p^2 - \varepsilon_c(x)E_pA_pe_p^2}{E_cI_0}$$

From which

$$\varepsilon_c(x) = \left[1 + \alpha_{EP}\frac{A_p}{A_0}\left(1 + \frac{e_p^2}{i^2}\right)\right]\frac{\sigma_{pc}(x)}{E_c} \qquad (5.49)$$

where $\varepsilon_c(x)$ = elastic compressive strain in concrete at the centroid of prestressing tendons due to the prestressing force after transfer.

$\sigma_{pc}(x)$ = normal stress of concrete at the centroid of prestressing tendons due to the prestressing force after transfer.

A_p = area of prestressing tendons.

A_0 = area of converted cross-section.

I_0 = moment of inertia of the converted section.

e_p = eccentricity of the prestressing tendons with respect to the centroid of the converted section.

i = radius of gyration of the converted section $i = \sqrt{\frac{I_0}{A_0}}$.

E_c = modulus of elasticity of concrete at transfer.

E_P = modulus of elasticity of prestressing tendon.

α_{EP} = ratio of modulus of elasticity $\alpha_{EP} = \frac{E_p}{E_c}$.

$\sigma_{pc}(x)$ can be obtained by

$$\sigma_{pc}(x) = \frac{N_{p0}}{A_0} + \frac{N_{p0}e_p^2}{I_0} \qquad (5.50)$$

where N_{P0} = effective stressing force prior to transfer.

The prestress loss due to elastic shortening can be attained from Eq. (5.49), as

$$\sigma_{l4}(x) = \alpha_{EP}\left[1 + \alpha_{EP}\frac{A_p}{A_0}\left(1 + \frac{e_p^2}{i^2}\right)\right]\sigma_{pc}(x) \tag{5.51}$$

The following alternative equation may be taken for simplified analysis:

$$\sigma_{l4}(x) = \alpha_{EP}\sigma_{pc}(x) \tag{5.52}$$

5.2.5.2 $\sigma_{l4}(x)$ in the post-tensioned members

If all prestressing tendons in the cross-section of a post-tensioned concrete member are stretched at once, the tendons do not shorten at transfer. If the prestressing tendons are stretched in batches, the previously stressed tendons will shorten when the subsequent tendons are stretched, thus, prestress loss due to elastic shortening occurs in the previously stressed tendons. Obviously, the first stressed tendons bear the highest prestress loss and the last has no prestress loss.

When the $(i + m)$-th batch of tendons is stretched, the compressive strain in concrete at the centroid of the ith batch of tendons will be produced, as

$$\varepsilon_{i,i+m}(x) = \frac{\sigma_{i,i+m}(x)}{E_c} \tag{5.53}$$

where $\sigma_{i,i+m}(x)$ = compressive stress of concrete at the centroid of the ith batch of tendons due to stretching the $(i + m)$-th batch of tendons.

Hence, the prestress loss of the ith batch of tendons due to stretching the subsequent m batches of tendons is obtained as

$$\sigma_{l4}(x) = \sum_{m=1}^{n-i} E_p\varepsilon_{i,i+m} = \sum_{m=1}^{n-i} E_p\frac{\sigma_{i,i+m}}{E_c} = \alpha_{Ep}\sum_{m=1}^{n-i}\sigma_{i,i+m} = \alpha_{Ep}\sigma_{pc}(x) \tag{5.54}$$

where $\sigma_{l4}(x)$ = prestress loss due to elastic shortening.

n = number of identical prestressing tendons.

$n - i$ = number of subsequently stretched identical prestressing tendons.

$\sigma_{pc}(x)$ = normal stress of concrete at the centroid of prestressing tendons due to stretching subsequent prestressing tendons.

Generally, different profiles of curved prestressing tendons are arranged longitudinally in a post-tensioned structure, bringing the complexity to calculate $\sigma_{l4}(x)$ by Eq. (5.54) by hand. To simplify the analysis, the average of the prestress losses of all prestressing tendons in a cross-section is taken as the characteristic value of the loss, hence Eq. (5.54) can be rewritten as

$$\sigma_{l4}(x) = \frac{1}{n} \sum_{i=1}^{n} (i-1)\alpha_{Ep}\Delta\sigma_c(x) = \alpha_{Ep}\Delta\sigma_c(x)\frac{1}{n}\sum_{i=1}^{n}(i-1) = \frac{n-1}{2}\alpha_{Ep}\Delta\sigma_c(x)$$

$$(5.55)$$

where $\Delta\sigma_c(x) =$ normal stress of concrete at the centroid of prestressing tendons caused by each batch prestressing tendons stretched subsequently, and can be calculated approximately by

$$\Delta\sigma_c(x) = \frac{1}{n}\sigma_{pc}(x)$$

$\sigma_{pc}(x) =$ total normal stress of concrete at the centroid of all prestressing tendons generated by all prestressing tendons.

Eq. (5.55) is often written in another form:

$$\sigma_{l4}(x) = \frac{n-1}{2n}\alpha_{Ep}\sigma_{pc}(x) \qquad (5.56)$$

In the post-tensioned structures, $\sigma_{pc}(x)$ and $\Delta\sigma_c(x)$ are different at different cross-sections, so $\sigma_{l4}(x)$ should be calculated at the typical sections, such as the sections with maximum moment, the section at mid-span, and one-fourth effective span for the flexural structures.

To reduce the prestress loss due to elastic shortening, the previous tendons can be stretched again if it is feasible with the construction technology.

5.2.6 Effective stress in prestressing tendons immediately after transfer

For pretensioned members, the prestressing tendons are stretched first and temporarily anchored on the abutments, then the concrete is cast. Once the concrete is hardened sufficiently, the temporary anchorage is released and the prestressing force is transferred to the concrete. As shown in Fig. 5.2B, the effective stress in prestressing tendons immediately after transfer can be obtained as

$$\sigma_{pe}(x, t_1) = \sigma_{con} - [\sigma_{l2}(x) + \sigma_{l3}(x) + \sigma_{l4}(x) + \xi_1\sigma_{l5}(x, t_1)] \qquad (5.57)$$

where $\sigma_{l5}(x, t_1) =$ relaxation loss occurring at the time period of $(t_1 - t_0)$, calculated by Eq. (2.35).

$t_0 =$ time that the tendons are anchored temporarily.

$t_1 =$ time that the prestressing force is transferred to the concrete.

$\xi_1 =$ proportion of relaxation loss that has occurred before transfer to the total relaxation loss.

For post-tensioned members, the tendons are stretched and anchored when the concrete has hardened sufficiently. As shown in Fig. 5.2A, the effective stress in prestressing tendons immediately after transfer can be obtained as

$$\sigma_{pe}(x, t_1) = \sigma_{con} - [\sigma_{l1}(x) + \sigma_{l2}(x) + \sigma_{l4}(x)] \qquad (5.58)$$

where $t_1 =$ time that the last batch of prestressing tendons is stretched.

5.3 Long-term effective stress in prestressing tendons

In a prestressed concrete structure, shrinkage of concrete gives rise to shortening of the structure, and creep of concrete makes the compressive strain in concrete increase with time, with both resulting in prestress loss in prestressing tendons. The decreased tensile stress in tendons due to shrinkage and creep reduces the development rate of relaxation, thereby the relaxation should be discounted by the effect of shrinkage and creep. On the other hand, the decreased tensile stress in tendons due to relaxation reduces the compressive stress in concrete, so the change in compressive strain due to creep and the corresponding prestress loss also need to be abated.

Generally, the reinforcing steels are arranged in prestressed concrete structures. Reinforcing and prestressing steels or prestressing carbon fibers do not experience creep in the natural environment during their service period, so the deformation due to shrinkage and creep of concrete is restrained by the reinforcements. Considering that the reinforcements are not distributed uniformly in the cross-section, the centroid of all reinforcements does not coincide with the centroid of the concrete section in most cases, therefore, the reinforcement constraint on the shrinkage and creep of concrete is uneven in the cross-section, resulting in stress redistribution between the reinforcement and the concrete within a section. Hence, the effect of reinforcement eccentricity should be taken into account in the calculation of the constraint.

As discussed above, the long-term prestress losses due to concrete shrinkage and creep and tendon relaxation are the result of a comprehensive time-dependent effect. A unified approach to calculate the long-term prestress loss is introduced briefly first, then the separated approaches provided in the codes are discussed.

5.3.1 Unified approach to calculate long-term prestress loss

Consider a prestressed concrete structure in which the prestressing force is transferred to the concrete at time t_1. Assume that the prestressing tendons and the surrounding concrete have been bonded completely. If the reinforcement constraint on concrete deformation due to shrinkage is disregarded, the prestress loss due to shrinkage is attained as

$$\sigma_{l,sh}(x, t) = E_p \varepsilon_{p,sh}(t, t_1) = E_p \varepsilon_{sh}(t, t_1) \qquad (5.59)$$

where $\varepsilon_{p,sh}(t, t_1)$ = strain change in prestressing tendons due to shrinkage of concrete during the time period $t - t_1$.

$\varepsilon_{sh}(t, t_0)$ = shrinkage strain of concrete during the time period $t - t_1$.

t_1 = initial time to calculate shrinkage.

t = time under consideration.

If the reinforcement constraint on concrete deformation due to creep of concrete is disregarded, the prestress loss due to creep is attained as

$$\sigma_{l,cr}(x, t) = E_p\varepsilon_{p,cr}(t, t_1) = E_p\varepsilon_{cr}(t, t_1) = E_p\frac{\sigma_{pc}(t_1)}{E_c}\varphi(t, t_1) = \alpha_{EP}\sigma_{pc}(t_1)\varphi(t, t_1)$$

(5.60)

where $\sigma_{pc}(t_1)$ = concrete stress at the centroid of prestressing tendons caused by the prestressing force at the time of loading t_1, for flexural members the stress caused by the self-weight of the member should be taken into account.

$\varepsilon_{p,cr}(t, t_1)$ = strain change in prestressing tendons due to creep of concrete during the time period $t - t_1$.

$\varepsilon_{cr}(t, t_1)$ = strain change in concrete due to creep at the centroid of prestressing tendons during the time period $t - t_1$.

$\varphi(t, t_1)$ = creep coefficient at time t for the load is applied at time t_1.

α_{EP} = ratio of modulus of elasticity, $\alpha_{EP} = \frac{E_p}{E_c}$.

E_p = modulus of elasticity of prestressing tendon.

E_c = modulus of elasticity of concrete.

Hence, the prestress loss in prestressing tendons due to creep and shrinkage of concrete without considering the influence of reinforcement restraint can be attained from Eqs. (5.59) and (5.60):

$$\sigma_{l,sh}(x, t) + \sigma_{l,cr}(x, t) = E_p\varepsilon_{sh}(t, t_1) + \alpha_{EP}\sigma_{pc}(t_1)\varphi(t, t_1)$$

(5.61)

Based on the internal force equilibrium and deformation compatibility of the cross-section, the reinforcement constraint reduction coefficient for the prestress loss can be derived as

$$\gamma_{sp}(x, t, t_1) = \frac{1}{1 + \left(\alpha_{ES}\frac{A_s}{A_c} + \alpha_{EP}\frac{A_p}{A_c}\right)\left(1 + \frac{e_{ps}^2}{i^2}\right)[1 + \chi(t, t_1)\varphi(t, t_1)]}$$

(5.62)

where $\gamma_{sp}(x, t, t_1)$ = constraint reduction coefficient by all reinforcements at time t when the concrete stress $\sigma_{pc}(t_1)$ is applied at time t_1.

$\chi(t, t_1)$ = aging coefficient of concrete at the time of loading t_1 to the calculating time t, varies from 0.6 to 0.9, generally, 0.82 is adopted.

α_{ES} = ratio of modulus of elasticity, $\alpha_{ES} = \frac{E_s}{E_c}$.

A_c = area of the net concrete section.

I_c = moment of inertia of the net concrete section.

i = radius of gyration of the net concrete section, $i = \sqrt{\frac{I_c}{A_c}}$.

e_{ps} = distance from the centroid of all reinforcements to the centroid of the concrete section.

From Eqs. (5.61) and (5.62), the prestress loss due to creep and shrinkage of concrete considering the reinforcement restraint influence gives:

$$\sigma_{l6,sh}(x,t) + \sigma_{l6,cr}(x,t) = \gamma_{sp}(x,t)\left[E_p\varepsilon_{sh}(t,t_1) + \alpha_{EP}\sigma_{pc}(t_1)\varphi(t,t_1)\right] \qquad (5.63)$$

For railway and highway prestressed concrete beams, there is a considerable period of time between the transfer and the placement of the deck system, during which the stress in concrete changes due to its shrinkage and creep. This stress change and the newly generated stress in concrete due to the imposed weight of the deck system will give rise to prestress loss in the tendons due to shrinkage and creep of the concrete in-service period, in this case, Eq. (5.63) is rewritten as

$$\sigma_{l6,sh}(x,t) + \sigma_{l6,cr}(x,t)$$

$$= \gamma_{sp}(x,t,t_1)E_p\varepsilon_{sh}(t,t_1) + \gamma_{sp}(x,t_2,t_1)\alpha_{EP}\sigma_{pc}(t_1)\varphi(t_2,t_1)$$

$$+ \gamma_{sp}(x,t,t_2)\alpha_{EP}\Delta\sigma_{pc}(t_2)\varphi(t,t_2) \qquad (5.64)$$

$$\Delta\sigma_{pc}(t_2) = \sigma_{pc}(t_2) + \Delta\sigma_{pc,t}(t_2,t_1) \qquad (5.65)$$

where $\sigma_{pc}(t_2)$ = stress change in concrete at the centroid of the prestressing tendons due to the weight of the imposed deck system.

$\Delta\sigma_{pc,t}(t_2,t_1)$ = stress change in concrete at the centroid of the prestressing tendons due to the time-dependent effect of shrinkage and creep between transfer and the placement of the deck system.

$\gamma_{sp}(x,t_2,t_1)$ = constraint reduction coefficient by all reinforcements at time t_2 when the concrete stress $\sigma_{pc}(t_1)$ is applied at time t_1, calculated by Eq. (5.62).

$\gamma_{sp}(x,t,t_2)$ = constrain reduction coefficient by all reinforcements at time t when the concrete stress $\Delta\sigma_{pc,t}(t_2,t_1)$ is applied at time t_2, calculated by Eq. (5.62).

$\varphi(t_2,t_1)$ = creep coefficient at time t_2 for load is applied at time t_1.

$\varphi(t,t_2)$ = creep coefficient at time t for load is applied at time t_2.

The prestress loss due to tendon relaxation considering the reduced influence by shrinkage and creep of concrete can be expressed as

$$\sigma_{l5}(x,t) = \gamma_r(t,t_0)\sigma_r(t,t_1) \qquad (5.66)$$

where $\sigma_r(t,t_1)$ = natural relaxation of the prestressing tendons during the time period $t - t_1$, which can be calculated by Eqs. (2.36)–(2.38).

$\gamma_r(t, t_1)$ = reduction coefficient of relaxation in prestressing tendons accounting for the shrinkage and creep of concrete.

The magnitude of $\gamma_r(t, t_1)$ relates to the initial stress in prestressing tendons, type of tendons, characteristics of shrinkage and creep of concrete, reinforcement ratio, and eccentricity, etc. For a commonly used post-tensioned structure, the initial stress in prestressing tendons and the eccentricity of reinforcement vary along the span, so it is very complicated to obtain the value of $\gamma_r(t, t_1)$ accurately. Generally, it can be taken as equal to 0.60–0.80.

For the structures in which the time period between the transfer and the placement of the deck system is short and the time-dependent effect in this period is disregarded, a unified formula to calculate the long-term prestress loss considering the mutual reduction effect between concrete creep and shrinkage and tendon relaxation can be obtained from Eqs. (5.63) and (5.66) (Hu and Chen, 2003), as

$$\sigma_{l5}(x, t) + \sigma_{l6}(x, t) = \gamma_r(t, t_1)\sigma_r(t, t_1)$$
$$+ \gamma_{sp}(x, t)\left\{\alpha_{EP}\sigma_{pc}(t_1)\varphi(t, t_1) + E_p\varepsilon_{sh}(t, t_1) - \frac{\gamma_r(t, t_1)\sigma_r(t, t_1)}{E_c}\frac{A_p}{A_c}[1 + \chi(t, t_1)\varphi(t, t_1)]\right\}$$

$$(5.67)$$

Rewrite Eq. (5.67) as

$$\sigma_{l5}(x, t) + \sigma_{l6,sh}(x, t) + \sigma_{l6,cr}(x, t)$$
$$= \gamma_r(t, t_1)\sigma_r(t, t_1)\left\{1 - \gamma_{sp}(x, t)\frac{A_p}{A_c}[1 + \chi(t, t_1)\varphi(t, t_1)]\right\} \qquad (5.68)$$
$$+ \gamma_{sp}(x, t)\left[E_p\varepsilon_{sh}(t, t_1) + \alpha_{EP}\sigma_{pc}(t_1)\varphi(t, t_1)\right]$$

where $\sigma_{pc}(t_1)$ = concrete stress at the centroid of the prestressing tendons due to the prestressing force, self-weight of the member, and subsequently applied weight of the deck system at the time of loading t_1.

For the structures in which the time period between the transfer and the placement of the deck system is long and the time-dependent effect in this period should be considered, Eqs. (5.64)–(5.66) give

$$\sigma_{l5}(x, t) + \sigma_{l6,sh}(x, t) + \sigma_{l6,cr}(x, t)$$

$$= \gamma_r(t, t_1)\sigma_r(t, t_1)\left\{1 - \gamma_{sp}(x, t)\frac{A_p}{A_c}[1 + \chi(t, t_1)\varphi(t, t_1)]\right\}$$
$$+ \gamma_{sp}(x, t, t_1)E_p\varepsilon_{sh}(t, t_1) + \gamma_{sp}(x, t_2, t_1)\alpha_{EP}\sigma_{pc}(t_1)\varphi(t_2, t_1)$$
$$+ \gamma_{sp}(x, t, t_2)\alpha_{EP}\Delta\sigma_{pc}(t_2)\varphi(t, t_2) \qquad (5.69)$$

where $\sigma_{pc}(t_1) =$ concrete stress at the centroid of the prestressing tendons due to the prestressing force at the time of transfer t_1.

In some design codes, such as EN1992, the long-term prestress loss is calculated by a uniform formula such as Eqs. (5.68) or (5.69), while the prestress losses due to shrinkage of concrete, creep of concrete, and relaxation of tendons are commonly calculated separately in other codes. For those structures that are sensitive to long-term deflection and crack resistance, such as high-speed railway girders with a ballastless track system, highway long-span continuous girders, step-by-step time-dependent analysis with high calculation accuracy should be taken to compute the long-term prestress losses, or Eq. (5.69) is used. For general structures, the approximately simplified formulas provided in the design codes based on Eqs. (5.68) or (5.69) present an agreeable prediction accuracy.

5.3.2 Calculation of long-term prestress loss specified in the codes

5.3.2.1 Prestress loss due to relaxation specified in the codes

Under normal circumstances, the ultimate value of relaxation loss is $0.03 - 0.08\sigma_{con}$ in the service period, which is commonly adopted in the design codes for general prestressed concrete structures.

In the Q/CR 9300-2018 and JTG 3362-2018 codes, the ultimate value of relaxation loss of prestressing threaded steel bars may be taken as follows.

One-time tensioning:

$$\sigma_{l5}(t \rightarrow \infty) = 0.05\sigma_{con} \tag{5.70}$$

Overtensioning:

$$\sigma_{l5}(t \rightarrow \infty) = 0.035\sigma_{con} \tag{5.71}$$

In the GB50010-2010 code, the prestress loss of prestressing threaded steel bars due to relaxation can be taken as

$$\sigma_{l5}(t \rightarrow \infty) = 0.03\sigma_{con} \tag{5.72}$$

And the prestress loss for medium-strength prestressed steel wires can be taken as

$$\sigma_{l5}(t \rightarrow \infty) = 0.08\sigma_{con} \tag{5.73}$$

The ultimate value of relaxation loss of wires and strands can be calculated from Table 2.25. If the time-dependent analysis is required, the interval values can be determined by the ratio coefficient in Table 2.26.

5.3.2.2 Prestress loss due to shrinkage and creep specified in the codes

When the experimental data on long-term prestress losses are used to amend Eqs. (5.63) and (5.64), the simplified formulas are attained.

In the JTG 3362-2018 code, the prestress loss due to shrinkage and creep of concrete may be taken as

$$\sigma_{l6}(x, t) = 0.9 \frac{\alpha_{EP}\sigma_{pc}(t_1)\varphi(t, t_1) + E_p\varepsilon_{sh}(t, t_1)}{1 + 15\rho\rho_{ps}} \tag{5.74}$$

$$\sigma'_{l6}(x, t) = 0.9 \frac{\alpha_{EP}\sigma'_{pc}(t_1)\varphi(t, t_1) + E_p\varepsilon_{sh}(t, t_1)}{1 + 15\rho'\rho'_{ps}} \tag{5.75}$$

where $\sigma_{l6}(x, t)$ = prestress loss due to shrinkage and creep of concrete at the centroid of all reinforcements in the tension zone.

$\sigma'_{l6}(x, t)$ = prestress loss due to shrinkage and creep of concrete at the centroid of all reinforcements in the compression zone.

σ_{pc} = normal stress of concrete at the centroid of all reinforcements in the tension zone due to the prestressing force and the self-weight of the member.

σ'_{pc} = normal stress of concrete at the centroid of all reinforcements in the compression zone due to the prestressing force and the self-weight of the member.

e_p = distance from the centroid of the concrete section to the centroid of the prestressing tendons in the tension zone.

e'_p = distance from the centroid of the concrete section to the centroid of the prestressing tendons in the compression zone.

e_s = distance from the centroid of the concrete section to the centroid of the reinforcing steels in the tension zone.

e'_s = distance from the centroid of the concrete section to the centroid of the reinforcing steels in the compression zone.

e_{ps} = distance from the centroid of the concrete section to the centroid of all reinforcements in the tension zone, $e_{ps} = \frac{A_p e_p + A_s e_s}{A_p + A_s}$.

e'_{ps} = distance from the centroid of the concrete section to the centroid of all reinforcements in the compression zone, $e'_{ps} = \frac{A'_p e'_p + A'_s e'_s}{A'_p + A'_s}$.

ρ = reinforcement ratio in tension zone, $\rho = \frac{A_p + A_s}{A_c}$.

ρ' = reinforcement ratio in compression zone, $\rho' = \frac{A'_p + A'_s}{A_c}$.

i = radius of gyration, properties of the converted section and net section are used for pretensioned and post-tensioned members, respectively

$$i = \sqrt{\frac{I}{A}}, \quad \rho_{ps} = 1 + \frac{e_{ps}^2}{i^2}, \quad \rho'_{ps} = 1 + \frac{e'^2_{ps}}{i^2}$$

In the Q/CR 9300-2018 code, the prestress loss due to shrinkage and creep of concrete can be estimated by

$$\sigma_{l6}(x, t \to \infty) = \frac{0.8\alpha_{EP}\sigma_{pc}(t_0)\varphi(\infty, t_1) + E_p\varepsilon_{sh}(\infty, t_1)}{1 + \left[1 + \frac{\varphi(\infty, t_1)}{2}\right]\rho\rho_{ps}} \qquad (5.76)$$

where $\sigma_{l6}(x, t \to \infty)$ = ultimate value of prestress loss due to shrinkage and creep of concrete at the centroid of prestressing tendons.

σ_{pc} = normal stress of concrete at the centroid of prestressing tendons due to the prestressing force and the self-weight of the member.

ρ = reinforcement ratio, $\rho = \frac{A_p + A_p' + A_s + A_s'}{A_c}$.

e_{ps} = distance from the centroid of the concrete section to the centroid of all reinforcements, $e_{ps} = \frac{A_p e_p + A_s e_s + A_p' e_p' + A_s' e_s'}{A_p + A_p' + A_s + A_s'}$.

i = radius of gyration, $i = \sqrt{\frac{I}{A}}$, $\rho_{ps} = 1 + \frac{e_{ps}^2}{i^2}$.

The GB50010-2010 code stipulates that the prestress loss due to creep and shrinkage of concrete can be calculated by Eqs. (5.74) and (5.75) for important structures or special structures, while the following formulas can be used for general prestressed concrete members.

Pretensioned members:

$$\sigma_{l6}(x, t \to \infty) = \frac{60 + 340\frac{\sigma_{pc}}{f_{cu}'}}{1 + 15\rho} \qquad (5.77)$$

$$\sigma_{l6}'(x, t \to \infty) = \frac{60 + 340\frac{\sigma_{pc}'}{f_{cu}'}}{1 + 15\rho'} \qquad (5.78)$$

Post-tensioned members:

$$\sigma_{l6}(x, t \to \infty) = \frac{55 + 300\frac{\sigma_{pc}}{f_{cu}'}}{1 + 15\rho} \qquad (5.79)$$

$$\sigma_{l6}'(x, t \to \infty) = \frac{55 + 300\frac{\sigma_{pc}'}{f_{cu}'}}{1 + 15\rho'} \qquad (5.80)$$

where σ_{pc} = normal stress of concrete at the centroid of prestressing tendons in the tension zone due to the prestressing force and the self-weight of the member.

σ'_{pc} = normal stress of concrete at the centroid of prestressing tendons in the compression zone due to the prestressing force and the self-weight of the member.

f'_{cu} = compressive strength of cubic concrete at transfer.

ρ = prestressing tendon ratio in the tension zone.

ρ' = prestressing tendon ratio in the compression zone.

When the structure works in an environment with an average annual relative humidity less than 40%, the obtained value from Eqs. (5.74)–(5.77) should be increased by 30%.

5.4 Estimation of effective stress in simplified analysis and preliminary design

For important prestressed concrete structures and those structures that are sensitive to long-term deflection (or camber) and crack resistance, such as the prestressed concrete box-girders in the high-speed railway with a ballastless track system and the prestressed concrete continuous girders with long spans in the highway, it is necessary to fulfill the step-by-step time-dependent analysis from stretching the tendons to the service period. These detailed calculation processes generally can be achieved on the computer.

The approach to estimating the effective stress in prestressing tendons provided in the codes can be used for general prestressed concrete structures. As discussed above, all prestress losses can be classified into two categories, instantaneous losses, $\sigma_{l,I}$, and long-term losses, $\sigma_{l,II}$. $\sigma_{l,I}$ occurs during construction, and is also called the first stage loss; $\sigma_{l,II}$ occurs in the service period, and is also called the second stage loss, which is broadly adopted in the codes to simplify the structural analysis under the prestressing force and design load, as briefly listed in Table 5.4.

For pretensioned members:

$$\sigma_{l,I} = \sigma_{l2}(x) + \sigma_{l3}(x) + \sigma_{l4}(x) + \xi_1 \sigma_{l5}(x, t \to \infty) \tag{5.81}$$

$$\sigma_{l,II} = \xi_2 \sigma_{l5}(x, t \to \infty) + \sigma_{l6}(x, t) \tag{5.82}$$

For post-tensioned members:

$$\sigma_{l,I} = \sigma_{l1}(x) + \sigma_{l2}(x) + \sigma_{l4}(x) \tag{5.83}$$

Table 5.4 Calculation of effective stress in prestressing tendons in the codes.

Time under consideration	Prestress losses	Effective stress in prestressing tendons
Immediately after transfer	$\sigma_{l,I}$	$\sigma_{con} - \sigma_{l,I}$
Service period	$\sigma_{l,I} + \sigma_{l,II}$	$\sigma_{con} - (\sigma_{l,I} + \sigma_{l,II})$

$$\sigma_{l,II} = \sigma_{l5}(x, t) + \sigma_{l6}(x, t) \tag{5.84}$$

In the Q/CR 9300-2018 and JTG 3362-2018 codes, $\xi_1 = \xi_2 = 0.5$ is taken in Eqs. (5.81) and (5.82). In the GB50010-2010 code, $\xi_1 = 1.0$ and $\xi_2 = 0$ are taken. The GB50010-2010 code stipulates that when the total prestress losses calculated from Eqs. (5.81)−(5.84) are less than the following values, the following values should be adopted: 100 MPa for pretensioned members and 80 MPa for post-tensioned members.

If the prestressing tendons with polylines or curves are arranged in the pretensioned members, frictional loss will occur between the steering gears and the prestressing tendons at stretching, in this case, $\sigma_{l1}(x)$ should be considered in Eq. (5.81).

In the preliminary design, the effective stress must be estimated first to determine the amount and layout of the prestressing tendons according to the design load and the preliminary proposed structural system and sectional dimensions. Consider the fact that $\sigma_{l1}(x)$, $\sigma_{l2}(x)$, and $\sigma_{l6}(x, t)$, etc. cannot be calculated before the preliminary design is completed, that is, Eq. (5.1) or Table 5.4 cannot be used to predict the effective stress in prestressing tendons. To facilitate the preliminary design, some codes or manuals provide the estimated value of prestress loss for each item, such as PCI and AASHTO, etc. These values are determined mostly for the members with a standardized design using conventional materials under typical ambient conditions, it is commonly observed that there is an obvious discrepancy between these values and the actual data in a statically indeterminate prestressed concrete structure with long prestressing tendons. Actually, it is more meaningful to give the ratio of final effective stress to initial jacking stress (control stress at stretching) than to estimate the prestress loss by each item in the preliminary design, called the coefficient of prestressing efficiency, expressed by

$$\lambda_{pe} = \frac{\sigma_{pe}(x, t \rightarrow \infty)}{\sigma_{con}} \tag{5.85}$$

where λ_{pe} = coefficient of prestressing efficiency, defined as the ratio of final effective stress to initial jacking stress.

$\sigma_{pe}(x, t \rightarrow \infty)$ = effective stress in the prestressing tendons at the final time when all prestress losses have taken place.

σ_{con} = control stress in prestressing tendons at the jacking point prior to transfer.

Generally, at the midspan section of the simply supported prestressed concrete T-girders and box girders with a span of 10−40 m by standardized design, made using C50 concrete and low-relaxation strands ($7\phi5$) with a characteristic tensile strength of 1860 MPa, widely used in railways and highways in China, λ_{pe} could be taken as equal to 0.7−0.8 for pretensioned girders and 0.65−0.75 for post-tensioned girders, respectively. At the midspan section and support section of the prestressed concrete continuous box girders with long spans, λ_{pe} could be taken as equal to 0.6−0.7.

5.5 Example 5.1

A railway post-tensioned simply supported box girder with an effective span of 24.0 m is made of C50 concrete and 14(7φ5) strands with a characteristic tensile strength of 1860 MPa. Two typical strands, N4 and N6, are embedded in the corrugated metal ducts, as shown in Fig. 5.8, and are stretched at both ends by a jacking stress of 1250 MPa. The measured retraction of the tendons is 5.8 mm and the compacted deformation of the anchorage is 1.0 mm due to the anchorage set. The tested coefficient of friction between the prestressing tendons and the duct wall is 0.265 and the coefficient of wobble friction is 0.003. Calculate the effective stresses in prestressing tendons according to the approach provided in the JTG 3362-2018 code.

5.6 Solution

5.6.1 Parameters

$\sigma_{con} = 1250$ MPa, $\mu = 0.265$, $k = 0.003$

5.6.2 Strand N4

$$R_2 = 10 \text{ m}, \ x_B = 2.373 \cos 25° = 2.151 \text{ m}$$

$$x_C = x_B + 2R \sin\left(\frac{L}{R} \cdot \frac{180}{2\pi}\right) \cos 12.5° = 2.151 + 2 \times 10 \times \sin\left(\frac{4.363}{10} \cdot \frac{180}{2\pi}\right)$$

$$\times \cos 12.5°$$

$$= 6.377 \text{ m}$$

Assuming $x_B < l_f < x_C$, according to Eq. (5.29):

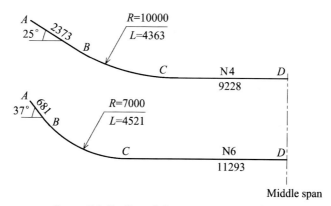

Figure 5.8 Profiles of the prestressing tendons.

$$l_f = \sqrt{\dfrac{\dfrac{\Delta l E_p}{\sigma_{con}} + \dfrac{\mu}{R_2}x_B^2}{k + \dfrac{\mu}{R_2}}} = \sqrt{\dfrac{\dfrac{6.8 \times 10^{-3} \times 1.95 \times 10^5}{1250} + \dfrac{0.265}{10} \times 2.151^2}{0.003 + \dfrac{0.265}{10}}} = 6.334 \text{ m}$$

Since $x_B = 2.151$ m $< l_f = 6.334$ m $< x_C = 6.377$ m, the assumed value of l_f is valid.

The prestress loss at the end of the anchorage device due to the anchorage set can be obtained as

$$\sigma_{l2}^0 = \sigma_{con}\left[2\kappa l_f + \dfrac{2\mu}{R_2}\left(l_f - x_B\right)\right]$$

$$= \sigma_{con} \times \left[2 \times 0.003 \times 6.334 + \dfrac{2 \times 0.265}{10} \times (6.334 - 2.151)\right] = 0.260\sigma_{con}$$

$$= 325 \text{ MPa}$$

The tendon stress at the end of the anchorage device at the anchorage set is

$$\sigma_{p,0} = \sigma_{con} - \sigma_{l2}^0 = 0.740\sigma_{con} = 925 \text{ MPa}$$

And the tendon stress at the fixed point can be obtained from Eq. (5.31), as

$$\sigma_{p,s} = \sigma_{con}\left[1 - \kappa l_f - \dfrac{\mu}{R_2}\left(l_f - x_B\right)\right]$$

$$= \sigma_{con} \times \left[1 - 0.003 \times 6.334 - \dfrac{0.265}{10} \times (6.334 - 2.151)\right] = 0.870\sigma_{con}$$

$$= 1087.5 \text{ MPa}$$

5.6.3 Strand N6

$$R_2 = 7.0 \text{ m}, \quad x_B = 0.681 \cos 37° = 0.544 \text{ m}$$

$$x_C = x_B + 2R\sin\left(\dfrac{L}{R} \cdot \dfrac{180}{2\pi}\right)\cos 12.5° = 0.544 + 2 \times 7 \times \sin\left(\dfrac{4.521}{7} \cdot \dfrac{180}{2\pi}\right)$$

$$\times \cos 18.5°$$

$$= 4.757 \text{ m}$$

Assuming $x_B < l_f < x_C$, according to Eq. (5.29):

$$l_f = \sqrt{\frac{\dfrac{\Delta l E_p}{\sigma_{con}} + \dfrac{\mu}{R_2}x_B^2}{k + \dfrac{\mu}{R_2}}} = \sqrt{\frac{\dfrac{6.8 \times 10^{-3} \times 1.95 \times 10^5}{1250} + \dfrac{0.265}{7} \times 0.544^2}{0.003 + \dfrac{0.265}{7}}} = 5.122 \text{ m}$$

Since $l_f = 5.122$ m $> x_C = 4.757$ m, the assumed value of l_f is invalid, so l_f should be calculated again by Eq. (5.32):

$$l_f = \sqrt{\frac{1}{\kappa}\left[\frac{\Delta l E_p}{\sigma_{con}} - \frac{\mu}{R_2}\left(x_C^2 - x_B^2\right)\right]}$$

$$= \sqrt{\frac{1}{0.003} \times \left[\frac{6.8 \times 10^{-3} \times 1.95 \times 10^5}{1250} - \frac{0.265}{7} \times \left(4.757^2 - 0.544^2\right)\right]} = 8.472 \text{ m}$$

Since $x_C = 4.757$ m $< l_f = 8.472$ m $< l_p/2 = 12.0$ m, the assumed value of l_f is valid.

The prestress loss at the end of the anchorage device due to the anchorage set can be obtained as

$$\sigma_{l2}^0 = \sigma_{con}\left[2\kappa l_f + \frac{2\mu}{R_2}(x_C - x_B)\right]$$

$$= \sigma_{con} \times \left[2 \times 0.003 \times 0.872 + \frac{2 \times 0.265}{7} \times (4.757 - 0.544)\right] = 0.324\sigma_{con}$$

$$= 405 \text{ MPa}$$

The tendon stress at the end of the anchorage device at the anchorage set is

$$\sigma_{p,0} = \sigma_{con} - \sigma_{l2}^0 = \sigma_{con} - 0.324\sigma_{con} = 0.676\sigma_{con} = 845 \text{ MPa}$$

And the tendon stress at the fixed point can be obtained from Eq. (5.34), as

$$\sigma_{p,s} = \sigma_{con}\left\{1 - \left[\kappa l_f + \frac{\mu}{R_2}(x_C - x_B)\right]\right\}$$

$$= \left[1 - 0.003 \times 0.872 - \frac{0.265}{7} \times (4.757 - 0.544)\right] = 0.838\sigma_{con} = 1047.5 \text{ MPa}$$

5.7 Example 5.2

A highway post-tensioned T-girder with a length of 24.90 m is simply supported over a span of 24.20 m, as shown in Figs. 5.9 and 5.10. Concrete C40 and four batches of $7\varphi5$ strands with a characteristic tensile strength of 1860 MPa are used. The distance between the two anchoring points of the strands is $l_a = 24.60$ m, where one side of the strands is

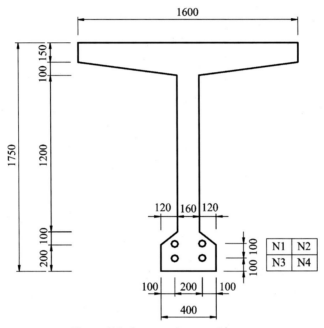

Figure 5.9 Cross-section at midspan.

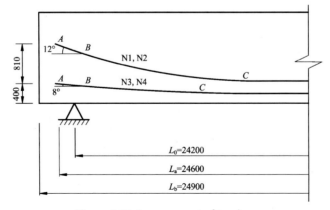

Figure 5.10 Arrangement of tendons.

fixed by an anchorage and the other side is stretched by a jack. The strands N1−N4 are designed as the typical profile, as shown in Fig. 5.6, where $l_{AB} = l_{EF} = 0.5$ m. N1 and N2 are bent by $\theta = 12°$ and N3 and N4 are bent by $\theta = 8°$ totally in curve BC. The corrugated metal ducts with a diameter of 60 mm are embedded in concrete before casting, and the strands are stretched in the order of $N3 \rightarrow N4 \rightarrow N1 \rightarrow N2$ by a jacking stress $0.75f_{ptk}$ as the strength of concrete reaches the design value. The annual average relative humidity at the bridge site is 55%. The load intensities due to the self-weight of the girder and the weight of the deck system are 15.71 kN/m and 4.4 kN/m, respectively. Calculate the prestress loss at the midspan section under the prestressing force and the weight of the structure in 10 years according to the JTG 3362-2018 code.

5.8 Solution

5.8.1 Material properties

$f_{cd} = 18.4$ MPa, $f_{cu,k} = 40$ MPa, $f_{cmo} = 40$ MPa

$\quad f_{ptk} = 1860$ MPa, $f_{pd} = 1260$ MPa, $A_{p,0} = 690.9$ mm^2

$\quad \sigma_{con} = 0.75f_{ptk} = 0.75 \times 1860 = 1395.0$ MPa, $\mu = 0.20$, $k = 0.0015$

$\quad A_n = 616690$ mm^2, $I_n = 2.16663 \times 10^{11}$ mm^4, $I_0 = 2.33505 \times 10^{11}$ mm^4

$\quad E_p = 1.95 \times 10^5$ MPa, $E_c = 3.25 \times 10^4$ MPa

$$\alpha_{EP} = \frac{E_p}{E_c} = \frac{1.95 \times 10^5}{3.25 \times 10^4} = 6.0$$

5.8.2 Geometric parameters of the strands

$$a_p = \frac{2 \times 100 + 2 \times 210}{4} = 155 \text{ mm}$$

$$h_0 = h - a_p = 1750 - 155 = 1595 \text{ mm}$$

In this example, the radius in the segment BC of the strands is not provided, therefore it is necessary to obtain the radius first, then to obtain the other geometric parameters of the strands. Fig. 5.11 shows the geometric relation in the curved segment BC.

In Fig. 5.11, let d_v be the vertical distance between points B and C, l_1 be the horizontal projected length of the curve BC, and l_2 be the length of segment CD. In Fig. 5.11, $KC = OC - OK$, giving

$$d_v = r - r \cos \theta$$

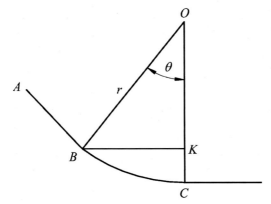

Figure 5.11 Geometric relation in the curved segment BC.

Thereby the radius in the segment BC is obtained:

$$r = \frac{d_v}{1 - \cos \theta}$$

For N1 and N2:

$$d_v = 1210 - 210 - 500 \times \sin 12° = 896 \text{ mm}$$

For N3 and N4:

$$d_v = 400 - 100 - 500 \times \sin 8° = 230 \text{ mm}$$

Considering the relations $r = d_v/(1 - \cos \theta)$, $l_1 = r \sin \theta$ and $l_2 = l_a - 2l_1$, the geometric parameters of the strands can be obtained from Figs. 5.9 and 5.10, as listed in Table 5.5.

5.8.3 Prestress loss due to friction

$$\sigma_{l1} = \sigma_{con}\left[1 - e^{-(\mu\theta + kx)}\right]$$

At midspan:

$$x = \frac{l_a}{2} = \frac{24.60}{2} = 12.30 \text{ m}$$

Table 5.5 Geometric parameters of the strands.

Strands	$\theta/°$	d_v/mm	$r = d_v/(1 - \cos \theta)$/mm	$l_1 = r \sin \theta$/mm	$l_2 = l_a - 2l_1$/mm
N1, N2	12	896	41,002	8524	7552
N3, N4	8	230	23,633	3289	18,022

$\theta = 12° = 0.209$ rad (N1, N2), $\theta = 8° = 0.140$ rad (N3, N4).

For N1 and N2:

$$\sigma_{l1} = \sigma_{con}\left[1 - e^{-(\mu\theta+kx)}\right] = 1395 \times \left[1 - e^{-(0.20\times0.209+0.0015\times12.30)}\right] = 81.57 \text{ MPa}$$

For N3 and N4:

$$\sigma_{l1} = \sigma_{con}\left[1 - e^{-(\mu\theta+kx)}\right] = 1395 \times \left[1 - e^{-(0.20\times0.140+0.0015\times12.30)}\right] = 63.32 \text{ MPa}$$

5.8.4 Prestress loss due to the anchorage set

The retraction of the tendons due to the anchorage set is taken as 6 mm from Table 5.3.

The influence length of retraction due to the anchorage set can be calculated by Eqs. (5.37) and (5.38), as

$$l_f = \sqrt{\frac{\sum \Delta l E_p}{\Delta\sigma_d}}, \ \Delta\sigma_d = \frac{\sigma_{l1,m}}{l_s}$$

For N1 and N2:

$x = l_a = 24600$ mm, $\theta = 12^0 = 0.209$ rad

$$\sigma_{l1,m} = \sigma_{con}\left[1 - e^{-(\mu\theta+kx)}\right] = 1395 \times \left[1 - e^{-(0.20\times0.209+0.0015\times24.60)}\right]$$
$$= 105.58 \text{ MPa}$$

$$\Delta\sigma_d = \frac{\sigma_{l1,m}}{l_a} = \frac{105.58}{24600} = 4.292 \times 10^{-3} \text{ MPa/mm}$$

$$l_f = \sqrt{\frac{6 \times 1.95 \times 10^5}{4.292 \times 10^{-3}}} = 16510 \text{ mm } < l = 24600 \text{ mm}$$

$$\Delta\sigma = 2\Delta\sigma_d l_f = 2 \times 4.292 \times 10^{-3} \times 16510 = 141.72 \text{ MPa}$$

$$\sigma_{l2} = \Delta\sigma\frac{l_f - x}{l_f} = 141.72 \times \frac{16510 - 12300}{16510} = 36.14 \text{ MPa}$$

For N3 and N4:

$x = l_a = 24600$ mm, $\theta = 8° = 0.140$ rad

$$\sigma_{l1,m} = 1395 \times \left[1 - e^{-(0.20\times0.140+0.0015\times24.60)}\right] = 87.66 \text{ MPa}$$

$$\Delta\sigma_d = \frac{\sigma_{l1,m}}{l_a} = \frac{87.66}{24600} = 3.563 \times 10^{-3} \text{ MPa}$$

$$l_f = \sqrt{\frac{6 \times 1.95 \times 10^5}{3.563 \times 10^{-3}}} = 18121 \text{ mm}$$

$$\Delta\sigma = 2\Delta\sigma_d l_f = 2 \times 3.563 \times 10^{-3} \times 18121 = 129.13 \text{ MPa}$$

$$\sigma_{l2} = \Delta\sigma \frac{l_f - x}{l_f} = 129.13 \times \frac{18121 - 12300}{18121} = 41.48 \text{ MPa}$$

5.8.5 Prestress loss due to elastic shortening

For N1 and N2:

$$\sigma_{pe} = \sigma_{con} - \sigma_{l1}(x) - \sigma_{l2}(x) = 1395 - 81.57 - 36.14 = 1277.29 \text{ MPa}$$

$$N_p = \sigma_{pe} A_p = 1277.29 \times 2 \times 690.9 = 1764959 \text{ N}$$

For N3 and N4:

$$\sigma_{pe} = \sigma_{con} - \sigma_{l1}(x) - \sigma_{l2}(x) = 1395 - 63.32 - 41.48 = 1290.20 \text{ MPa}$$

$$N_p = \sigma_{pe} A_p = 1290.20 \times 2 \times 690.9 = 1782798 \text{ N}$$

Hence

$$e_{pn} = \frac{\sigma_{pe} A_p \gamma_{pn}}{N_p} = \frac{\sum N_{pi} \gamma_{pni}}{\sum N_{pi}} = \frac{1764959 \times (1176 - 210) + 1782798 \times (1176 - 100)}{1764959 + 1782798} = 1021 \text{ mm}$$

$$\Delta\sigma_{pc} = \frac{1}{4}\left(\frac{N_p}{A_n} + \frac{N_p e_{pn}}{I_n}\gamma_n\right)$$

$$= \frac{1}{4} \times \left(\frac{1764959 + 1782798}{616690} + \frac{(1764959 + 1782798) \times 1021}{2.16663 \times 10^{11}} \times (1176 - 155)\right) = 5.706 \text{ MPa}$$

$$\sigma_{l4} = \frac{m-1}{2}\alpha_{EP}\Delta\sigma_{pc} = \frac{4-1}{2} \times 6 \times 5.706 = 51.35 \text{ MPa}$$

5.8.6 Prestress loss due to relaxation of strands

For N1 and N2:

$$\sigma_{pe} = 1225.94 \text{ MPa}$$

$$\sigma_{l5} = \psi_\varsigma\left(0.52\frac{\sigma_{pe}}{f_{pk}} - 0.26\right)\sigma_{pe}$$

$$= 1 \times 0.3 \times \left(0.52 \times \frac{1225.94}{1860} - 0.26\right) \times 1225.94 = 30.43 \text{ MPa}$$

For N3 and N4:

$$\sigma_{pe} = 1258.85 \text{ MPa}$$

$$\sigma_{l5} = \psi_{\varsigma}\left(0.52\frac{\sigma_{pe}}{f_{ptk}} - 0.26\right)\sigma_{pe}$$

$$= 1 \times 0.3 \times \left(0.52 \times \frac{1258.85}{1860} - 0.26\right) \times 1258.85 = 34.72 \text{ MPa}$$

5.8.7 Prestress loss due to shrinkage and creep of concrete

(1) Shrinkage strain and creep coefficient.

The shrinkage strain can obtain from Eqs. (2.27)–(2.31). Some of the parameters used in the calculations are as follows:

$$t_s = 7 \text{ } d, \text{ } t = 3650 + 28 = 3678 \text{ } d, \text{ } \beta_{sc} = 5.0, \text{ } f_{cm} = 0.8f_{cu,k} + 8 = 0.8 \times 40 + 8$$
$$= 40 \text{ MPa}$$

The gross cross-sectional area can be obtained from Fig. 5.9, as

$$A = 1600 \times 150 + \frac{100}{2} \times (160 + 1600) + 1200 \times 160 + \frac{100}{2} \times (160 + 400) + 200$$
$$\times 400$$
$$= 628000 \text{ mm}^2$$

The perimeter of the section is

$$u = 1600 + 150 \times 2 + 2 \times \sqrt{100^2 + 720^2} + 1200 \times 2 + 2 \times \sqrt{100^2 + 120^2} + 2$$
$$\times 200 + 400$$
$$= 6866 \text{ mm}$$

The notional size of the girder is

$$h = \frac{2A}{u} = \frac{2 \times 628000}{6866} = 182.9 \text{ mm}$$

Then

$$\beta_{RH} = 1.55\left[1 - \left(\frac{RH}{RH_0}\right)^3\right] = 1.55\left[1 - \left(\frac{55}{100}\right)^3\right] = 1.292$$

$$\beta_s(t - t_s) = \left[\frac{\dfrac{t - t_s}{t_1}}{350\left(\dfrac{h}{h_0}\right)^2 + \dfrac{t - t_s}{t_1}}\right]^{0.5} = \left[\frac{3678 - 7}{350\left(\dfrac{182.9}{100}\right)^2 + (3678 - 7)}\right]^{0.5} = 0.871$$

$$\varepsilon_{sh}(f_{cm}) = \left[160 + 10\beta_{sc}\left(9 - \frac{f_{cm}}{f_{cmo}}\right)\right] \times 10^{-6} = \left[160 + 10 \times 5\left(9 - \frac{40}{40}\right)\right] \times 10^{-6}$$
$$= 560 \times 10^{-6}$$

$$\varepsilon_{sho} = \varepsilon_{sh}(f_{cm})\beta_{RH} = 560 \times 10^{-6} \times 1.292 = 723.52 \times 10^{-6}$$

$$\varepsilon_{sh}(t, t_s) = \varepsilon_{sho}\beta_s(t - t_s) = 723.52 \times 10^{-6} \times 0.871 = 0.63 \times 10^{-3}$$

The creep coefficient can be calculated from Eqs. (2.16)–(2.22) or obtained from Table 2.8.

$$\beta_H = 150\left[1 + \left(1.2\frac{RH}{RH_0}\right)^{18}\right]\frac{h}{h_0} + 250 = 150\left[1 + \left(1.2 \times \frac{55}{100}\right)^{18}\right] \times \frac{182.9}{100}$$
$$+ 250$$
$$= 524.5$$

$$\beta_c(t - t_0) = \left[\frac{(t - t_0)/t_1}{\beta_H + (t - t_0)/t_1}\right]^{0.3} = \left[\frac{3678 - 7}{524.5 + 3678 - 7}\right]^{0.3} = 0.875$$

According to $h = 182.9$ mm and $RH = 55\%$, $\varphi(t, t_0) = 2.33$ is obtained from Table 2.8.

(2) Normal stress of concrete due to the effective stressing force and the weight of the structure.

At the end of the first stage, σ_{l1}, σ_{l2}, and σ_{l4} have taken place, so the effective stress in the strands at the beginning of creep and shrinkage is

$$\sigma_{pe} = \sigma_{con} - \sigma_{l,I} = \sigma_{con} - \sigma_{l1} - \sigma_{l2} - \sigma_{l4}$$

For N1 and N2:

$$\sigma_{pe} = 1225.94 \text{ MPa}$$

$$N_p = \sigma_{pe} \times 2 \times A_{p,0} = 1225.94 \times 2 \times 690.9 = 1694004 \text{ N}$$

For N3 and N4:

$$\sigma_{pe} = 1258.85 \text{ MPa}$$

$$N_p = \sigma_{pe} \times 2 \times A_{p,0} = 1238.85 \times 2 \times 690.9 = 1711843 \text{ N}$$

Then

$$e_{pn} = \frac{\sigma_{pe} A_p \gamma_{pn}}{N_p} = \frac{\sum N_{pi} \gamma_{pni}}{\sum N_{pi}}$$

$$= \frac{1694004 \times (1176 - 210) + 1711843 \times (1176 - 100)}{1694004 + 1711843} = 1021 \text{ mm}$$

$$\sigma_{pc,p} = \frac{N_p}{A_n} + \frac{N_p e_{pn}}{I_n} y_n$$

$$= \frac{1694004 + 1711843}{616690} + \frac{(1694004 + 1711843) \times 1021}{2.16663 \times 10^{11}} \times (1176 + 155)$$

$$= 21.91 \text{ MPa}$$

The bending moments at the midspan due to the weight of the girder and deck system are

$$M_{d1} = \frac{q_g l_0^2}{8} = \frac{15.71 \times 24.2^2}{8} = 1150 \text{ kN·m}$$

$$M_{d2} = \frac{q_{deck} l_0^2}{8} = \frac{4.4 \times 24.2^2}{8} = 322 \text{ kN·m}$$

In the service period, the properties of the converted section are used to calculate the stress, hence

$$\sigma_{cd,p} = -\frac{(M_{d1} + M_{d2}) \times y_0}{I_0} = -\frac{(1150 + 322) \times 10^6 \times 1150}{2.33505 \times 10^{11}} = -7.25 \text{ MPa}$$

Therefore, normal stress of concrete at the centroid of all steels due to the effective stressing force and the weight of the girder and deck system is obtained as

$$\sigma_{pc} = \sigma_{pc,p} + \sigma_{cd,p} = 21.91 - 7.25 = 14.66 \text{ MPa}$$

(3) Prestress loss due to shrinkage and creep of concrete.

There are no prestressing steels in the compression zone, so the prestress loss due to shrinkage and creep of concrete is given by

$$\sigma_{l6}(t) = \frac{0.9 \times \left[E_p\varepsilon_{cs}(t, t_0) + \alpha_{EP}\sigma_{pc}\varphi(t, t_0)\right]}{1 + 15\rho\rho_{ps}}$$

where σ_{pc} is the normal stress of concrete at the centroid of all steels due to the prestressing force and the weight of the girder and deck system.

$$\rho = \frac{A_p + A_s}{A_n} = \frac{4 \times 690.9}{616690} = 0.00448$$

$$e_{ps} = e_{pn} = 1021 \text{ mm}$$

$$i^2 = \frac{I_n}{A_n} = \frac{2.16663 \times 10^{11}}{616690} = 3.513 \times 10^5 \text{ mm}^2$$

$$\rho_{ps} = 1 + \frac{e_{ps}^2}{i^2} = 1 + \frac{1021^2}{3.513 \times 10^5} = 3.967$$

Therefore:

$$\sigma_{l6} = \frac{0.9\left[E_p\varepsilon_{cs}(t, t_0) + \alpha_{EP}\sigma_{pc}\varphi(t, t_0)\right]}{1 + 15\rho\rho_{Ps}}$$

$$= \frac{0.9 \times \left[1.95 \times 10^5 \times 0.63 \times 10^{-3} + 6 \times 14.66 \times 2.33\right]}{1 + 15 \times 0.00448 \times 3.967} = 232.92 \text{ MPa}$$

5.8.8 Total prestress loss at midspan section

The total prestress losses at midspan section occur after 10 years.
    For N1 and N2:

$$\sigma_l = \sigma_{l1} + \sigma_{l2} + \sigma_{l4} + \sigma_{l5} + \sigma_{l6} = 81.57 + 36.14 + 51.35 + 30.43 + 232.92$$
$$= 432.41 \text{ MPa}$$

For N3 and N4:

$$\sigma_l = \sigma_{l1} + \sigma_{l2} + \sigma_{l4} + \sigma_{l5} + \sigma_{l6} = 63.32 + 41.48 + 51.35 + 34.72 + 232.92$$
$$= 423.79 \text{ MPa}$$

5.9 Example 5.3

Consider the pretensioned hollow slab shown in Fig. 5.12 with a total length of 4.0 m and an effective span of 3.8 m. C40 concrete and ϕ^P4 ($n = 12$) straight steel wires are adopted. The distance between two abutments for temporarily anchoring wires is

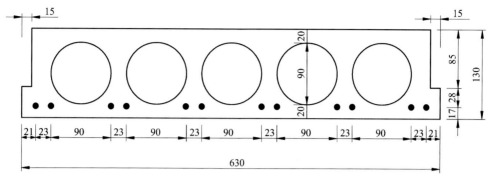

Figure 5.12 Section dimension.

80 m and the retraction of the tendons is 5 mm due to the anchorage set. During heat treatment curing, the temperature increases by 20°C, and the temporary anchoring is released when the strength of the concrete reaches 85% of the characteristic compressive strength. Calculate the prestress loss of the slab according to the GB50010-2010 code. Reinforcing steels are disregarded in the calculation.

5.10 Solution

5.10.1 Material properties

$f_{ck} = 26.8 \, \text{MPa}, f_{tk} = 2.39 \, \text{MPa}, f_{cd} = 19.1 \, \text{MPa}, f_{ptk} = 1570 \, \text{MPa}, f_{pd} = 1250 \, \text{MPa},$
  $E_c = 3.25 \times 10^4 \, \text{MPa}, E_p = 2.05 \times 10^5 \, \text{MPa}$

$$\alpha_{EP} = \frac{E_p}{E_c} = \frac{2.05 \times 10^5}{3.25 \times 10^4} = 6.31$$

5.10.2 Convert a hollow section into an I-section

Convert the hollows in the slab into a rectangular section with the same area, centroid, and moment of inertia. The equivalent section for a hollow with a width of b_1 and depth of h_1 can be obtained by

$$A_{hole} = \frac{\pi d^2}{4} = b_1 h_1$$

$$I_{hole} = \frac{\pi d^4}{64} = \frac{1}{12} b_1 h_1^3$$

From which

$$b_1 = \frac{\pi}{2\sqrt{3}} d = \frac{3.1416}{2\sqrt{3}} 90 = 82 \, \text{mm}$$

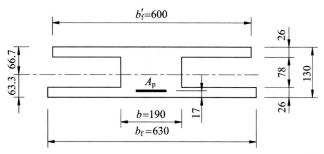

Figure 5.13 Converted section.

$$h_1 = \sqrt{3}d/2 = \sqrt{3} \times 90/2 = 78 \text{ mm}$$

Accordingly, the converted I-shaped section is achieved, as shown in Fig. 5.13:

$$b_f = 630 \text{ mm}; \ b_f' = 630 - 2 \times 15 = 600 \text{ mm}$$

$$b = b_f' - nb_1 = 600 - 5 \times 82 = 190 \text{ mm}$$

$$h_f = 20 + \frac{90 - 78}{2} = 26 \text{ mm}$$

$$h_{middle} = h - 2h_f = 130 - 2 \times 26 = 78 \text{ mm}$$

5.10.3 Geometrical properties of the section

The area of the converted section is

$$A_0 = 600 \times 26 + 78 \times 190 + 26 \times 630 + (6.31 - 1) \times 150.80 = 47601 \text{ mm}^2$$

The second moment of the converted section to the bottom fiber is

$$S_0 = 600 \times 26 \times \left(130 - \frac{26}{2}\right) + 78 \times 190 \times \left(26 + \frac{78}{2}\right)$$

$$+ 26 \times 630 \times \frac{26}{2} + (6.31 - 1) \times 150.80 \times 17$$

$$= 3015053 \text{ mm}^3$$

The distance from the centroid of the converted section to the bottom fiber is

$$y_0 = \frac{S_0}{A_0} = \frac{3015053}{47601} = 63.3 \text{ mm}$$

The distance from the centroid of the converted section to the top fiber is

$$y_2 = 130 - y_0 = 130 - 63.3 = 66.7 \text{ mm}$$

The distance from the centroid of the prestressing tendons to the centroid of the converted section is

$$e_{p0} = y_0 - a_p = 63.3 - 17 = 46.3 \text{ mm}$$

Therefore, the moment of inertia of the converted section is

$$I_0 = \frac{1}{3} \times (630 - 190) \times 26^3 + \frac{1}{3} \times 190 \times 130^3 + (600 - 190) \times 26 \times (130 - 26/2)^2$$

$$+ \frac{1}{12} \times (600 - 190) \times 26^3 + (6.31 - 1) \times 150.8 \times 17^2 - 47601 \times 63.3^2 \text{ mm}^4$$

$$= 1.028 \times 10^8 \text{ mm}^4$$

5.10.4 Prestress losses at the first stage

$$\sigma_{con} = 0.75 f_{ptk} = 0.75 \times 1570 = 1177.5 \text{ MPa}$$

Prestress loss due to anchorage set:

$$\sigma_{l2} = \frac{\Delta l}{l_p} E_p = \frac{5}{80 \times 10^3} \times 2.05 \times 10^5 = 12.8 \text{ MPa}$$

Prestress loss due to heat treatment curing:

$$\sigma_{l3} = 2\Delta t = 2 \times 20 = 40 \text{ MPa}$$

Prestress loss due to relaxation of steel wire:

$$\psi = 1.0$$

$$\sigma_{l5} = 0.4\psi \left(\frac{\sigma_{con}}{f_{ptk}} - 0.5 \right) \sigma_{con}$$

$$= 0.4 \times 1.0 \times (0.75 - 0.5) \times 1177.5 = 117.75 \text{ MPa}$$

In the GB50010-2010 code, $\xi_1 = 1.0$ and $\xi_2 = 0$ in Eqs. (5.81) and (5.82), hence, the prestress loss at the first stage is

$$\sigma_{l,I} = \sigma_{l2} + \sigma_{l3} + \sigma_{l5} = 12.8 + 40 + 117.75 = 170.55 \text{ MPa}$$

5.10.5 Prestress loss at the second stage

This hollow slab is a general prestressed member, so Eqs. (5.77) and (5.78) are used to calculate the prestress loss due to shrinkage and creep of concrete.

The effective stressing force in prestressing tendons at the end of the first stage and the tendon eccentricity are as follows:

$$N_{p0I} = (\sigma_{con} - \sigma_{lI})A_p = (1177.5 - 170.55) \times 150.8 = 151848 \text{ N}$$

$$e_{p0I} = 46.3 \text{ mm}$$

The normal stress of concrete at the centroid of the prestressing tendons due to the effective stressing force:

$$\sigma_{pc} = \frac{N_{p0I}}{A_0} + \frac{N_{p0I}e_{p0I}^2}{I_0} = \frac{151848}{47601} + \frac{151848 \times 46.3^2}{1.028 \times 10^8} = 6.36 \text{ MPa}$$

Prestressing steel ratio:

$$\rho_p = \frac{A_p}{A_0} = \frac{150.8}{47601} = 0.00317$$

Thus, the prestress loss due to shrinkage and creep of concrete is obtained:

$$\sigma_{l6} = \frac{60 + 340\dfrac{\sigma_{pc}}{f_{cu}'}}{1 + 15\rho} = \frac{60 + 340 \times 0.212}{1 + 15 \times 0.00317} = 126.08 \text{ MPa}$$

5.10.6 Total prestress losses in the service period

$$\sigma_l = \sigma_{l,I} + \sigma_{l,II} = 170.55 + 126.08 = 296.63 \text{ MPa}$$

References

[1] Hu Di, Chen Z. Experimental research on shrinkage & creep and deflection in prestressed concrete bridges. Civil Eng J 2003;36(8):79–85.
[2] Hu Di. The steel restraint influence coefficient method to analyze time-dependent effect in prestressed concrete bridges. Eng Mech 2006;23(6):120–6.
[3] Hu Di, Chen Z. Computation of stress losses in prestressed steel due to anchorage set considering the function of reverse-friction in PC members. China J Highw Transp 2004;17(1):34–8.
[4] Hu Di, Chen Z. Automatically step-up method to analyze creep in prestressed concrete bridges. Eng Mech 2004;21(5):41–5. 47.
[5] Hu Di. Basic principles of prestressed concrete structure design. 2nd ed. Beijing: China Railway Publishing House; 2019.
[6] Hu Di. Theory of Creep effect on concrete structures. Beijing: Science Press; 2015.
[7] Bažant ZP. Prediction of concrete creep effects using age-adjusted effective modulus method. ACI J 1972;69:212–7.
[8] GB 50010: 2015. Code for design of concrete structures. Beijing: China Construction Industry Press; 2015.
[9] Q/CR 9300:2018. Code for design on railway bridge and culvert. Beijing: China Railway Publishing House; 2018.

[10] JTG 3362: 2018. Specifications for design of highway reinforced concrete and prestressed concrete bridges and culverts. Beijing: People's Communications Press; 2018.

[11] FIB. Model code 2010 — final draft, volume 1. Fib bulletin N.65. Ernst & Sohn; 2012. fib International Federation for Structural Concrete: fib Model Code for concrete Structures 2010.Ernst & Sohn, Berlin, 2013.

[12] Eurocode 2:Design of concrete structures. European standard, prEN 1992-1-1. European Committee for Standardization; 2002.

[13] AASHTO LFRD Bridge design specifications. 8th ed. Washington D.C.: American Association of State and Highway Transportation Officials; 2017.

CHAPTER 6

Effects of the prestressing force on structures

Contents

6.1 Equivalent loads of the prestressing force

To analyze the effects of prestressing force on a prestressed concrete structure, one feasible approach is to transform the prestressing force into external loads exerted on the concrete structures and carry out the analysis accordingly. The external loads are called the equivalent loads, and the structural analysis based on the equivalent loads is called the equivalent load method.

For post-tensioned structures, the anchorage compresses the supporting concrete due to the concentrated prestressing force. For prestressing tendons with a segment of arc or parabola, the radial pressure on concrete on the concave side is generated due to the prestressing force; for prestressing tendons with broken lines, a concentrated force on the concrete on the concave side at the turning point is generated due to the prestressing force. Therefore, the equivalent loads are related to the profile of the prestressing tendon and the anchoring method.

Considering that the stress over the entire length of the bonded prestressing tendons is different due to friction loss and set loss (Fig. 5.5), the theoretical distribution of the radial pressure is complex and difficult to obtain. Thus, it is commonly assumed that the effective stress in the prestressing tendons is constant along their entire length in a simplified analysis. The equivalent loads on concrete structures caused by prestressing tendons with typical profiles, such as a straight line, broken lines, and a parabola, in the post-tensioned structures, are discussed in this chapter.

Analysis and Design of Prestressed Concrete
ISBN 978-0-12-824425-8, https://doi.org/10.1016/B978-0-12-824425-8.00006-3

6.1.1 Equivalent loads of the straight prestressing tendon

Consider a simply supported prestressed concrete beam arranged with an eccentric straight prestressing tendon anchored at both ends, as shown in Fig. 6.1A. The effective prestressing force after all the prestress losses have taken place is given by

$$N_p = \sigma_{pe} A_p \tag{6.1}$$

where σ_{pe} = effective stress of the prestressing tendon after all the prestress losses have taken place.

A_p = cross-sectional area of the prestressing tendon.

To counterbalance the tensile force in the prestressed tendon, the concrete under anchorage bears compression along the direction of the tendon, namely, the equivalent load of the prestressed force acting on the concrete under an anchorage is a pair of compression of N_p (directs to inside the beam; Fig. 6.1C). In addition, the bending moment in the concrete cross-section is also generated due to the eccentric prestressing force.

Hence, the equivalent loads of a straight prestressing tendon in a post-tensioned member include the following parts.

At the left tendon end:

$$N_{L,x} = N_p \tag{6.2a}$$

At the right tendon end:

$$N_{R,x} = -N_p \tag{6.2b}$$

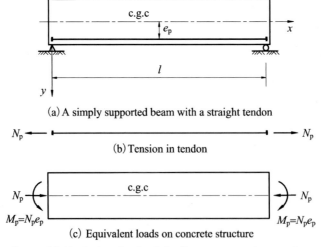

(a) A simply supported beam with a straight tendon

(b) Tension in tendon

(c) Equivalent loads on concrete structure

Figure 6.1 Equivalent loads of the linear prestressing tendon.

Along the entire length of tendon:

$$M_p = N_p e_p \tag{6.2c}$$

where e_p = eccentricity, the distance from the centroid of the tendon to the center of gravity of the concrete section.

6.1.2 Equivalent loads of the broken-line prestressing tendon

As shown in Fig. 6.2A, a broken-line prestressing tendon is anchored at the centroid of the concrete section at both ends of a simply supported beam. The bending angles of the left and right segments of the tendon are θ_1 and θ_2, respectively, and the eccentricity at the turning point is e_p. The effective stress in the prestressing tendon is σ_{pe} and the cross-sectional area is A_p, so the effective prestressing force is $N_p = \sigma_{pe} A_p$.

As shown by the analysis in Fig. 6.1, the equivalent load under an anchorage due to the prestressing force is a compression acting on the concrete along the direction of the tendon, so the equivalent loads at the tendon ends can be decomposed into two forces perpendicular to each other, as follows.

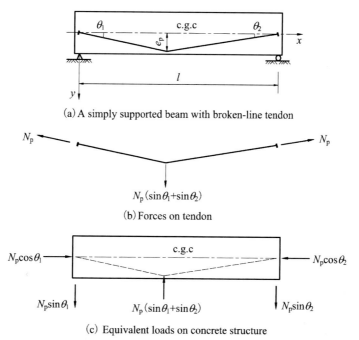

(a) A simply supported beam with broken-line tendon

(b) Forces on tendon

(c) Equivalent loads on concrete structure

Figure 6.2 Equivalent loads by a broken-line tendon.

At the left tendon end:

$$N_{L,x} = N_p \cos \theta_1 \tag{6.3a}$$

$$N_{L,y} = N_p \sin \theta_1 \tag{6.3b}$$

At the right tendon end:

$$N_{R,x} = -N_p \cos \theta_2 \tag{6.3c}$$

$$N_{R,y} = N_p \sin \theta_2 \tag{6.3d}$$

At the turning point, the tendon exerts an upward compression on the concrete:

$$C_{turning,y} = -N_p(\sin \theta_1 + \sin \theta_2) \tag{6.3e}$$

Along the entire length of the left segment:

$$M_{L,p} = \lambda_x e_p N_p \cos \theta_1 \tag{6.3f}$$

Along the entire length of the right segment:

$$M_{R,p} = \lambda_x e_p N_p \cos \theta_2 \tag{6.3g}$$

where λ_x = ratio of the eccentricity at the calculating point to e_p at the turning point.

Therefore, the equivalent loads exerted on the concrete due to the prestressing force provided by a broken-line prestressing tendon include $N_{L,x}$, $N_{L,y}$, $N_{R,x}$, $N_{R,y}$, $C_{turning,y}$, $M_{L,p}$, and $M_{R,p}$ as given in Eqs. (6.3a)–(6.3g).

6.1.3 Equivalent loads of the parabolic prestressing tendon

Consider a simply supported beam arranged with a parabolic prestressing tendon anchored at the centroid of the concrete section at both ends, as shown in Fig. 6.3A. The effective prestressing force is $N_p = \sigma_{pe} A_p$.

Assume the parabola can be expressed as

$$e_{px}(x) = \frac{4e_p}{l^2}x^2 - \frac{4e_p}{l}x \tag{6.4}$$

where $e_{px}(x)$ = eccentricity of the parabolic tendon at any point under consideration.

e_p = eccentricity of the parabolic tendon at midspan.

l = effective span of the simply supported beam.

The bending moment at any section directly produced by the prestressing force is

$$M_p(x) = N_p \cos \theta(x) \cdot e_{px}(x) \tag{6.5}$$

where $\theta(x)$ = bending angle of the prestressing tendon at any point.

Separating the tendon and anchorages from the concrete beam, as shown in Fig. 6.3B, the tendon bears a radial stress w_p provided by the concrete to counterbalance the inclined

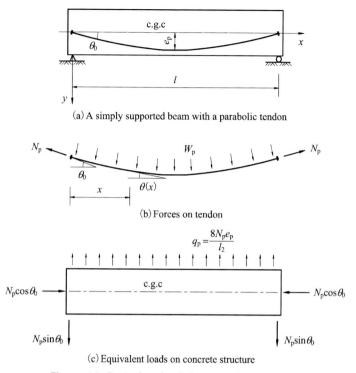

(a) A simply supported beam with a parabolic tendon

(b) Forces on tendon

(c) Equivalent loads on concrete structure

Figure 6.3 Equivalent loads by a parabolic tendon.

tensile forces at two tendon ends. On the contrary, the concrete bears the radial compression, which can be derived by

$$q_p(x) = \frac{d^2 M_p(x)}{dx^2} = -N_p \frac{d^2\left[N_p \cos\theta(x) \cdot e_{px}(x)\right]}{dx^2} \tag{6.6}$$

In a real prestressed concrete structure, the sag of the prestressing tendon is relatively small compared to the effective span, so $\cos\theta(x) \approx 1.0$ can be taken in a simplified analysis. Substituting Eq. (6.3) into Eq. (6.5) yields

$$q_p(x) = -\frac{8N_p e_p}{l^2} \tag{6.7}$$

Eq. (6.7) shows that the equivalent load of the parabolic prestressing tendon along its entire length is a uniformly distributed load exerted on the concrete on the concave side.

At the two tendon ends, the equivalent loads can be obtained from Eqs. (6.3a)–(6.3d) where $\theta_1 = \theta_2 = \theta_0$, as follows.

At the left tendon end:

$$N_{L,0} = N_p \cos \theta_0 \tag{6.8a}$$

$$N_{L,0} = N_p \sin \theta_0 \tag{6.8b}$$

At the right tendon end:

$$N_{R,l} = -N_p \cos \theta_0 \tag{6.8c}$$

$$N_{R,l} = N_p \sin \theta_0 \tag{6.8d}$$

Hence, the equivalent loads of the parabolic prestressing tendon include a uniformly distributed load along the entire length of the tendon given by Eq. (6.7), with concentrated horizontal and vertical loads at two ends given by Eqs. (6.8a)–(6.8d).

When numerous distinct prestressing tendons are arranged in a structure, the total equivalent loads can be obtained by superposition of the equivalent loads of each tendon. Once the equivalent loads due to the prestressing force are obtained, the structural analysis can be carried out by applying the equivalent loads as external loads on the structure.

6.2 Primary internal forces produced by the prestressing force

The internal forces directly obtained by applying compression provided by the prestressing force on the section are called the primary internal forces caused by the prestressing force.

Consider a simply supported beam arranged with a parabolic prestressing tendon, as shown in Fig. 6.4A. Set the eccentricity of the prestressing tendon as $e_p(x)$, the bending angle as $\theta_p(x)$, and the effective prestressing force as N_p. Take away the right portion of

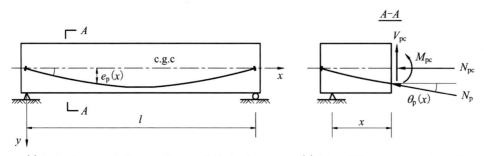

(a) A simply supported beam with a parabolic tendon   (b) Internal forces on concrete section

Figure 6.4 Diagram of the primary internal forces.

the beam at section $A - A$, as shown in Fig. 6.4B, the internal forces produced by the prestressing force at the centroid of the $A - A$ section are obtained as follows

$$N_{pc}(x) = -N_p \cos \theta_p(x) \tag{6.9}$$

$$V_{pc}(x) = -N_p \sin \theta_p(x) \tag{6.10}$$

$$M_{pc}(x) = N_p e_p(x) \cos \theta_p(x) \tag{6.11}$$

The forces given by Eqs. (6.9)−(6.11) are the primary internal forces produced by the prestressing force.

If the sag or eccentricity $e_p(x)$ of the prestressing tendon is relatively small compared to the effective span, $\sin \theta_p(x) \approx \theta_p(x)$ and $\cos \theta_p(x) \approx 1$ can be taken, then

$$N_{pc}(x) \approx -N_p \tag{6.12}$$

$$V_{pc}(x) \approx -N_p \theta_p(x) \tag{6.13}$$

$$M_{pc}(x) \approx N_p e_p(x) \tag{6.14}$$

6.3 Secondary internal forces produced by the prestressing force

In a statically indeterminate structure, the deformation caused by the eccentric prestressing force is restrained by the redundant constraints, then the reactions are generated in the surplus restrains resulting in additional internal forces in the sections. These kinds of additional internal forces are called secondary internal forces. Hence, the total internal forces produced by the prestressing force are the sum of the primary and the secondary internal forces.

Fig. 6.5 shows the calculation process for internal forces in a two-span post-tensioned continuous beam with a linear prestressing tendon, where the support B can bear both compression and tension.

If the support B is removed, the beam will generate a camber at B when the prestressing tendons are stretched and anchored, as shown in Fig. 6.5B. Therefore, a reaction at the support B is produced to restrict this camber, as shown in Fig. 6.5C and D.

The primary internal forces in a section produced by the prestressing force are

$$N_{p,0} = N_p \tag{6.15}$$

$$M_{p,0} = N_p e_p \tag{6.16}$$

The reaction at the support B R_b, called the secondary reaction produced by the prestressing force, can be derived by the force method. At the support B, the following equation is obtained:

$$\delta_{11} R_b + \Delta_{1b} = 0 \tag{6.17}$$

where $\delta_{11} =$ displacement at the support B in the basic structure due to unit force.

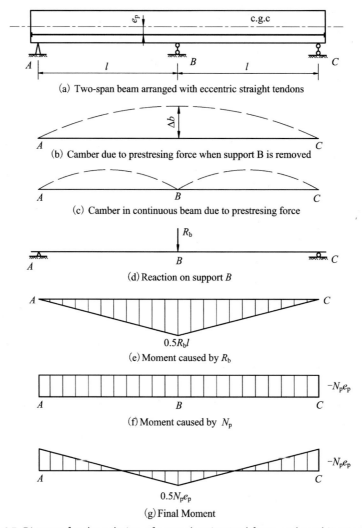

Figure 6.5 Diagram for the solution of secondary internal force and total internal force.

Δ_{1b} = displacement at the support B in the basic structure due to R_b.

Let $R_b = 1$ in Fig. 6.5E, δ_{11} is obtained as

$$\delta_{11} = \frac{1}{EI} \left(\frac{1}{2} \times l \times \frac{l}{2} \right) \left(\frac{2}{3} \times \frac{l}{2} \right) \times 2 = \frac{l^3}{6EI} \tag{6.18}$$

From Fig. 6.5E and F, Δ_{1b} is obtained as

$$\Delta_{1b} = \frac{1}{EI}\left(\frac{1}{2} \times l \times \frac{l}{2}\right) \times N_p e_p \times 2 = \frac{N_p e_p l^2}{2EI} \tag{6.19}$$

Substituting Eqs. (6.18) and (6.19) into Eq. (6.17), the secondary reaction at the support B produced by the prestressing force is achieved:

$$R_b = -\frac{3N_p e_p}{l} \tag{6.20}$$

Then the secondary internal force (moment) caused by the prestressing force is given by

$$M_{p,s}(x) = \frac{R_b}{2}x = -\frac{3N_p e_p}{2l}x \tag{6.21}$$

Hence, the total bending moment at any section caused by the prestressing force is derived:

$$\begin{aligned} M(x) &= M_{p,0} + M_{p,s}(x) \\ &= N_p e_p - \frac{3N_p e_p}{2l}x \end{aligned} \tag{6.22}$$

The distribution of the total bending moment due to the prestressing force is shown in Fig. 6.5G.

6.4 Linear transformation and concordant tendon

6.4.1 C-line

The line connecting the point of thrust at any section over the entire length of the structure is called the C-line. Obviously, in the statically determinate structures, the C-line coincides with the line of thrust caused by the prestressing force; the prestressing force and the internal forces of the section constitute a self-equilibrium system. In the statically indeterminate structures, the C-line usually does not coincide with the line of thrust caused by the prestressing force, since the prestressing force is counterbalanced by the primary and secondary internal forces. The location of the C-line in the statically indeterminate structure depends on the structural system and the external loads. For the continuous beam shown in Fig. 6.5A, the eccentricity of the C-line, $y_p(x)$, the distance from the C-line to the center of gravity of the section under the action of the prestressing force, can be obtained from Eq. (6.22), as

$$y_p(x) = \frac{M(x)}{N_p} = e_p\left(1 - \frac{3x}{2l}\right) \tag{6.23}$$

where $M(x)$ = moment due to the prestressing force.

When external loads are exerted on the continuous beam, the prestressing tendon will have an incremental force ΔN_p, then the eccentricity of the C-line turns into

$$y_p(x) = \frac{M(x) + M_L(x)}{N_p + \Delta N_p} = e_p\left(1 - \frac{3x}{2l}\right)\frac{N_p}{N_p + \Delta N_p} + \frac{M_L(x)}{N_p + \Delta N_p} \tag{6.24}$$

where $M_L(x)$ = moment due to the external loads.

At service loads, the incremental stress in the prestressing tendon due to the external loads is relatively small compared with the effective stress, so ΔN_p in Eq. (6.24) can be disregarded in simplified analysis, hence

$$y_p(x) \approx \frac{M(x) + M_L(x)}{N_p} = e_p\left(1 - \frac{3x}{2l}\right) + \frac{M_L(x)}{N_p} \tag{6.25}$$

Eq. (6.25) indicates that the position of the C-line varies with the change of external loads. According to Eq. (6.25), the increase of the bending moment produced by the external loads can be counterbalanced by enlarging the C-line eccentricity, while the value of eccentricity is limited since it must be inside the section; this means that the magnitude of external loads is also limited for a prestressed concrete structure.

6.4.2 Concordant tendon and linear transformation principle

In the statically indeterminate structures, the prestressing tendon whose centroid coincides with the C-line is called the concordant tendon. Obviously, the concordant tendon does not produce reactions on intermediate supports and secondary internal forces in any section in a statically indeterminate structure. Consequently, the total moment at any section produced by the prestressing force in a statically indeterminate structure with concordant tendons equals the primary moment.

The position of the C-line can be obtained by Eq. (6.23) in the continuous beam shown in Fig. 6.5A, which is illustrated in Fig. 6.6B. Apparently, the prestressing tendon whose profile matches the C-line is the concordant tendon, as shown in Fig. 6.6B. Although Fig. 6.6A and B have different layouts of the prestressing tendon, these two continuous beams bear the same bending moment due to the prestressing force.

Comparison of Fig. 6.6A and B shows that the eccentricity of the prestressing tendon at two end supports is the same, while it is different at the intermediate support, and the profiles of segments AB and BC at each span are the same (straight line). This

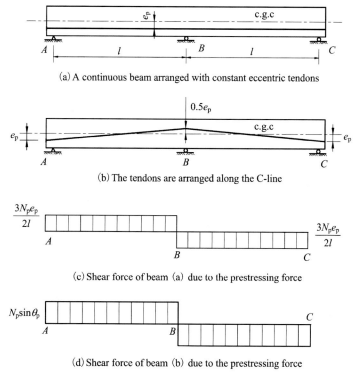

(a) A continuous beam arranged with constant eccentric tendons

(b) The tendons are arranged along the C-line

(c) Shear force of beam (a) due to the prestressing force

(d) Shear force of beam (b) due to the prestressing force

Figure 6.6 Diagram of the linear transformation and shear force.

phenomenon leads to the principle of a linear transformation of the prestressing tendon: in the statically indeterminate structure, in the condition of keeping the profile (straight line, polyline, or curve) of the tendon in each span and the eccentricity at end supports unchanged, the change of eccentricity at the intermediate support does not change the C-line position.

By using the linear transformation principle, the location of the prestressing tendons can be adjusted without changing the C-line, which is helpful in the preliminary design. It should be noted that the shear force due to the prestressing force changes after linear transformation, as shown in Fig. 6.6C and D. Actually, a concordant tendon is rarely completely adopted in the design, because the reasonable profile of the prestressing tendons depends on obtaining an ideal design to meet the requirements of ultimate limit state and serviceability limit state in the case of economical sectional dimension and tendon amount.

6.5 Redistribution of the prestress-caused moment due to creep

During the construction of a long prestressed concrete continuous beam with numerous spans, such as the long continuous beam in a highway or railway bridge over a river or canyon which cannot be erected by the full scaffold method, transformation of the structural system is inevitable. For instance, a continuous beam with three spans can be achieved by connecting three simply supported beams at the ends (Fig. 6.7), or by connecting the ends of two T-structures along with two end parts when the segmental cantilever method is used in building (Fig. 6.8). In this case, the prestressing force exerted on the previous stage structures before connecting will generate a bending moment in the final structure due to creep.

In order to describe the effect of transformation of the structural system on the moment redistribution due to creep, a two-span post-tensioned continuous beam with structural system transformation by two construction stages is analyzed. In Fig. 6.9, two simply supported beams with the first-stage tendons are fabricated in the factory and erected first (Fig. 6.9A), then they are connected by casting concrete at the end gap and the second-stage tendons are stretched and anchored at concrete age t_0. The moment at any section caused by the second-stage tendons in the second stage (final structure) can be calculated by the method discussed in Section 6.4, so the following

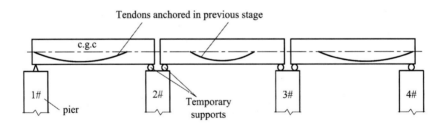

(a) Three simply supported beams before connecting

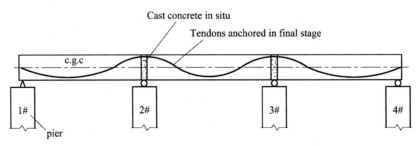

(b) A completed prestressed concrete continuous beam

Figure 6.7 Connect several simply supported beams to a continuous beam.

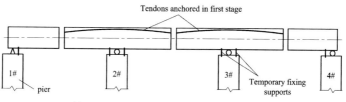

(a) Two T-structures along with two end parts before connecting

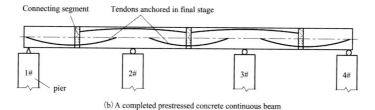

(b) A completed prestressed concrete continuous beam

Figure 6.8 Connect several T-structures to a continuous beam.

analyzes the effect of concrete creep on the first-stage prestressing force to the bending moment in the final structure. To simplify the analysis, the restrain on creep effect by the reinforcements is disregarded.

In an axial concrete member under the prestressing force, creep of concrete leads to shortening of the beam, as follows

$$\Delta_c(t, t_0) = \varepsilon_e(t_0) \cdot \varphi(t, t_0) \cdot l = \Delta_e(t_0) \cdot \varphi(t, t_0) \tag{6.26}$$

where $\varphi(t, t_0)$ = creep coefficient of concrete.
$\varepsilon_e(t_0)$ = elastic strain of concrete due to the prestressing force.
l = effective length of the member.
$\Delta_e(t_0)$ = elastic shortening caused by the prestressing force.
t = concrete age at the moment under consideration.
t_0 = concrete age at applying the prestressing force.

Differentiating Eq. (6.26) with respect to time yields

$$d\Delta_c(t, t_0) = \Delta_e(t_0)d\varphi(t, t_0) \tag{6.27}$$

Similar to Eq. (6.27), the change of bending curvature in a flexural member due to creep in the time period dt can be expressed as

$$d\phi_c(t, t_0) = \phi_e(t_0)d\varphi(t, t_0) \tag{6.28}$$

where $\phi_e(t_0)$ = elastic bending curvature due to the prestressing force.
$\phi_c(t, t_0)$ = bending curvature at time t under consideration.

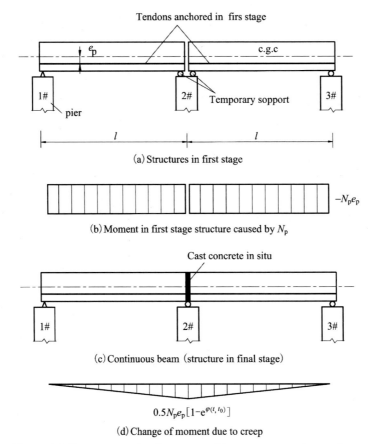

(a) Structures in first stage

(b) Moment in first stage structure caused by N_p

(c) Continuous beam (structure in final stage)

(d) Change of moment due to creep

Figure 6.9 Connect two simply supported beams to a continuous beam.

Set ϕ_{1p} as the bending curvature at the specific section K in the final structure just after two simply supported beams are connected at concrete age t_0 (Fig. 6.9C), which is caused by the first-stage prestressing force; ϕ_{11} as the bending curvature at the section K in the final structure, which is caused by a unit force applied on the section K; M_{1t} as the change of bending moment due to creep at the section K in the final structure at concrete age t ($t > t_0$). In the time period dt, the change of bending curvature at the section K consists of following three parts:

(1) Due to moment and change of creep, $\phi_{11}M_{1t}d\varphi(t, t_0)$.

(2) Due to change of moment, $\phi_{11}dM_{1t}$.

(3) Due to initial curvature and change of creep, $\phi_{1p}d\varphi(t, t_0)$.

Hence, the following equation is attained according to the deformation compatibility in the section:

$$\phi_{11}M_{1t}d\varphi(t, t_0) + \phi_{11}dM_{1t} + \phi_{1p}d\varphi(t, t_0) = 0 \tag{6.29}$$

In the continuous beam (final structure) under the prestressing force provided by the first-stage tendons, as shown in Fig. 6.9C, the moment M_k at section K can be calculated by

$$\phi_{11}M_k + \phi_{1p} = 0 \tag{6.30}$$

From which

$$M_k = -\frac{\phi_{1p}}{\phi_{11}} \tag{6.31}$$

Substituting Eq. (6.31) into Eq. (6.29) yields

$$\frac{dM_{1t}}{d\varphi(t, t_0)} = -M_{1t} + M_k \tag{6.32}$$

From which

$$M_{1t} = M_k + \zeta e^{-\varphi(t,t_0)} \tag{6.33}$$

where $\zeta =$ a constant coefficient.

Consider that $M_{1t} = 0$ when $\varphi(t, t_0) = 0$, resulting in

$$M_{1t} = M_k[1 - e^{-\varphi(t,t_0)}] \tag{6.34}$$

Eq. (6.34) shows that the prestressing force provided by the first-stage tendons will generate a time-dependent moment in the final statically indeterminate structure due to creep of concrete when the structural system undergoes a transformation during construction.

If M_k is the moment at the section above pier 2#, obviously, $M_k = 0$ when two simply supported beams are connected at concrete age t_0, while $M_k = M_{1t}$ in the continuous beam at concrete age t ($t > t_0$), which can be obtained from Eq. (6.34) along with Eq. (6.21):

$$M_k = 0.5N_p e_p[1 - e^{-\varphi(t,t_0)}] \tag{6.35}$$

And the moment change along the entire length of the continuous beam at concrete age t ($t > t_0$) (Fig. 6.9C) due to creep of concrete is

$$\Delta M(x, t, t_0) = \overline{M}_k(x) \cdot 0.5N_p e_p[1 - e^{-\varphi(t,t_0)}] \tag{6.36}$$

where $\overline{M}_k(x) =$ bending moment along the entire length of the beam when a unit moment is applied at the section above pier 2# in the basic structure (two simply supported beams).

The change of moment given by Eq. (6.36) due to creep in the final continuous beam produced by the first-stage prestressing force is shown in Fig. 6.9D when the structural system undergoes a transformation. This kind of change of prestress-caused internal forces in a section due to creep of concrete is called the secondary internal forces due to creep.

Therefore, the total moment in the continuous beam at concrete age t, caused by the first-stage prestressing force, is obtained as

$$M(x, t, t_0) = -N_p e_p + \overline{M}_k(x) \cdot 0.5 N_p e_p [1 - e^{-\varphi(t,t_0)}] \tag{6.37}$$

Generally, the total moment $M(x, t, t_0)$ produced by the previous-stage prestressing force in the final statically indeterminate structure with a transformation can be calculated by

$$M_{pt} = M_{1pt} + \overline{M}_k(x) M_k [1 - e^{-\varphi(t,t_0)}] \tag{6.38}$$

where $M_{1pt} =$ bending moment in the previous structure caused by the previous-stage prestressing force.

Set $M_{2pt} = M_{1pt} + \overline{M}_k(x) M_k$ as the bending moment in the final structure caused by the previous-stage prestressing force, and Eq. (6.38) turns into

$$M_{pt} = M_{1pt} + (M_{2pt} - M_{1pt})[1 - e^{-\varphi(t,t_0)}] \tag{6.39}$$

where $M_{pt} =$ total moment in the continuous beam at concrete age t, caused by the previous-stage prestressing force.

$M_{1pt} =$ bending moment in the previous structure caused by the previous-stage prestressing force.

$M_{2pt} =$ bending moment in the final structure caused by the previous-stage prestressing force.

References

[1] Lin TY, Burns NH. Design of prestressed concrete structures. 3rd ed. New York: John Wiley and Sons; 1981.
[2] Hu Di. Basic principles of prestressed concrete structure design. 2nd ed. Beijing: China Railway Publishing House; 2019.
[3] Naaman AE. Prestressed concrete structure analysis and design. 2nd ed. New York: McGraw Hill; 2004.

CHAPTER 7

Stress analysis of prestressed concrete flexural members

Contents

Analysis and Design of Prestressed Concrete
ISBN 978-0-12-824425-8, https://doi.org/10.1016/B978-0-12-824425-8.00007-5

7.1 Flexural behavior of a prestressed concrete flexural member

When the external load exerted on a prestressed concrete flexural member increases from zero to the magnitude that causes the member to fail, the behavior can be distinguished into three stages, an approximate elastic working stage before cracking, the working stage after cracking, and the failure stage. The first two stages are the basis of the analysis and design of serviceability limit state while the last stage is the basis of analysis and design of the ultimate limit state.

For a typical precast prestressed concrete railway or highway beam, the whole process can be described as a transfer of the prestressing force to the concrete, transportation, and erection of the beam, in service before cracking and in service after cracking if cracking is allowed in the service period. The following outlines the bending behavior of a precast beam in all stages.

7.1.1 Stage of transfer of the prestressing to the concrete

When concrete attains sufficient strength, the tendons are stretched and anchored to form a post-tensioned concrete member; while the jacking force is released over some time at two ends of the concrete with sufficient strength to form a pretensioned concrete member. This process is called the stage of transfer of the prestressing to the concrete. Generally, the prestressing force is designed to counterbalance the bending moment due to self-weight and the subsequently applied dead load and live load, so the prestressing tendons in the flexural section must be arranged eccentrically. Thus, the member has camber at transfer, and consequently it turns into a simply supported beam subjected to the eccentric prestressing force and self-weight, as shown in Fig. 7.1A. In this stage, the first stage loss, as compiled in Table 5.5 or given by Eqs. (5.81) and (5.83), has taken place. The stresses in a typical section are shown in Fig. 7.1A.

For a prestressed concrete beam, whether cracking is allowed or not at service loads, cracking is not allowed during construction, so the whole concrete section should be in full compression, or tensile stress below the tensile strength (or modulus of rupture) is permitted at the top edge.

7.1.2 Stages of transportation and erection

The beam is erected after being transported from the prefabricated factory to the construction site. The positions of the temporary supports of the beam during transportation and the hanging points during erection move to the middle of the span (Fig. 7.1B), so the bending moment in the beam in this stage is different from that in the stage of transfer. In addition, the beam is in motion during transportation and erection, so augmentation of self-weight due to impact should be taken into account.

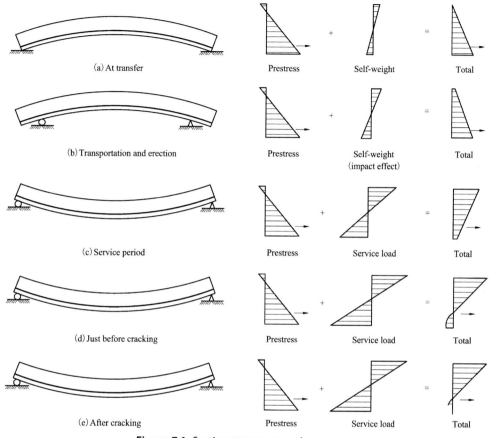

Figure 7.1 Section stresses at various stages.

7.1.3 Stage of service before cracking

In the service period, the simply supported beam is subjected to the effective prestressing force and the combined action of dead load and live load, and the stress analysis in the section can be achieved by the elastic theory before cracking. In this stage, it is assumed that all prestress losses have occurred, and the whole concrete section should be in full compression; if tensile stress occurs, its value should be less than the tensile strength (or modulus of rupture), as shown in Fig. 7.1C and D.

7.1.4 Stage of service after cracking

For beams allowed to crack in the service period, cracking develops when the tensile stress of concrete in the extreme tension fiber reaches the tensile strength (or modulus of rupture). The moment corresponding to the critical state of cracking is called the cracking moment. After cracking, the effective sectional area and moment of inertia reduce; stress analysis should be based on the cracked section. If the maximum crack width is within a limited value, generally, the behavior of the cracked prestressed concrete beam still works approximately elastically.

If the external load increases till the prestressed concrete flexural member fails, the variation of the deformation and concrete stress in the section at midspan with bending moments is as shown in Fig. 7.2.

In Fig. 7.2, each point corresponds to a specific state as follows:

① At transfer of the prestressing force to the concrete;
② At applying the second-stage dead load, such as the weight of the track system in a railway bridge;
③ Full section under uniform compression (no deflection);

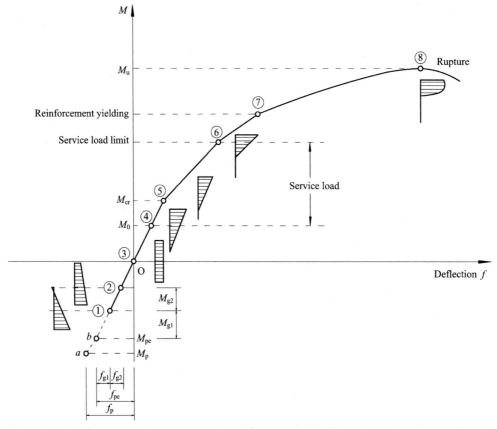

Figure 7.2 Bending moment versus vertical deformation of a flexural member. f_{g1} - deflection caused by self-weigh; f_{g2} - deflection caused by second-stage dead load; f_{pe} - camber caused by the effective prestressing force; f_p - camber caused by the jacking force; M_0 - decompression moment; M_{cr} - cracking moment; M_u - ultimate moment.

④ Decompression state (concrete stress at the edge of tension zone is zero);

⑤ Cracking limit;

⑥ Service load limit;

⑦ Outermost layer reinforcement yielding;

⑧ Ultimate.

In practice, the service load (design load) of the prestressed concrete flexural members is designed in the region of points ④ and ⑥. If the service load is much lower than the cracking moment, the design is uneconomical since the material strength cannot be utilized effectively; meanwhile, the deflection is too large and the bearing capacity of the member is insufficient or the safety margin is insufficient if the service load exceeds the service load limit.

As discussed above, both the upper side and bottom side in a section of a prestressed concrete flexural member go through tension and compression from the transfer stage to the service period. At transfer, the prestressing force is maximum while the design load is minimum (only the self-weight of the member is applied), in this case, the opposite edge of the eccentric tendons bears small compressive stress or tensile stress, so the concrete stress in this edge fiber should be checked. During transportation and erection, the position of supports moves closer to the midspan, the top edge above the support possibly bears tension under the action of the prestressing force, and the weight of the cantilever at the end along with the effect of impact due to movement, where the concrete stress still needs to be checked. In the service period, the reinforcement stress in the tension zone, the concrete stresses in the tension and compression zones in the midspan section, and the principal stress of concrete in the oblique sections should be checked to guarantee sufficient crack resistance and safety. Generally, it is necessary to compute the maximum tensile or compressive stress of concrete, maximum tensile stresses of prestressing tendons, and reinforcing steels of a prestressed concrete flexural member at each stage.

7.2 Stress analysis before cracking

7.2.1 Analysis of normal stress in the uncracked section

To facilitate the use of elastic mechanics formulas and superposition principle in stress analysis of the prestressed concrete flexural members, the original section with zero strain is taken as the reference section in the calculation.

The adopted areas of prestressing tendons (A_p, A_p') and reinforcing steels (A_s, A_s') in calculation of normal stress, unless otherwise specified, refer to the cross-sectional areas of the reinforcements perpendicular to the normal direction of the net or converted concrete section.

7.2.1.1 Normal stress at transfer

Generally, a prestressed concrete flexural member with eccentric prestressing tendons exhibits a camber at transfer. At the midspan of a member subjected to the prestressing force and the self-weight, the opposite edge of the eccentric tendons bears small compressive stress or tensile stress, while huge compression occurs at the other edge (Fig. 7.1A). The excessive tensile stress of concrete results in normal cracking, which is not allowed during construction, since this kind of crack not only destroys the integrity of the section but also reduces the durability of the section as water leaks deep into the cracks and is retained for a long time once the water comes into the cracking region. Also, the excessive compressive stress of concrete results in huge transverse tensile strain in concrete, which may give rise to longitudinal cracking. Therefore, both the maximum compressive stress and maximum tensile stress of concrete should be calculated and verified at transfer.

For pretensioned members, the tendons and concrete have been bonded together at transfer, so the properties of the converted section of concrete and all reinforcements are adopted in the calculation. For post-tensioned members, the tendons and concrete have not been bonded together yet at transfer, so the properties of the net section (consisting of concrete and reinforcing steels) are adopted in the calculation. At transfer, the first stage prestress loss has occurred.

The effective stress in prestressing tendons at transfer is given by

$$\sigma_{pe} = \sigma_{con} - \sigma_{l,I} \tag{7.1}$$

$$\sigma'_{pe} = \sigma'_{con} - \sigma'_{l,I} \tag{7.2}$$

where σ_{pe} = effective stress in prestressing tendons in the tension zone.

σ'_{pe} = effective stress in prestressing tendons in the compression zone.

σ_{con} = control stress in prestressing tendons in the tension zone when stretching.

σ'_{con} = control stress in prestressing tendons in the compression zone when stretching.

$\sigma_{l,I}$ = first stage prestress loss in prestressing tendons in the tension zone, calculated by Eqs. (5.81) and (5.83).

$\sigma'_{l,I}$ = first stage prestress loss of prestressing tendons in the compression zone, calculated by Eqs. (5.81) and (5.83).

The normal stress of concrete caused by the effective prestressing force is given by

$$\sigma_{pc} = \frac{N_p}{A_i} \pm \frac{N_p e_i}{I_i} y_i \tag{7.3}$$

where A_i = cross-sectional area, converted section (A_0) and net section (A_n) are taken for pretensioned and post-tensioned members, respectively.

I_i = moment of inertia, converted section (I_0) and net section (I_n) are taken for pretensioned and post-tensioned members, respectively.

e_i = distance from the point of prestressing force to the centroid of the converted section or net section in pretensioned (e_0) or post-tensioned members (e_n), respectively.

y_i = distance from calculating fiber to the centroid of the converted section or net section in pretensioned (y_0) or post-tensioned members (y_n), respectively.

N_p = effective prestressing force.

For statically indeterminate structures such as post-tensioned continuous beams, the influence of the secondary bending moment should be taken into account when tendons are stretched. In this case, the normal stress of concrete caused by the prestressing force is given by

$$\sigma_{pc} = \frac{N_p}{A_i} \pm \frac{N_p e_i}{I_i} y_i \pm \frac{M_{p2}}{I_i} y_i \tag{7.4}$$

Where M_{p2} = the secondary moment generated by stretching the tendons in the post-tensioned statically indeterminate structures.

For pretensioned members, the effective prestressing force and its eccentricity in Eq. (7.3) are given by

$$N_p = \sigma_{p0} A_p + \sigma'_{p0} A'_p \tag{7.5}$$

$$e_0 = \frac{\sigma_{p0} A_p y_{p0} - \sigma'_{p0} A'_p y'_{p0}}{N_p} \tag{7.6}$$

where σ_{p0} = stress in prestressing tendons in the tension zone, $\sigma_{p0} = \sigma_{con} - \sigma_{l,I} + \sigma_{l4}$.

σ'_{p0} = stress in prestressing tendons in the compression zone, $\sigma'_{p0} = \sigma'_{con} - \sigma'_{l,I} + \sigma'_{l4}$.

σ_{l4} = prestress loss in prestressing tendons in the tension zone due to elastic shortening.

σ'_{l4} = prestress loss in prestressing tendons in the compression zone due to elastic shortening.

A_p = area of prestressing tendons in the tension zone.

A'_p = area of prestressing tendons in the compression zone.

y_{p0} = distance from the centroid of prestressing tendons in the tension zone to the centroid of the converted section.

y'_{p0} = distance from the centroid of prestressing tendons in the compression zone to the centroid of the converted section.

For post-tensioned members, the effective prestressing force and its eccentricity in Eq. (7.3) are given by

$$N_p = \sigma_{pe} A_p + \sigma'_{pe} A'_p \tag{7.7}$$

$$e_n = \frac{\sigma_{pe} A_p \gamma_{pn} - \sigma'_{pe} A'_p \gamma'_{pn}}{N_p} \tag{7.8}$$

where γ_{pn} = distance from the centroid of prestressing tendons in the tension zone to the centroid of the net section.

γ'_{pn} = distance from the centroid of prestressing tendons in the compression zone to the centroid of the net section.

If the member exhibits camber at transfer, the self-weight of the member should be taken into account in the calculation. The normal stress of concrete due to the self-weight is given by

$$\sigma_{gc} = \pm \frac{M_g}{I_i} \gamma_i \tag{7.9}$$

where M_g = bending moment due to the self-weight of the member.

From Eqs. (7.3), (7.4), and (7.9), the normal stress of concrete at transfer is obtained as

$$\sigma_c = \sigma_{pc} + \sigma_{gc} \tag{7.10}$$

For post-tensioned flexural members, the tension force in tendons at stretching equals the pressure shown in the oil gauge, so the actual stresses in the prestressing tendons are equal to the effective stresses as given by Eqs. (7.1) and (7.2) at transfer, whether the member exhibits a camber or not. For those pretensioned members that do not exhibit a camber at transfer, the actual stresses in prestressing tendons are the effective stresses as given by Eqs. (7.1) and (7.2), while the actual stresses in the prestressing tendons should take into account the effect of the self-weight of the member for those pretensioned members that exhibit a camber at transfer, that is

$$\sigma_p = \sigma_{pe} + \alpha_{EP} \sigma_{pc} \tag{7.11}$$

$$\sigma'_p = \sigma'_{pe} + \alpha_{EP} \sigma'_{pc} \tag{7.12}$$

where σ_p = stress in prestressing tendons in the tension zone at transfer.

σ'_p = stress in prestressing tendons in the compression zone at transfer.

σ_{pc} = concrete stress at the centroid of prestressing tendons in the tension zone at transfer.

σ'_{pc} = concrete stress at the centroid of prestressing tendons in the compression zone at transfer.

α_{EP} = ratio of elastic modulus of prestressing tendon to concrete.

During the construction or the storage of the constructed members, the maximum normal stress usually occurs in the sections close to the midspan or the supports.

In this stage, the verification of normal stress of concrete stipulated in the codes is always expressed as follows.

For normal compressive stress of concrete:

$$\sigma_{cc} \leqslant \alpha f_{ck} \tag{7.13}$$

For normal tensile stress of concrete:

$$\sigma_{ct} \leqslant \beta f_{ctk} \tag{7.14}$$

where f_{ck} = characteristic strength of concrete in compression.

f_{ctk} = characteristic tensile strength of concrete.

α, β = coefficients, generally vary from 0.70 to 1.2, the values compiled in Table 7.1 are prescribed in the *Code for design on railway bridge and culvert* (Q/CR 9300-2018), *Specifications for design of highway reinforced concrete and prestressed concrete bridges and culverts* (JTG 3362-2018), and *Codes for design of concrete structures* (GB 50010-2010).

Generally, the stress in prestressing tendons at transfer is smaller than that at stretching, so it is not necessary to check the stress of prestressing tendons in most cases. Meanwhile, if the members are stored in the prefabrication factory or site for a long time after construction, cracking at the top side of the section or excessive camber of the members subjected to the prestressing force and self-weight is probable, in this case, the stress in prestressing tendons should be controlled. The Q/CR 9300-2018 code stipulates that the stress in prestressing tendons at transfer should satisfy

$$\sigma_p \leqslant 0.65 f_{ptk} \tag{7.15}$$

where σ_p = stress in prestressing tendons at transfer.

f_{ptk} = characteristic tensile strength of prestressing tendons.

Table 7.1 Values of α and β.

Name	Railway members (Q/CR 9300-2018)	Highway members (JTG 3362-2018)	Civil and industrial members (GB 50010-2010)
α	0.70 (C40–C45) 0.75 (C50–C60) 0.80 (temporary over-tension)	0.70	0.80
β	0.70	1.15 ($\rho_{ps} \geqslant 0.4\%$) 0.70 ($\rho_{ps} \geqslant 0.2\%$) Lear interpolation ($0.2\% \leqslant \rho_{ps} \leqslant 0.4\%$)	1.00 (cracking is allowed in the tension zone) 1.20 (in the vicinity of the end of simply supported members)

Note: ρ_{ps} is the reinforcement ration in the pre-tensioned zone.

7.2.1.2 Normal stress during transportation and erection

During transportation, the temporary supports for prestressed concrete members possibly move closer to midspan to keep the member stable, as shown in Fig. 7.3. Hence, the effective span of the member is shorter than the design value and the cantilever length at the ends is longer than it should be. In this condition, the compressive stress of concrete at the top side of the midspan section caused by the self-weight reduces due to the shortening of the effective span, and the stress in concrete at the top edge of the section above the temporary support changes from compression caused by the self-weight in the original beam to tension caused by the cantilevered weight in the new structure. On the other hand, the concrete stress at the bottom in these two sections also alters. Thus, the concrete stresses in these two sections during transportation should be checked. The situation is similar to the above during erection.

The normal stress of concrete at the midspan section during transportation due to the self-weight of the member is obtained as

$$\sigma_{gc,m} = \pm \frac{\lambda M_{g,m} y_{0,m}}{I_{0,m}} \tag{7.16}$$

where $M_{g,m}$ = bending moment at midspan during transportation calculated based on the effective span of L_1 produced by the self-weight of the member.

λ = impact coefficient, which is greater than 1.0.

$I_{0,m}$ = moment of inertia of section converted to concrete.

$y_{0,m}$ = distance from the centroid of the converted section to the top or bottom fiber.

The normal stress of concrete in the section above the temporary support during transportation due to cantilever weight is obtained as

$$\sigma_{gc,f} = \mp \frac{\lambda M_{g,f} y_{0,f}}{I_{0,f}} \tag{7.17}$$

where $M_{g,f}$ = bending moment at the section above the temporary support during transportation calculated based on the cantilever with a length of L_c.

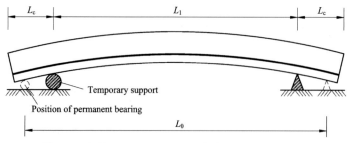

Figure 7.3 Temporary supports during transportation.

$I_{0,f}$ = moment of inertia of section converted to concrete.

$y_{0,f}$ = distance from the centroid of the converted section to the top fiber.

Thus, the normal stress of concrete in the top or bottom extreme fiber of the section is given by

$$\sigma_{c,m} = \sigma_{pc,m} \pm \sigma_{gc,m} \tag{7.18}$$

$$\sigma_{c,f} = \sigma_{pc,f} \mp \sigma_{gc,f} \tag{7.19}$$

where $\sigma_{c,m}$ = normal stress of concrete at the midspan section during transportation.

$\sigma_{c,f}$ = normal stress of concrete at the section above the temporary support during transportation.

$\sigma_{pc,m}$ = normal stress of concrete at the midspan section due to the effective prestressing force during transportation.

$\sigma_{pc,f}$ = normal stress of concrete at the section above the temporary support due to the effective prestressing force during transportation.

In the Q/CR 9300-2018 code, 1.5 is taken as the impact coefficient λ during transportation, and 1.2 is taken during erection.

For members in which cracking is not allowed, the concrete stresses given by Eqs. (7.18) and (7.19) should satisfy

$$\sigma_{c,m,c} = \sigma_{pc,m} + \sigma_{gc,m} \leqslant 0.8 f_{ctk} \tag{7.20}$$

$$\sigma_{c,m,t} = \sigma_{pc,m} - \sigma_{gc,m} \leqslant 0.8 f_{ck} \tag{7.21}$$

$$\sigma_{c,f,c} = \sigma_{pc,f} - \sigma_{gc,f} \leqslant 0.8 f_{ctk} \tag{7.22}$$

$$\sigma_{c,f,t} = \sigma_{pc,f} + \sigma_{gc,f} \leqslant 0.8 f_{ck} \tag{7.23}$$

where $\sigma_{c,m,c}$ = normal stress of concrete in the top extreme fiber of the midspan section during transportation.

$\sigma_{c,m,t}$ = normal stress of concrete in the bottom extreme fiber of the midspan section during transportation.

$\sigma_{c,f,c}$ = normal stress of concrete in the top extreme fiber of the section above the temporary support during transportation.

$\sigma_{c,f,t}$ = normal stress of concrete in the bottom extreme fiber of the section above the temporary support during transportation.

When the bridge-erecting machine is used to install the member, the maximum normal stress of concrete should be calculated according to the actual construction method; the corresponding values of α and β in Eqs. (7.20)–(7.24) should be adopted from the codes.

7.2.1.3 *Normal stress in the service period*

For a prestressed concrete flexural member subjected to the prestressing force and permanent actions (such as self-weight of the member and subsequently applied weight of track system, etc.) in the service period, the creep and shrinkage of concrete and the relaxation of prestressing tendons do not only cause the prestress losses and alter the value of the prestressing force but also lead to stress redistribution among the concrete, prestressing tendons, and reinforcing steels in a section. What is more complicated is that the different ages of stretching prestressing tendons and applying permanent actions results in a different time-dependent effect. Hence, at a specific time in the service period, stresses of concrete, prestressing tendons, and reinforcing steels in a section consist of the following parts: stress produced by the effective prestressing force, stresses produced by the permanent actions, stresses produced by the variable actions (or live loads), and the incremental stress due to the time-dependent effect. Therefore, concrete creep and shrinkage and tendon relaxation along with the age of stretching prestressing tendons and applying permanent actions should be considered comprehensively to carry out the time-dependent stress analysis for a prestressed concrete flexural member. Obviously, this is a complicated process and difficult to achieve by hand. For those structures that are sensitive to time-dependent effects, the stress and deflection should be controlled strictly, such as the prestressed concrete flexural box girders in high-speed railway and the continuous prestressed concrete beams with long spans in highway, step-by-step time-dependent analysis based on a computer must be adopted. However, for general structures, the simplified approach provided in the codes will give an agreeable result.

In present codes, the service load is the result of a specified combination of actions, such as the characteristic combination and quasi-permanent combination of actions, consequently, the commonly adopted approach of simplified stress analysis provided in the codes is to calculate the section stresses produced by the effective prestressing force and by the action combination separately. The latter does not consider the influence of concrete creep and shrinkage and tendon relaxation; the corresponding prestress loss is taken into account in the effective prestressing force according to the stage of the calculation. In stress analysis, when the original section without stress is taken as the reference section, the stress change in reinforcing steels due to the prestress loss and the stress redistribution in the section due to time-dependent effect needs to be considered in the effective prestressing force.

To simplify the calculation, it is assumed that the strain change at the centroid of the reinforcing steels due to concrete creep and shrinkage is approximately equal to the strain change at the centroid of the prestressing tendons nearby. Fig. 7.4 shows the relations of strain change in prestressing tendons and the reinforcing steels due to concrete creep and shrinkage since transfer to the service period, giving

$$\Delta e_s \approx \Delta e_p, \ \Delta \varepsilon_s' \approx \Delta \varepsilon_p'$$

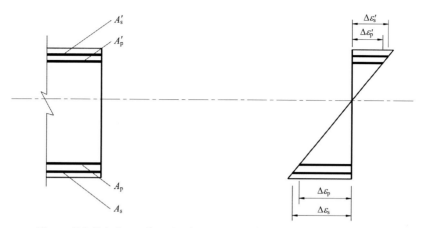

Figure 7.4 Relations of strain change in tendons and reinforcing steels.

where Δe_s = strain change in reinforcing steels in the tension zone due to concrete creep and shrinkage since the transfer.

Δe_p = strain change in prestressing tendons in the tension zone due to concrete creep and shrinkage since the transfer.

$\Delta \varepsilon_s'$ = strain change in reinforcing steels in the compression zone due to concrete creep and shrinkage since the transfer.

$\Delta \varepsilon_p'$ = strain change in prestressing tendons in the compression zone due to concrete creep and shrinkage since the transfer.

Considering that the elastic modulus of reinforcing steels is close to that of prestressing steels, $E_s \approx E_p$, the stress changes in prestressing tendons and reinforcing steels due to concrete creep and shrinkage since the transfer to the service period are obtained from the relations of strain change:

$$\Delta \sigma_s = E_s \Delta e_s \approx E_p \Delta e_p = \Delta \sigma_p$$

$$\Delta \sigma_s' = E_s \Delta \varepsilon_s' \approx E_p \Delta \varepsilon_p' = \Delta \sigma_p'$$

The stress changes in prestressing tendons due to concrete creep and shrinkage are given by Eqs. (5.74)−(5.80), so

$$\Delta \sigma_s \approx \Delta \sigma_p = \sigma_{l6} \tag{7.24a}$$

$$\Delta \sigma_s' \approx \Delta \sigma_p' = \sigma_{l6}' \tag{7.24b}$$

where σ_{l6} = prestress loss in prestressing tendons in the tension zone due to concrete shrinkage and creep.

σ'_{l6} = prestress loss in prestressing tendons in the compression zone due to concrete shrinkage and creep.

When the original section before applying any load is taken as the reference section in analysis, the resultant (compression) in the reinforcing steels due to concrete creep and shrinkage since transfer to the service period should be taken into account, as

$$\Delta N_s = -A_s \sigma_{l6} \qquad (7.25a)$$

$$\Delta N'_s = -A'_s \sigma'_{l6} \qquad (7.25b)$$

where ΔN_s = compression change in reinforcing steels in the tension zone due to concrete creep and shrinkage since transfer to the service period.

$\Delta N'_s$ = compression change in reinforcing steels in the compression zone due to concrete creep and shrinkage since transfer to the service period.

The resultant provided by the prestressing tendons considering the compression change in reinforcing steels due to concrete creep and shrinkage in the service period can be called the generalized prestressing force. It is also often abbreviated to the prestressing force without causing any misunderstanding in some cases.

Based on the above assumptions, the generalized prestressing force and its eccentricity in the section in the service period can be obtained from Fig. 7.5A and B.

Pretensioned members

For pretensioned members in the service period, the tendons and concrete have been bonded together, so the converted section of concrete and reinforcements is used in the calculation.

Considering the effect of concrete creep and shrinkage, the generalized prestressing force and its eccentricity can be obtained from Fig. 7.3A, as

$$N_{p0} = \sigma_{p0}A_p + \sigma'_{p0}A'_p - \sigma_{l6}A_s - \sigma'_{l6}A'_s \qquad (7.26)$$

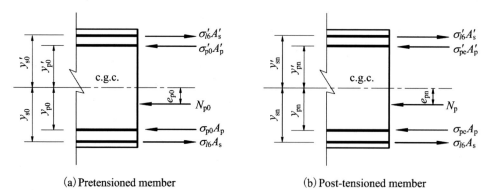

(a) Pretensioned member   (b) Post-tensioned member

Figure 7.5 Diagram for calculating the generalized prestressing force and its eccentricity.

$$e_{p0} = \frac{\sigma_{p0}A_p\gamma_{p0} - \sigma'_{p0}A'_p\gamma'_{p0} - \sigma_{l6}A_s\gamma_{s0} + \sigma'_{l6}A'_s\gamma'_{s0}}{N_{p0}} \qquad (7.27)$$

where σ_{p0} = stress in the prestressing tendons in the tension zone, $\sigma_{p0} = \sigma_{con} - \sigma_{l,I} - \sigma_{l,II} + \sigma_{l4}$.

σ'_{p0} = stress in prestressing tendons in the compression zone, $\sigma'_{p0} = \sigma'_{con} - \sigma'_{l,I} - \sigma'_{l,II} + \sigma'_{l4}$.

A_p = area of prestressing tendons in the tension zone.

A'_p = area of prestressing tendons in the compression zone.

A_s = area of reinforcing steels in the tension zone.

A'_s = area of reinforcing steels in the compression zone.

γ_{p0} = distance from the centroid of the prestressing tendons in the tension zone to the centroid of the converted section.

γ'_{p0} = distance from the centroid of the prestressing tendons in the compression zone to the centroid of the converted section.

γ_{s0} = distance from the centroid of the reinforcing steels in the tension zone to the centroid of the converted section.

γ'_{s0} = distance from the centroid of the reinforcing steels in the compression zone to the centroid of the converted section.

σ_{l4} = prestress loss in prestressing tendons in the tension zone due to elastic shortening.

σ'_{l4} = prestress loss in prestressing tendons in the compression zone due to elastic shortening.

σ_{l6} = prestress loss in prestressing tendons in the tension zone due to concrete creep and shrinkage.

σ'_{l6} = prestress loss in prestressing tendons in the compression zone due to concrete creep and shrinkage.

$\sigma_{l,I}$, $\sigma_{l,II}$, $\sigma'_{l,I}$, and $\sigma'_{l,II}$ are the prestress losses of prestressing tendons in different stages in the tension zone and compression zones, given in Eqs. (5.81)–(5.84).

The normal stress of concrete produced by the effective prestressing force in the service period is

$$\sigma_{pc} = \frac{N_{p0}}{A_0} \mp \frac{N_{p0}e_{p0}}{I_0}\, y_0 \qquad (7.28)$$

The normal stress of concrete caused by design load in the service period is

$$\sigma_{cc} = \pm \frac{M_{design}}{I_0}\, y_0 \qquad (7.29)$$

where M_{design} = design bending moment, corresponding to a specified combination of actions.

Hence, the total normal stress of concrete in the service period is obtained as

$$\sigma_c = \sigma_{pc} + \sigma_{cc}$$

$$= \frac{N_{p0}}{A_0} \mp \frac{N_{p0}e_{p0}}{I_0} y_0 \pm \frac{M_{design}}{I_0} y_0 \tag{7.30}$$

where N_{p0} = generalized prestressing force.

e_{p0} = eccentricity of N_{p0} with respect to the centroid of the converted section.

y_0 = distance from calculating fiber to the centroid of the converted section.

A_0 = area of section converted to concrete.

I_0 = moment of inertia of the converted section.

It is assumed that all prestress losses have occurred in the service period, the effective stress in prestressing tendons is given by

$$\sigma_{pe} = \sigma_{con} - \sigma_{l,I} - \sigma_{l,II} \tag{7.31}$$

$$\sigma'_{pe} = \sigma'_{con} - \sigma'_{l,I} - \sigma'_{l,II} \tag{7.32}$$

The total stress in prestressing tendons in the tension zone in the service period is given by

$$\sigma_p = \sigma_{pe} + \alpha_{EP}\sigma_{c,p} = \sigma_{pe} + \alpha_{EP}\left(-\frac{N_{p0}}{A_0} - \frac{N_{p0}e_{p0}}{I_0} y_{p0} + \frac{M_{design}}{I_0} y_{p0} \right) \tag{7.33}$$

where y_{p0} = distance from the centroid of the prestressing tendons to the centroid of the converted section.

$\sigma_{c,p}$ = normal stress of concrete at the centroid of the prestressing tendons, which can be obtained using Eq. (7.30).

α_{EP} = ratio of elastic modulus of prestressing tendon to concrete.

The stress in reinforcing steels in the tension zone in the service period is given by

$$\sigma_s = -\sigma_{l6} + \alpha_{ES}\sigma_{c,s} = -\sigma_{l6} + \alpha_{ES}\left(-\frac{N_{p0}}{A_0} - \frac{N_{p0}e_{p0}}{I_0} y_{s0} + \frac{M_{design}}{I_0} y_{s0} \right) \tag{7.34}$$

where y_{s0} = distance from the centroid of the reinforcing steels to the centroid of the converted section in the tension zone.

α_{ES} = ratio of elastic modulus of reinforcing steel to concrete.

$\sigma_{c,s}$ = normal stress of concrete at the centroid of the reinforcing steels, which can be obtained from Eq. (7.30).

Post-tensioned members

For post-tensioned members, the prestressing force acts on the net section consisting of concrete and reinforcing steels at transfer. In the service period, the prestress losses that occur after transfer along with the time-dependent stress redistribution results in a change in section stress, which must be calculated based on the converted section in which the boned prestressing tendons are taken into account. Therefore, the concrete stress produced by the generalized prestressing force can be obtained approximately as

$$\sigma_{pc} = \frac{N_{p1}}{A_n} \mp \frac{N_{p1}e_{pn1}}{I_n} y_n - \left(\frac{\Delta N_p}{A_0} \mp \frac{\Delta N_p e_{p02}}{I_0} y_0 \right) \tag{7.35}$$

$$N_{p1} = A_p \left(\sigma_{pe} - \sigma_{l6} \right) + A_p' \left(\sigma_{pe}' - \sigma_{l6}' \right) \tag{7.36}$$

$$\Delta N_p = \left(A_p + A_s \right) \sigma_{l6} + \left(A_p' + A_s' \right) \sigma_{l6}' \tag{7.37}$$

where N_{p1} = effective prestressing force at transfer.

$\quad \Delta N_p$ = change of the prestressing force due to prestress loss and stress redistribution.

$\quad e_{pn1}$ = eccentricity of N_{p1} with respect to the centroid of the net section.

$\quad e_{p02}$ = eccentricity of ΔN_p with respect to the centroid of the converted section.

$\quad y_n$ = distance from the calculating fiber to the centroid of the net section.

$\quad y_0$ = distance from the calculating fiber to the centroid of the converted section.

$\quad A_n$ = area of net section.

$\quad I_n$ = moment of inertia of the net section.

$\quad A_0$ = area of the converted section.

$\quad I_0$ = moment of inertia of the converted section.

The effective stress in prestressing tendons in the service period is given by

$$\sigma_{pe} = \sigma_{con} - \sigma_{l,I} - \sigma_{l,II} \tag{7.38a}$$

$$\sigma_{pe}' = \sigma_{con}' - \sigma_{l,I}' - \sigma_{l,II}' \tag{7.38b}$$

where σ_{pe} = effective stresses in prestressing tendons in the tension zone.

$\quad \sigma_{pe}'$ = effective stresses in prestressing tendons in the compression zone.

$\quad \sigma_{l,I}$ = prestress losses in the first stage in the tension zone.

$\quad \sigma_{l,I}'$ = prestress losses in the first stage in the compression zone.

$\quad \sigma_{l,II}$ = prestress losses of prestressing tendons in the second stage in the tension zone.

$\quad \sigma_{lII}'$ = prestress losses of prestressing tendons in the second stage in the compression zone.

Therefore, the normal stress of concrete under the action of the generalized prestressing force and design load in the service period is obtained as

$$\sigma_c = \sigma_{pc} + \sigma_{cc}$$

$$= \frac{N_{p1}}{A_n} \mp \frac{N_{p1}e_{pn}}{I_n} y_n - \left(\frac{\Delta N_p}{A_0} \mp \frac{\Delta N_p e_{p0}}{I_0}\right) y_0 \pm \frac{M_{design}}{I_0} y_0 \qquad (7.39)$$

where M_{design} = design bending moment, corresponding to a specified combination of actions.

The total stress in prestressing tendons in the tension zone in the service period is given by

$$\sigma_p = \sigma_{pe} + \alpha_{EP}\sigma_{c,p}$$

$$= \sigma_{pe} + \alpha_{EP}\left(-\frac{\Delta N_p}{A_0} - \frac{\Delta N_p e_{p02}}{I_0} y_{p0} + \frac{M_{design} - M_d}{I_0} y_{p0}\right) \qquad (7.40)$$

where $\sigma_{c,p}$ = concrete stress at the centroid of prestressing tendons in the tension zone due to prestress losses and applied loads after transfer.

M_g = bending moment due to the self-weight of the member.

y_{p0} = distance from the centroid of prestressing tendons to the centroid of the converted section.

α_{EP} = ratio of elastic modulus of prestressing tendon to that of concrete.

The stress in reinforcing steels in the tension zone in the service period is given by

$$\sigma_s = -\sigma_{l6} + \alpha_{ES}\sigma_{c,s}$$

$$= -\sigma_{l6} + \alpha_{ES}\left[-\frac{N_{p1}}{A_n} - \frac{N_{p1}e_{pn}}{I_n} y_{sn} + \left(\frac{\Delta N_p}{A_0} - \frac{\Delta N_p e_{p0}}{I_0} y_{s0}\right) + \frac{M_{design}}{I_0} y_{s0}\right]$$

$$(7.41)$$

where $\sigma_{c,s}$ = concrete stress at the centroid of reinforcing steels under the action of the effective prestressing force and design loads.

y_{sn} = distance from the centroid of reinforcing steels in the tension zone to the centroid of the net section.

y_{s0} = distance from the centroid of reinforcing steels in the tension zone to the centroid of the converted section.

α_{ES} = ratio of elastic modulus of reinforcing steel to concrete.

Eqs. (7.34)−(7.41) show the simplified approach of stress analysis provided in the Q/CR 9300-2018 code.

To further simplify the stress analysis of a post-tensioned flexural member in the service period, as provided in the JTG 3362-2018 and GB 50010-2010 codes, the properties of the net section are also used to calculate the stress change due to prestress losses after transfer. Thus, the generalized prestressing force and its eccentricity in the service period can be obtained from Fig. 7.5B, as

$$N_p = \sigma_{pe}A_p + \sigma'_{pe}A'_p - \sigma_{l6}A_s - \sigma'_{l6}A'_s \tag{7.42}$$

$$e_{pn} = \frac{\sigma_{pe}A_p y_{pn} - \sigma'_{pe}A'_p y'_{pn} - \sigma_{l6}A_s y_{sn} + \sigma'_{l6}A'_s y'_{sn}}{N_p} \tag{7.43}$$

where N_p = generalized prestressing force.

e_{pn} = eccentricity of the generalized prestressing force with respect to the centroid of the net section.

The effective stresses in prestressing tendons in the tension and compression zones in the service period, σ_{pe} and σ'_{pe}, are given by Eq. (7.38).

The normal stress of concrete produced by the effective prestressing force and design load in the service period is

$$\sigma_c = \frac{N_p}{A_n} \mp \frac{N_p e_{pn} y_n}{I_n} \pm \frac{M_{design}}{I_0} y_0 \tag{7.44}$$

where M_{design} = design bending moment, corresponding to a specified combination of actions.

y_n = distance from the calculating fiber to the centroid of the net section.

A_n = area of net section.

I_n = moment of inertia of the net section.

A_0 = area of the converted section.

I_0 = moment of inertia of the converted section.

The normal stresses in prestressing tendons and reinforcing steels in the tension zone in the service period are obtained as

$$\sigma_p = \sigma_{pe} + \alpha_{EP} \frac{M_{design} - M_g}{I_0} y_{p0} \tag{7.45}$$

$$\sigma_s = -\sigma_{L6} + \alpha_{ES} \left(-\frac{N_p}{A_n} - \frac{N_p e_{pn}}{I_n} y_{sn} + \frac{M_{design}}{I_0} y_{s0} \right) \tag{7.46}$$

where σ_{pe} = effective stress in prestressing tendons, given by Eq. (7.38a).

M_g = bending moment due to the self-weight of the member.

Stress limitation in the service period

For railway prestressed concrete flexural members, the Q/CR 9300-2018 code prescribes that the normal stress of concrete in the service period should be controlled to a limited value. For the members in which tensile stress is not allowed, the maximum normal tensile stress of concrete should satisfy

$$\sigma_{ct} \leqslant 0 \tag{7.47a}$$

For the members in which tensile stress is allowed while cracking is not allowed, the maximum normal tensile stress of concrete should satisfy

$$\sigma_{ct} \leqslant 0.7 f_{ctk} \tag{7.47b}$$

Under the combination of characteristic values of permanent action with frequent basic variables, the maximum normal compressive stress of concrete should satisfy

$$\sigma_c \leqslant 0.5 f_{ck} \tag{7.47c}$$

Under the combination of characteristic values of permanent action with frequent basic variables and other quasi-permanent values of other variables, the maximum normal compressive stress of concrete should satisfy

$$\sigma_c \leqslant 0.55 f_{ck} \tag{7.47d}$$

where σ_{ct} = normal tensile stress of concrete, calculated by Eqs. (7.30), (7.39), or (7.44).
σ_c = normal compressive stress of concrete, calculated by Eqs. (7.30), (7.39), or (7.44).
f_{ctk} = characteristic tensile strength of concrete.
f_{ck} = characteristic compressive strength of concrete.

For highway prestressed concrete flexural members in which cracking is not allowed in the service period, the JTG 3362-2018 code prescribes that the maximum normal compressive stress of concrete should satisfy

$$\sigma_c = \sigma_{kc} + \sigma_{pt} \leqslant 0.5 f_{ck} \tag{7.48}$$

The maximum tensile stress in prestressing tendons should satisfy

$$\text{Wires and strands } \sigma_p \leqslant 0.65 f_{ptk} \tag{7.49a}$$

$$\text{Prestressing steel bars } \sigma_p \leqslant 0.75 f_{ptk} \tag{7.49b}$$

where σ_{kc} = normal compressive stress of concrete produced by the design load (the characteristic combination of actions).
σ_{pt} = normal compressive stress of concrete produced by the generalized prestressing force.
f_{ptk} = characteristic tensile strength of prestressing tendons.

7.2.2 Shear stress and principal stresses before cracking

In a prestressed concrete flexural member subjected to the prestressing force and design load, both shear force and moment inevitably exist at the sections between the support and the midspan. Thus, the longitudinal prestressing tendons close to supports are required to bend to resist the shear force, sometimes transverse prestressing tendons are also needed in the webs of the box girder for long-span structures. Take an infinitesimal element dA located close to one-fourth of the span in a simply supported beam, as shown in Fig. 7.6. In the infinitesimal element, there are longitudinal normal stresses caused by the moment due to the longitudinal prestressing force and the design load, transverse normal stresses caused by the transverse compression due to the bent longitudinal prestressing force and transverse prestressing force, and shear stresses around the element.

According to Mohr's theory, the principal stresses in the element under bidirectional stresses can be obtained as

$$\sigma_{tp} = \frac{\sigma_{cx} + \sigma_{cy}}{2} - \sqrt{\left(\frac{\sigma_{cx} - \sigma_{cy}}{2}\right)^2 + \tau_c^2} \tag{7.50a}$$

$$\sigma_{cp} = \frac{\sigma_{cx} + \sigma_{cy}}{2} + \sqrt{\left(\frac{\sigma_{cx} - \sigma_{cy}}{2}\right)^2 + \tau_c^2} \tag{7.50b}$$

And the angle between the tensile principal stress and the beam's axis is given by

$$\text{tg} 2\alpha_0 = -\frac{\tau_c}{\sigma_{cx} - \sigma_{cy}} \tag{7.50c}$$

where σ_{tp} = tensile principal stress of concrete in an oblique section.

σ_{cp} = compressive principal stress of concrete in an oblique section.

α_0 = angle between the tensile principal stress and the beam's axis.

τ_c = shear stress of concrete due to the transverse prestressing force and design load.

σ_{cx} = longitudinal normal stress of concrete due to the longitudinal prestressing force and design load.

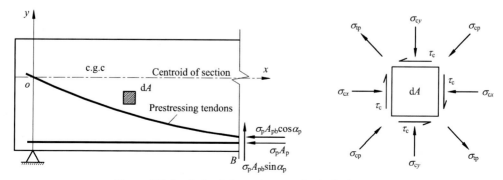

Figure 7.6 Analysis of shear stress and principal stresses.

σ_{cy} = transverse normal stress of concrete produced by the transverse prestressing tendons and the bent prestressing tendons.

Assuming point O is the intersection of the support centerline and the centroid line of the concrete section, B is the bottom edge point at midspan, as shown in Fig. 7.6. The inclined angle for tensile principal stresses depends on the magnitudes of the nominal stresses and shear stresses, while the tensile principal stress must be in the OB direction, meaning that the diagonal tension occurs along the OB direction in the bending-shearing region. Considering the low tensile strength of concrete, it is possible to develop oblique cracking due to excessive tensile principal stress in the case of normal stress under the tensile strength of concrete, and the propagation of cracking up and downward in webs may lead to diagonal tension failure. In the oblique section perpendicular to the direction of tensile principal stress, excessive compressive principal stress may result in crushing in the web. Therefore, it is necessary to check the tensile and compressive principal stresses in a flexural member.

7.2.2.1 Calculation of shear stress
The transverse internal force in the section generated by the longitudinal bent prestressing tendons and transverse prestressing tendons is opposite to the shear force due to external transverse loads; the shear stress along any section in a pretensioned flexural member can be calculated by

$$\tau_c = \frac{\left(-V_p + V_{design}\right) S_0}{bI_0} \tag{7.51}$$

$$V_p = \sum A_{pb}\sigma_{pb} \sin \theta_p = \sum A_{pb} \sin \alpha_p \left[\sigma_{con} - \left(\sigma_{l,I} + \sigma_{l,II}\right) + \sigma_{l4}\right] \tag{7.52}$$

where τ_c = shear stress of concrete at the level under consideration.

V_p = transverse force in the section produced by the prestressing force.

V_{design} = design value of shear force due to a specified combination of actions.

σ_{pb} = effective stress in bent-up prestressing tendons.

A_{pb} = area of the bent-up prestressing tendons.

θ_p = angle between the bent-up prestressing tendons and the beam's axis.

I_0 = moment of inertia of the converted section.

S_0 = first static moment of the converted area about the neutral axis of the portion of the section outside the shear plane considered.

b = width of the section at the level of the shear plane considered.

In a post-tensioned flexural member, the shear stress of concrete can be calculated by

$$\tau_c = \frac{\left(-V_p + V_g\right) S_n}{bI_n} + \frac{\left(\Delta V_p + V_{design} - V_g\right) S_0}{bI_0} \tag{7.53}$$

$$V_{p1} = \sum A_{pb}\left(\sigma_{con} - \sigma_{l,I}\right) \sin \theta_p \tag{7.54}$$

$$\Delta V_p = \sum \sigma_{l,II} A_{pb} \sin \alpha_p \qquad (7.55)$$

where V_{p1} = transverse force in the section produced by the prestressing force.

$\quad \Delta V_p$ = reduction of transverse force due to prestress loss.

$\quad V_g$ = shear forces due to the self-weight of the member.

$\quad V_{design}$ = design value of shear force due to a specified combination of actions in which the self-weight of the member is included.

$\quad I_n$ = moment of inertia of the net section.

$\quad S_n$ = first static moment of the net area about the neutral axis of the portion of the section outside the shear plane considered.

For the prestressed concrete flexural members with variable depth, the influence of depth change should be taken into account in the calculation of shear stress.

The formula of concrete shear stress can be shortened to

$$\tau_c = K_{f1}\tau - \tau_{pv} = K_{f1}\frac{VS_0}{bI_0} - \frac{S_i \sum A_{pb}\sigma_{pb} \sin \theta_p}{bI_i} \qquad (7.56)$$

where τ_c = shear stress of concrete at the level under consideration.

$\quad \tau$ = shear stress of concrete due to design load at the level under consideration.

$\quad \tau_{pv}$ = shear stress of concrete due to bent prestressing tendons at the calculating fiber.

$\quad V$ = shear force due to design load.

$\quad I_i$ = moment of inertia of the net or converted section; the converted section is adopted for pretensioned members and the net section is adopted for post-tensioned members.

$\quad S_i$ = first static moment of the net area about the neutral axis of the portion of the section outside the shear plane considered; the converted section is adopted for pretensioned members and the net section is adopted for post-tensioned members.

$\quad K_{f1}$ = coefficient, generally it is taken as equal to 1.0.

The Q/CR 9300-2018 code prescribes that the coefficient K_{f1} is taken as equal to 1.2 when the characteristic permanent actions are combined with the frequent variables; for the concrete structures that are made under the required condition of process manufacturing, K_{f1} should be increased by 10%.

7.2.2.2 Calculation of principal stresses

The tensile and compressive principal stresses of concrete in an oblique section of the prestressed concrete flexural members can be calculated by Eq. (7.50), where the transverse compressive stress of concrete is given by

$$\sigma_{cy} = \kappa \frac{n A_{pv}\sigma_{pv}}{bs_v} \qquad (7.57)$$

where σ_{cy} = transverse compressive stress of concrete produced by the transverse prestressing tendons and bent-up prestressing tendons.

κ = reduction coefficient considering the actual situation of construction quality, κ is taken as equal to 0.6 in the JTG 3362-2018 code, κ is taken as equal to 1.0 in the Q/CR 9300-2018 code.

n = number of legs of transverse prestressing tendons at a section.

A_{pv} = area of single-leg transverse prestressing tendons.

σ_{pv} = effective stress in transverse prestressing tendons.

b = width of the section at the level of the shear plane considered.

s_v = spacing of transverse prestressing tendons.

In the service period, the maximum compressive principal stress of concrete in an oblique section should satisfy

$$\sigma_{cp} \leqslant 0.6 f_{ck} \qquad (7.58)$$

where f_{ck} = characteristic strength of concrete in compression.

The Q/CR 9300-2018 code prescribes that for the members allowing tensile stresses, the maximum tensile principal stress of concrete in an oblique section in the service period should satisfy

$$\sigma_{tp} \leqslant f_{ctk} \qquad (7.59)$$

where f_{ctk} = characteristic tensile strength of concrete.

The verification for principal stresses should be carried out in the following positions: along the length direction of the member, the sections where both the shear force and bending moment are relatively large, and the location where the section shape and the thickness of web change obviously; along the height direction of the section, the centroid of the section and the transition regions between the web and the upper and lower flanges should be verified.

7.3 Stress analysis after cracking

For partially prestressed concrete members in which cracking is allowed, the cracking moment at the critical section should be obtained first to judge whether it occurs at service loads. Once cracking develops, the fatigue strength resistance and durability reduce. Stress analysis at the cracked section is not only the base for computing the crack width to achieve crack control but also the premise for fatigue control under repeated loading.

Considering that a prestressed concrete flexural member in practical engineering may crack at the beginning of the service period (the moment live loads are applied), or crack when the applied service load reaches the design load after a time period, this variation in cracking time affects the cracking moment when the time-dependent effect due to concrete creep and shrinkage is taken into account. To simplify the analysis, it is commonly assumed in the codes that the member cracks at the beginning of the service and all prestress losses have occurred.

7.3.1 Cracking moment

The cracking moment of a section refers to the bending moment corresponding to the critical state of cracking. To reach the critical state of cracking in a prestressed concrete flexural member, the section stress undergoes two stages. The first is that the whole section works elastically when the precompressed extreme fiber of concrete in the tension zone becomes zero, which is commonly called the decompression state, as shown in Fig. 7.7A. The second illustrates the stage that the stress in the extreme fiber of concrete in the tension zone varies from zero to the ultimate tensile strength just before cracking, which is called the critical cracking state. In consideration of safety, it is commonly assumed in the codes that the cracking develops as the tensile stress in concrete exceeds the characteristic tensile strength, as shown in Fig. 7.7B. Hence, the cracking moment is composed of two parts, M_1 and M_2, where M_1 corresponds to the decompression state while M_2 represents the change till the critical cracking state is reached.

The moment corresponding to the decompression state can be obtained as

$$M_1 = \sigma_{pc} W_0 \tag{7.60}$$

where W_0 = section modulus of the converted section with respect to the bottom fiber.

σ_{pc} = precompressed stress of concrete produced by the prestressing force in the extreme fiber in the tension zone, given by Eq. (7.28) or (7.44).

The second stage can be divided into two substages. The first corresponds to the stage in which the stress in the extreme fiber of concrete in the tension zone varies from zero to the characteristic tensile strength, and the second represents the next stage from the stress in the extreme fiber of concrete in the tension zone just reaching the characteristic tensile strength to the critical state before cracking. Before cracking, a portion of concrete close to the tensile edge exhibits plastic deformation, as shown in Fig. 7.7B, which needs to be considered in computing M_2. To simplify, a coefficient is introduced to consider the influence of concrete plasticity, then

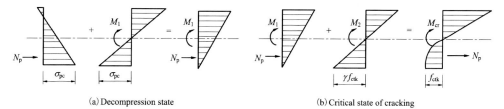

(a) Decompression state   (b) Critical state of cracking

Figure 7.7 Diagram of the cracking moment calculation.

$$M_2 = \gamma f_{ctk} W_0 \tag{7.61}$$

where γ = correction coefficient considering the influence of plasticity of part concrete.
f_{ctk} = characteristic tensile strength of concrete.
The cracking moment is derived from Eqs. (7.60) and (7.61):

$$M_{cr} = (\sigma_{pc} + \gamma f_{tk}) W_0 \tag{7.62}$$

For railway and highway prestressed concrete flexural members, stipulated by the Q/CR 9300-2018 and JTG 3362-2018 codes, the correction coefficient in Eq. (7.62) is given approximately by

$$\gamma = \frac{2S_0}{W_0} \tag{7.63}$$

where S_0 = static moment of the converted section area about the neutral axis.
Eq. (7.63) can be derived from Fig. 7.8.
In the critical cracking state, it is considered that the stress distributions in the compression and tension zones are approximately treated as rectangular patterns, as shown in Fig. 7.8B, which gives

$$M_2 = f_{ctk} A_c Z_c + f_{ctk} A_t Z_t = f_{ctk}(S_c + S_t) \tag{7.64}$$

where A_c = area of the compression zone.
A_t = area of the tension zone.
Z_c = distance from the resultant of compression to the bisector of the sectional area.
Z_t = distance from the resultant of tension to the bisector of the sectional area.
S_c = static moment of the compressive area above the bisector of the sectional area about the neutral axis.

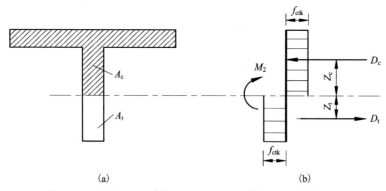

(a)   (b)

Figure 7.8 Diagram of the correction coefficient calculation.

S_t = static moment of the tensile area below the bisector of the sectional area about the neutral axis.

Substituting Eq. (7.61) into Eq. (7.64), yields

$$\gamma = \frac{S_c + S_t}{W_0} \tag{7.65}$$

If $S_c = S_t = S_0$ is adopted for further simplification in Eq. (7.65), Eq. (7.63) is obtained.

For prestressed concrete flexural members used in civil and industrial buildings, the GB 50010-2010 code prescribes that the correction coefficient in Eq. (7.61) should be calculated by

$$\gamma = \left(0.7 + \frac{120}{h}\right)\gamma_m \tag{7.66}$$

where γ_m = basic correction coefficient considering the influence of plasticity of part concrete, given by Table 7.2.

h = depth of section (mm), when $h < 400$, $h = 400$; when $h > 1600$, $h = 1600$; for circular and annular sections, $h = 2r$, where r is the radius of the circular section or the outer radius of the annular section.

7.3.2 Stress analysis in the cracked section

As cracking occurs, the crack width must be controlled to a limited value to guarantee the durability of the cracked sections. Research shows that the crack width is closely related to the tensile stresses in reinforcing steels and prestressing tendons at a crack, and the cracking reduces the rigidity of the section, resulting in a wider range of stress changes in reinforcing steels and prestressing tendons under repeated loading, which obviously affects the structural fatigue performance, therefore stresses in reinforcing steels and prestressing tendons at the cracked sections at service loads must be controlled. The depth of the compression zone shortens once cracked, the compressive stress of concrete at the cracked sections increases and it is also necessary to be controlled.

The stress analysis at the cracked sections can be achieved by the cut-and-trial method or the entire-section decompression method. In the cut-and-trial method, the stress in the cracked section is obtained by repeated cut-and-trial according to equilibrium condition and deformation coordination based on elastic analysis. In the entire-section decompression method, the generalized prestressing force exerted on the section is assumed to be counterbalanced by a virtual tension force first, in this condition there is zero concrete strain through the entire depth. Then this virtual tension and the design load are combined into an eccentric compression force acted on the section, and the stress analysis is fulfilled finally by using the same approach as that of a reinforced concrete

Table 7.2 Values of basic correction coefficient considering the influence of plasticity of concrete.

Item	1	2	3		4		5
Section shape	Rectangular section	T-section when the flange is in the tension zone	Symmetrical I-section or box section		Inverted T-section when the flange is in the tension zone		Circular and annular sections
			$b_f/b \leqslant 2$ h_f/b is an arbitrary value	$b_f/b > 2$ $h_f/b < 0.2$	$b_f/b \leq 2$ h_f/b is an arbitrary value	$b_f/b > 2$ $h_f/b < 0.2$	
γ_m	1.55	1.5 0	1.45	1.35	1.50	1.40	$1.6 - 0.24 r_1/r$

member subjected to large eccentric compression. Stress formulas can be obtained by the entire-section decompression method, so it is widely used.

7.3.2.1 General procedures for stress analysis of the cracked section

The following assumptions are adopted in the calculation:

(1) Plane sections remain plane after cracking, and the strain of concrete varies linearly through the depth of the section.

(2) The concrete and prestressing tendons and reinforcing steels are stressed within their elastic ranges.

(3) The contribution of the tensile strength due to concrete is neglected.

(4) The prestress losses have occurred completely.

Consider a typical cracked section of a simply supported prestressed concrete beam with prestressing tendons and reinforcing steel bars in the tension zone only, as shown in Fig. 7.9B. According to the idea of the decompression method, the stress in the cracked section is imagined to undergo the following three stages:

First state: the stress produced by the generalized prestressing force (N_p), the correspondingly generated strain in the section is represented by the line ① in Fig. 7.9A;

Second state: a virtual tension force N_{p2} is introduced which is in the location of N_p. Under the action of N_p and N_{p2}, the strain through the entire depth of the section is zero, which is represented by the line ② in Fig. 7.9A, this is the decompression state;

Third state: combining the design load, N_p and N_{p2} into a compression force with a large eccentricity on the section, as shown in Fig. 7.9C. Once this eccentric compression force is applied on the section, the stresses in reinforcements and concrete in the section

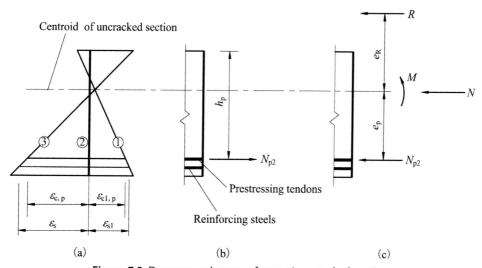

Figure 7.9 Decomposed states of stress in a cracked section.

can be attained, the corresponding strain in the section is represented by the line ③ in Fig. 7.9A. Certainly, the section cracks once the eccentric compression force is applied.

Stage 1: stress produced by the resultant force of prestressing tendons and reinforcing steels

For a pretensioned member, the stresses of concrete and reinforcing steels in the section produced by the generalized prestressing force are given by

$$\sigma_{c1} = \frac{N_p}{A_0}\left(1 + \frac{e_{p0}y_{c0}}{i_0^2}\right) = \varepsilon_{c1}E_c \tag{7.67}$$

$$\sigma_{s1} = \alpha_{ES}\frac{N_p}{A_0}\left(1 + \frac{e_{p0}y_{s0}}{i_0^2}\right) = \varepsilon_{s1}E_s \tag{7.68}$$

where N_p = generalized prestressing force, given by Eq. (7.26) where $A'_p = 0$ and. $A'_s = 0$.

e_{p0} = eccentricity of N_p, given by Eq. (7.27).

i_0 = radius of gyration of the converted section, $i_0^2 = I_0/A_0$.

y_{c0} = distance from the calculating concrete fiber to the centroid of the converted section.

y_{s0} = distance from the centroid of the reinforcing steels to the centroid of the converted section.

ε_{c1} = strain of concrete at the calculating fiber.

ε_{s1} = strain of reinforcing steels at the centroid of the reinforcing steels.

E_c = modulus of elasticity of concrete.

E_s = modulus of elasticity of reinforcing steel.

α_{ES} = ratio of modulus of reinforcing steel to concrete.

The normal compressive strain of concrete at the point of the N_p can be obtained from Eq. (7.67), as

$$\varepsilon_{c1,ps} = \frac{N_p}{E_c A_0}\left(1 + \frac{e_{p0}^2}{i_0^2}\right) \tag{7.69}$$

For a post-tensioned member, the stresses of concrete and reinforcing steels in the section produced by the generalized prestressing force are given by

$$\sigma_{c1} = \frac{N_p}{A_n}\left(1 + \frac{e_{pn}y_{cn}}{i_n^2}\right) = \varepsilon_{c1}E_c \tag{7.70}$$

$$\sigma_{s1} = \alpha_{ES}\frac{N_p}{A_n}\left(1 + \frac{e_{pn}y_{sn}}{i_n^2}\right) = \varepsilon_{s1}E_s \tag{7.71}$$

where N_p = resultant force of prestressing tendons and reinforcing steels, given by Eq. (7.42) where $A'_p = 0$ and. $A'_s = 0$.

e_{pn} = eccentricity of the resultant force of N_p, given by Eq. (7.43).

γ_{cn} = distance from the calculating concrete fiber to the centroid of the net section.

γ_{sn} = distance from the centroid of reinforcing steels to the centroid of the net section.

i_n = radius of gyration of the net section , $i_n^2 = I_n/A_n$.

The normal compressive strain of concrete at the point of the resultant N_p can be obtained from Eq. (7.70), as

$$\varepsilon_{c1,ps} = \frac{N_p}{E_c A_n}\left(1 + \frac{e_{pn}^2}{i_n^2}\right) \tag{7.72}$$

Stage 2: entire-section decompression state

To eliminate the stress in the section due to the resultant of prestressing tendons and reinforcing steels, a virtual tension ΔN_{p2} should be applied at the point of the resultant, which can be obtained from Eq. (7.69) or (7.72):

$$\Delta N_{p2} = \varepsilon_{c1,ps} E_p A_p + \varepsilon_{c1,ps} E_s A_s \tag{7.73}$$

Hence, the temporary total virtual tension force is given by

$$N_{p2} = N_p + \Delta N_{p2} = N_p + \varepsilon_{c1,ps} E_p A_p + \varepsilon_{c1,ps} E_s A_s \tag{7.74}$$

where N_p = generalized prestressing force.

$\varepsilon_{c1,ps}$ = normal compressive strain of concrete at the point of N_p, given by Eq. (7.69) or (7.72).

The stress change in prestressing tendons and reinforcing steels at their respective centroids due to ΔN_{p2} can be derived by

$$\sigma_{p2} = \alpha_{EP}\frac{\Delta N_{p2}}{A_0}\left(1 + \frac{e_{p0}\gamma_{p0}}{i_0^2}\right) \tag{7.75}$$

$$\sigma_{s2} = \alpha_{ES}\frac{\Delta N_{p2}}{A_0}\left(1 + \frac{e_{p0}\gamma_{s0}}{i_0^2}\right) \tag{7.76}$$

Stage 3: stress under the action of virtual tension and design load

In the service period, the design loads N and M are applied to the section. Combining the design loads and the temporary virtual tension force into an equivalent compression force R with an eccentricity of e_R, as

$$R = N_{p2} + N \tag{7.77}$$

$$e_R = \frac{M - e_p N_{p2}}{R} \tag{7.78}$$

where R = equivalent compression force with a large eccentricity.

$\quad e_R$ = distance from the equivalent compression force to the centroid of the converted section.

$\quad N_{p2}$ = temporary virtual tension force, given by Eq. (7.74).

$\quad M$ = bending moment in the cracked section produced by the combination of actions.

$\quad N$ = axial force in the cracked section produced by the combination of actions.

$\quad e_p$ = distance from N_{p2} to the centroid of the converted section.

When the compression R with a large eccentricity is acted on the section of a prestressed concrete member, the produced stress of concrete σ_{c3}, stress of prestressing tendons σ_{p3}, and stress of reinforcing steels σ_{s3} can be obtained by the same approach of solving the reinforced concrete member subjected to compression with a large eccentricity.

Hence, the final stresses of concrete, prestressing tendons, and reinforcing steels in the cracked section in a prestressed concrete flexural member are obtained as

$$\sigma_c = \sigma_{c3} \tag{7.79}$$

$$\sigma_s = \sigma_{s0} + \sigma_{s3} \tag{7.80}$$

$$\sigma_p = \sigma_{p1} + \sigma_{p2} + \sigma_{p3} \tag{7.81}$$

where σ_{s0} = additional stress in reinforcing steels due to concrete creep and shrinkage and tendon relaxation in the service period comparing to the reference section with zero concrete strain, approximately $\sigma_{s0} = -\sigma_{l6}$ is taken, where σ_{l6} is the prestress loss in prestressing tendons due to concrete creep and shrinkage.

$\quad \sigma_{p1}$ = stress in prestressing tendons at stage 1, $\sigma_{p1} = \sigma_{pe}$.

$\quad \sigma_{p2}$ = stress change in prestressing tendons at stage 2, given by Eq. (7.75).

$\quad \sigma_{pe}$ is the effective stress in prestressing tendons, $\sigma_{pe} = \sigma_{con} - \sigma_{l,I} - \sigma_{l,II}$, where $\sigma_{l,I}$ and $\sigma_{l,II}$ are the first and second stage prestress losses, which are calculated by Eqs. (5.81)–(5.84).

7.3.2.2 Stresses in the cracked T-section

For the flexural members with different types of sections, the expressions of σ_{c3}, σ_{p3}, and σ_{s3} are different. Considering a typical cracked T-section of a prestressed concrete flexural member, as shown in Fig. 7.10A, the design load and the generalized prestressing force in the service period are combined into an equivalent compression R with a large eccentricity of e_R, as shown in Fig. 7.10B.

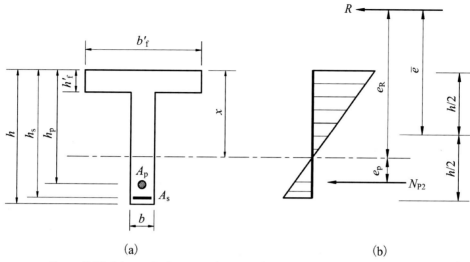

Figure 7.10 Schematic diagram of stress calculation in the cracked T-section.

Set the distance between the equivalent eccentric compression and half depth of the T-section as $\bar{e}$, then

$$\bar{e} = e_R + e_p - h_p + \frac{h}{2} \tag{7.82}$$

Considering $\Sigma M_R = 0$ at the line of eccentric compression, the following equation is obtained:

$$\frac{1}{2}\sigma_{c3}xb'_f\left(\bar{e} - \frac{h}{2} + \frac{x}{3}\right) - \frac{1}{2x}\sigma_{c3}\left(b'_f - b\right)\left(x - h'_f\right)\frac{x - h'_i}{x}\left(\bar{e} - \frac{h}{2} + h'_f + \frac{x - h'_f}{3}\right)$$

$$- \sigma_{p3}A_p\left(\bar{e} - \frac{h}{2} + h_p\right) - \sigma_{s3}A_s\left(\bar{e} - \frac{h}{2} + h_s\right) = 0 \tag{7.83}$$

From Fig. 7.9B, the following relations can obtain:

$$\sigma_{p3} = \alpha_{EP}\sigma_{c3}\frac{h_p - x}{x} \tag{7.84}$$

$$\sigma_{s3} = \alpha_{ES}\sigma_{c3}\frac{h_s - x}{x} \tag{7.85}$$

Substituting Eqs. (7.84) and (7.85) into Eq. (7.83), yields

$$x^3 + Ax^2 + Bx + C = 0 \tag{7.86}$$

where $A = 3\left(\bar{e} - \dfrac{h}{2}\right)$

$$B = \frac{3}{b}\left[\left(b_f' - b\right)\left(2\bar{e} - h + h_f'\right)h_f' + 2\alpha_{EP}A_p\left(\bar{e} - \frac{h}{2} + h_p\right) + 2\alpha_{ES}A_s\left(\bar{e} - \frac{h}{2} + h_s\right)\right]$$

$$C = -\frac{h_f'^2\left(b_f' - b\right)}{b}\left(3\bar{e} + 2h_f' - \frac{3}{2}h\right) - \frac{6}{b}\,\alpha_{EP}A_p h_p\left(\bar{e} - \frac{h}{2} + h_p\right)$$
$$- \frac{6}{b}\alpha_{ES}A_s h_s\left(\bar{e} - \frac{h}{2} + h_s\right)$$

If $x > h_f'$, then

$$A = 3\left(\bar{e} - \frac{h}{2}\right)$$

$$B = \frac{6}{b}\left[\alpha_{EP}A_p\left(\bar{e} - \frac{h}{2} + h_p\right) + \alpha_{ES}A_s\left(\bar{e} - \frac{h}{2} + h_s\right)\right]$$

$$C = -\frac{6}{b_f'}\left[\alpha_{EP}A_p h_p\left(\bar{e} - \frac{h}{2} + h_p\right) + \alpha_{ES}A_s h_s\left(\bar{e} - \frac{h}{2} + h_s\right)\right]$$

Thus, the depth of the compression zone can be calculated by Eq. (7.86), and the concrete stress due to eccentric compression is obtained as

$$\sigma_{c3} = \frac{Rx}{\dfrac{1}{2}\,b_f'x^2 - \dfrac{1}{2}\left(b_f' - b\right)\left(x - h_f'\right)^2 - \alpha_{ES}A_s(h_s - x) - \alpha_{EP}A_p\left(h_p - x\right)} \tag{7.87}$$

If $x \leqslant h_f'$, $b = b_f'$ is taken , and Eq. (7.87) turns into

$$\sigma_{c3} = \frac{Rx}{\dfrac{1}{2}\,b_f'x^2 - \alpha_{ES}A_s(h_s - x) - \alpha_{EP}A_p\left(h_p - x\right)} \tag{7.88}$$

As σ_{c3} is obtained, σ_{p3} is obtained from Eq. (7.84) and σ_{s3} is obtained from Eq. (7.85), respectively. Finally, the actual stresses of concrete, prestressing tendons, and reinforcing steels in the cracked section in a prestressed concrete flexural member are obtained from Eqs. (7.79)–(7.81).

7.3.2.3 Stress calculation in the cracked section of a railway structure (Q/CR 9300-2018)

The approach to stress calculation in the cracked section provided in the Q/CR 9300–2018 code is the same as that described above, and Eqs. (7.80) and (7.81) are simplified into

$$\sigma_s^c = \sigma_{s0} + \sigma_{s3} \tag{7.89}$$

$$\sigma_p^c = \sigma_{pe} + \sigma_{p2} + \sigma_{p3} \tag{7.90}$$

$$\sigma_{p2} = \frac{10\sigma_p A_p}{A}\left(1 + \frac{e_p^2}{i^2}\right) + \frac{10\sigma_{s0} A_s}{A}\left(1 + \frac{e_p e_s}{i^2}\right) \tag{7.91}$$

where σ_s^c = stress in reinforcing steels in the cracked section.

σ_p^c = stress in prestressing tendons in the cracked section.

σ_{s0} = additional stress in reinforcing steels due to concrete creep and shrinkage and tendon relaxation in the service period compared to the reference section with zero concrete strain, approximately $\sigma_{s0} = -\sigma_{l6}$ is taken, where σ_{l6} is the prestress loss in prestressing tendons due to concrete creep and shrinkage.

σ_{s3} = incremental stress in reinforcing steels in the cracked section from decompression state which can be calculated by Eq. (7.85) for a T-section.

σ_{pe} = effective stress in prestressing tendons.

σ_{p2} = stress change in prestressing tendons at decompression state, which can be calculated by Eq. (7.75).

σ_{p3} = incremental stress in prestressing tendons in the cracked section from decompression state which can be calculated by Eq. (7.84) for a T-section.

e_p = distance from the centroid of the prestressing tendons to the centroid of the section.

e_s = distance from the centroid of the reinforcing steels to the centroid of the section.

i = radius of gyration of section, properties of the converted and net sections are used for pretensioned and post-tensioned members, respectively.

A = cross-sectional area, the converted and net sections are used for pretensioned and post-tensioned members, respectively.

The stress in prestressing tendons (strands and wires) in the cracked section should satisfy

$$\sigma_p^c \leqslant 0.6 f_{ptk} \tag{7.92}$$

where σ_p^c = stress in prestressing tendons in the cracked section.

f_{ptk} = characteristic tensile strength of prestressing tendons.

7.3.2.4 Stress calculation in the cracked section of a highway structure (JTG 3362-2018 code)

In the state of decompression, the design load, M_k, and the resultant of prestressing tendons and reinforcing steels N_{p0}, is combined into an equivalent compression N_{p0} with a large eccentricity, expressed as

$$N_{p0}\left(e_N + h_{ps}\right) = M_k \pm M_{p2} \tag{7.93}$$

$$e_N = \frac{M_k \pm M_{p2}}{N_{p0}} - h_{ps} \tag{7.94}$$

where N_{p0} = generalized prestressing force, given by Eq. (7.26) or (7.42).

  M_k = bending moment in the cracked section produced by the characteristic combination of actions.

  M_{p2} = secondary bending moment in the post-tensioned statically indeterminate structures.

  e_N = distance from the point of equivalent compression to the extreme compression fiber.

  h_{ps} = distance from the point of the generalized prestressing force to the extreme compression fiber.

Under the action of equivalent compression N_{p0} with a large eccentricity e_N given in Eqs. (7.93) and (7.94), the depth of the compression zone can be calculated by Eq. (7.86), then the concrete stress in the extreme compression fiber in the cracked section caused by the equivalent eccentric compression is obtained as

$$\sigma_{cc} = \frac{N_{p0}}{A_{cr}} + \frac{N_{p0}e_{0N}c}{I_{cr}} \tag{7.95}$$

$$e_{0N} = e_N + c \tag{7.96}$$

$$h_{ps} = \frac{\sigma_{p0}A_p h_p - \sigma_{l6}A_s h_s + \sigma'_{p0}A'_p d'_p - \sigma'_{l6}A'_s d'_s}{N_{p0}} \tag{7.97}$$

And the stress change in prestressing tendons in the tension zone is given by

$$\sigma_p = \alpha_{EP}\left[\frac{N_{p0}}{A_{cr}} - \frac{N_{p0}e_{0N}(h_p - c)}{I_{cr}}\right] \tag{7.98}$$

Where N_{p0} = generalized prestressing force, given by Eq. (7.26) or (7.42).

  σ_{p0} = stress in prestressing tendons in the tension zone.

  σ'_{p0} = stress in prestressing tendons in the compression zone.

  A_p = area of prestressing tendons in the tension zone.

  A'_p = area of prestressing tendons in the compression zone.

  A_s = area of reinforcing steels in the tension zone.

  A'_s = area of reinforcing steels in the compression zone.

  σ_{l6} = prestress loss in the prestressing tendons in the tension zone due to concrete creep and shrinkage.

  σ'_{l6} = prestress loss in the prestressing tendons in the compression zone due to concrete creep and shrinkage.

  A_{cr} = converted area of the cracked section where the tensile concrete is neglected.

I_{cr} = moment of inertia of the cracked section.

e_{0N} = distance from the point of equivalent compression to the centroid of the cracked section.

c = distance from the extreme compression fiber to the centroid of the cracked section.

h_p = distance from the centroid of the prestressing tendons in the tension zone to the extreme compression fiber.

d'_p = distance from the centroid of the prestressing tendons in the compression zone to the extreme compression fiber.

h_s = distance from the centroid of the reinforcing steels in the tension zone to the extreme compression fiber.

d'_s = distance from the centroid of the reinforcing steels in the compression zone to the extreme compression fiber.

The maximum compressive stress of concrete in the cracked section should satisfy

$$\sigma_{cc} \leqslant 0.5 f_{ck} \tag{7.99}$$

The maximum tensile stress of prestressing strands or wires in the tension zone in the cracked section should satisfy

$$\sigma_p \leqslant 0.65 f_{ptk} \tag{7.100}$$

The maximum tensile stress of prestressing threaded bars in the tension zone in the cracked section should satisfy

$$\sigma_p \leqslant 0.75 f_{ptk} \tag{7.101}$$

where f_{ck} = characteristic strength of concrete in compression.

f_{ptk} = characteristic tensile strength of prestressing tendons.

7.4 Verification of crack resistance

For prestressed concrete flexural structures, generally, sufficient crack resistance can be achieved by limiting the values of normal stress of concrete in the normal sections and limiting the values of principal tensile or compressive stresses of concrete in the oblique sections. Accordingly, the critical sections are required to verify the crack resistance. Those sections that have maximum positive or negative bending moment during construction and at the service period, such as the midspan section and the support section of a continuous beam, are needed to verify the normal crack resistance. Those sections where both the shear force and bending moment are relatively large, and the web thickness changes suddenly, should be selected for verification of crack resistance of the oblique section. For a prestressed concrete flexural structure with a box girder section, the crack resistance in oblique sections in the web, bottom plate, and top plate of the critical regions should be verified.

When verifying the crack resistance of a pretensioned member in the vicinity of the anchorage region, the stress of the prestressing tendon is zero at the end of the member and reaches a maximum (σ_{pe}) at the end of the transfer length, and the values over the transfer length vary linearly.

7.4.1 Verification of crack resistance in the normal section

The verification of crack resistance of the normal section is essentially to limit the maximum normal tensile stress of concrete produced by the effective prestressing force and design load in the service period, expressed as

$$\sigma_{ct} - \kappa_1 \sigma_{pc} \leqslant \kappa_2 f_{ctk} \tag{7.102}$$

where σ_{ct} = maximum normal tensile stress of concrete produced by the design load.

σ_{pc} = normal compressive stress of concrete produced by the prestressing force.

f_{ctk} = characteristic tensile strength of concrete.

κ_1 = coefficient related to the level of crack control and construction methods, etc.

κ_2 = coefficients related to the level of crack control and type of action combination.

The values of κ_1 and κ_2 are provided in the codes, as compiled in Table 7.3.

Table 7.3 Values of κ_1 and κ_2.

Item	κ_1	κ_2
Railway members (Q/CR 9300-2018)	Tensile stress is not allowed: $1.0/\gamma_{kf}$ Tensile stress is not allowed while cracking is not allowed: 1.00	Tensile stress is not allowed: γ_0/γ_{kf} Tensile stress is not allowed while cracking is not allowed: 0.70
Highway members (JTG 3362-2018)	Fully prestressed concrete members: 0.85 (prefabricated members) 0.8 (longitudinal block members by segment casting or mortar joints) Class A members: 1.00	Fully prestressed concrete members: 0.00 Class A members: 0.70 (frequent combination of actions) 0.00 (quasi-permanent combination of actions)
Civil and industrial members (GB 50010-2010)	Cracking is not allowed: 1.00	Cracking is not allowed strictly: 0.00 Cracking is not allowed generally: 1.00 (frequent combination of actions) 1.00 (quasi-permanent combination of actions)

In Table 7.3, γ_0 is a correction coefficient of concrete plasticity at the tension zone which is given by Eq. (7.66). γ_{kf} is a comprehensive influence factor of crack resistance, and is taken to be equal to 1.2 when the characteristic values of permanent actions are combined with the frequent values of variables. The concrete structures that are made under the required condition of process manufacturing, γ_{kf} should be increased by 10%.

7.4.2 Verification of crack resistance in the oblique section

For the prestressed concrete flexural members allowing cracking, the cracks in the normal section are closed in most cases in the service period, especially when the live loads are absent. Considering that the oblique cracks are difficult to close automatically once cracking occurs, the provisions in the codes on crack resistance in the oblique section are more stringent.

Since excessive compressive principal stress reduces the crack resistance in the perpendicular direction, both the tensile and compressive principal stresses of concrete in the oblique section are required for verification. Generally, the principal stresses of concrete should satisfy

$$\sigma_{tp} \leqslant \gamma_1 f_{tk} \tag{7.103}$$

$$\sigma_{cp} \leqslant \gamma_2 f_{ck} \tag{7.104}$$

where σ_{tp} = maximum tensile principal stress of concrete in the service period.

σ_{cp} = maximum compressive principal stress of concrete in the service period.

f_{ctk} = characteristic tensile strength of concrete.

γ_1 = coefficient related to the level of crack control and construction methods, etc.

γ_2 = coefficients related to the level of crack control.

The values of γ_1 and γ_2 are provided in the codes, as compiled in Table 7.4.

Table 7.4 Values of γ_1 and γ_2.

Item	Railway members (Q/CR 9300-2018)	Highway members (JTG 3362-2018)	Civil and industrial members (GB 50010-2010)
γ_1	1.0 (tensile stress in concrete is not allowed) 0.7 (tensile stress in concrete and cracking are allowed)	Fully prestressed members: 0.60 (prefabricated) 0.40 (cast in situ, prefabricated assembly) Class A members: 0.70 (prefabricated) 0.50 (cast in situ, prefabricated assembly)	0.85 (cracking is not allowed strictly) 0.95 (cracking is not allowed generally)
γ_2	0.6 (tensile stress is not allowed)	—	0.60

7.5 Example 7.1

The cross-section at the midspan of a simply supported railway pretensioned beam is shown in Fig. 7.11. The beam is made of C50 concrete and two groups of bonded strands $10\phi^S15.2$ (10-7ϕ5) with a characteristic tensile strength of 1860 MPa. The bending moments due to dead load and live load are 4.38 kN m and 43 kN m, respectively. The control stress at the stretching tendon is 1302 MPa, and the sequentially occurred prestresses loss are $\sigma_{l2} = 92$ MPa (due to anchorage seating), $\sigma_{l3} = 52$ MPa (due to heat treatment curing), $\sigma_{l4} = 76$ MPa (due to shortening), $\sigma_{l5} = 94$ MPa (due to tendon relaxation), and $\sigma_{l6} = 56$ MPa (due to creep and shrinkage of concrete). Tensile stress is allowed while cracking is not allowed in this beam in the service period.

Calculate and verify the stresses of concrete and prestressing strands according to *Code for design on railway bridge and culvert* (TB10002-2017):

(1) Calculate and verify the stresses of concrete and prestressing strands at transfer.
(2) Calculate and verify the stresses of concrete and prestressing strands in the service period.
(3) Verify crack resistance in the normal section during the service period.

7.6 Solution

7.6.1 Parameters

$M_g = 4.38$ kN·m, $M_L = 43.0$ kN·m, $a_p = 5$ mm.
$f_{ptk} = 1860$ MPa, $\sigma_{con} = 1302$ MPa, $f_{ctk} = 3.10$ MPa, $f_{ck} = 33.5$ MPa.

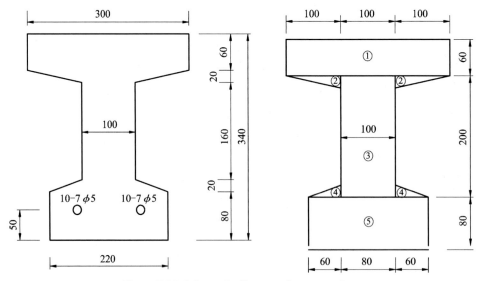

Figure 7.11 Schematic diagram of cross-section.

$E_p = 1.95 \times 10^5$ MPa, $E_c = 3.55 \times 10^4$ MPa, $\alpha_{EP} = \frac{E_p}{E_c} = 5.493$.

$\sigma_{l2} = 92$ MPa, $\sigma_{l3} = 52$ MPa, $\sigma_{l4} = 76$ MPa, $\sigma_{l5} = 94$ MPa, $\sigma_{l6} = 56$ MPa.

$A_{p0} = 10 \times 140 = 1400$ mm^2 (each $10\phi^S 15.2$).

$A_p = 2 \times 1400 = 2800$ mm^2 (two $10\phi^S 15.2$).

7.6.2 Geometrical properties

For the convenience of calculation, the section is divided into several portions (Fig. 7.11) to calculate the sectional geometrical properties, as compiled in Table 7.5.

From Table 7.5, the following parameters are obtained:

Area of converted section:

$$A_0 = 603.38 \text{ cm}^2$$

Distance from the centroid of the converted section to the top edge:

$$y_0' = \frac{9903.39}{603.38} = 16.413 \text{ cm}$$

Distance from the centroid of the converted section to the bottom edge:

$$y_0 = 34 - 16.413 = 17.587 \text{ cm}$$

Distance from the centroid of the strands to the centroid of the converted section:

$$e_0 = 12.587 \text{ cm}$$

Moment of inertia of the converted section:

$$I_0 = 69754.73 + 8152.45 = 77907.18 \text{ cm}^4$$

Section modulus of the converted section to the top edge:

$$W_0' = \frac{I_0}{y_0'} = \frac{77918.453}{16.413} = 4747.36 \text{ cm}^3$$

Section modulus of the converted section to the bottom edge:

$$W_0 = \frac{I_0}{y_0} = \frac{77918.453}{17.587} = 4430.46 \text{ cm}^3$$

Section modulus of the converted section to the centroid of the strands:

$$W_c = \frac{I_0}{e_0} = \frac{77918.453}{12.587} = 6190.39 \text{ cm}^3$$

Table 7.5 Calculation of geometrical properties.

Item	A_i/cm²	d/cm	A_d/cm³	y/cm	y^2/cm²	A_iy^2/cm⁴	I_i/cm⁴
①	$30 \times 6 = 180$	3	540	13.413	179.908	32,383.44	$1/12 \times 30 \times 6^3 = 540$
②	$2 \times 1/2 \times 10 \times 2 = 20$	6.67	133.4	9.743	94.926	1898.52	$2 \times 1/36 \times 10 \times 2^3 = 4.44$
③	$10 \times 20 = 200$	16	3200	0.413	0.171	34.20	$1/12 \times 10 \times 20^3 = 6666.7$
④	$2 \times 1/2 \times 6 \times 2 = 12$	25.33	303.96	−8.917	79.512	954.15	$2 \times 1/36 \times 6 \times 2^3 = 2.67$
⑤	$22 \times 8 = 176$	30	5280	−13.587	184.61	32,490.76	$1/12 \times 22 \times 8^3 = 938.67$
Σ	588		9457.36			67,761.66	8152.45
Strands	$4.493 \times 2.8 = 12.58$	29	364.83	−12.587	158.432	1993.07	
Σ	603.38		9822.19			69,754.73	8152.45

Notes: A_i = area of each portion.

d_i = distance from the centroid of A_i to the top edge of the beam.

A_d = second moment of each portion to the top edge of the beam.

y = distance from the centroid of A_i to the centroid of the converted section.

I_i = moment of inertia of each portion with respect to its centroid.

The static moment of the converted section area above (or below) the centroid of the converted section to the centroid line:

$$S_0' = S_0 = A_1 y_1 + A_2 y_2 + \frac{1}{2} \times 10 (y_0' - 6)^2$$

$$= 180 \times 13.413 + 20 \times 9.743 + \frac{1}{2} \times 10 \times (16.413 - 6)^2 = 3151.35 \text{ cm}^3$$

7.6.3 Stresses of concrete and prestressing strands at transfer

$$\sigma_{pe1} = \sigma_{con} - \sigma_{lI} = \sigma_{con} - \sigma_{l2} - \sigma_{l3} - \sigma_{l4} - 0.5\sigma_{l5} = 1302 - 92 - 52 - 76 - 0.5 \times 94$$
$$= 1035 \text{ MPa}$$

$$\sigma_{p0} = \sigma_{pe1} + \sigma_{l4} = 1035 + 76 = 1111.0 \text{ MPa}$$

$$N_{p0} = A_p \sigma_{p0} = 0.280 \times 1111 = 311.08 \text{ kN}$$

Stress of concrete in the top fiber:

$$\sigma_c' = \frac{N_{p0}}{A_0} - \frac{N_{p0} e_0}{W_0'} + \frac{M_g}{W_0'}$$

$$= \frac{311080}{60338} - \frac{311080 \times 125.87}{4747360} + \frac{4.38 \times 10^6}{4747360} = -2.165 \text{ MPa (tensile stress)}$$

Stress of concrete in the bottom fiber:

$$\sigma_c = \frac{N_{p0}}{A_0} + \frac{N_{p0} e_0}{W_0} - \frac{M_g}{W_0} = \frac{311080}{60338} + \frac{311080 \times 125.87}{4747360} - \frac{4.38 \times 10^6}{4747360} = 12.48 \text{ MPa}$$

Verifying the normal compressive stress of concrete at transfer:

$$\sigma_{cc} = 12.48 \text{ MPa} < \alpha f_{ck} = 0.75 \times 33.5 = 25.12 \text{ MPa}$$

Verifying the normal tensile stress of concrete at transfer:

$$\sigma_{ct} = 2.165 \text{ MPa} < \beta f_{ctk} = 0.7 \times 3.10 = 2.17 \text{ MPa}$$

Verifying the tensile stress of prestressing strands at transfer:

$$\sigma_p = \sigma_{pe1} = 1035 \text{ MPa} < 0.65 f_{ptk} = 0.65 \times 1860 = 1209 \text{ MPa}$$

Hence, the stresses of concrete and prestressing strands at transfer satisfy the requirements prescribed in the code.

7.6.4 Stresses of concrete and prestressing strands in the service period

$$\sigma_{pe} = \sigma_{con} - \sigma_I - \sigma_{II} = \sigma_{con} - \sigma_{l2} - \sigma_{l3} - \sigma_{l4} - \sigma_{l5} - \sigma_{l6}$$
$$= 1302 - 92 - 52 - 76 - 49 - 94 = 718 \text{ MPa}$$

$$\sigma_{p0}^* = \sigma_{pe} + \sigma_{l4} = 718 + 76 = 794 \text{ MPa}$$

$$N_p = A_p \sigma_{p0}^* = 0.280 \times 794 = 222.32 \text{ kN}$$

$$M = M_g + M_L = 4.38 + 43 = 47.38 \text{ kN.m}$$

Stress of concrete in the top fiber:

$$\sigma_c' = \frac{N_p}{A_0} - \frac{N_p e_0}{W_0'} + \frac{M}{W_0'} = \frac{222320}{60338} - \frac{222320 \times 125.87}{4747360} + \frac{47.38 \times 10^6}{4747360} = 7.77 \text{ MPa}$$

Stress of concrete in the bottom fiber:

$$\sigma_c = \frac{N_p}{A_0} + \frac{N_p e_0}{W_0} - \frac{M}{W_0} = \frac{222320}{60338} + \frac{222320 \times 125.87}{4747360} - \frac{47.38 \times 10^6}{4747360}$$
$$= -0.402 \text{ MPa}$$

$$\sigma_p = \sigma_{pe} + \alpha_{EP} \frac{M_l}{I_0} e_0 = \sigma_{pe} + \alpha_{EP} \frac{M_l}{W_c} = 718 + 5.429 \times \frac{43.0 \times 10^6}{6190390} = 756.1 \text{ MPa}$$

Verifying the normal compressive stress of concrete in the service period:

$$\sigma_c = 7.77 \text{ MPa} < 0.5 f_{ck} = 0.5 \times 33.5 = 16.75 \text{ MPa}$$

Verifying the normal tensile stress of concrete in the service period when tensile stress is allowed while cracking is not allowed:

$$\sigma_{ct} = 0.402 \text{ MPa} < 0.7 f_{ctk} = 0.7 \times 3.10 = 2.17 \text{ MPa}$$

Verifying the tensile stress of the prestressing strands in the service period:

$$\sigma_p = 756.1 \text{ MPa} < 0.65 f_{ptk} = 0.65 \times 1860 = 1209 \text{ MPa}$$

Hence, the stresses of concrete and prestressing strands in the service period satisfy the requirements prescribed in the code.

7.6.5 Verification of crack resistance of the normal section in the service period

$$\sigma_{ct} = \frac{M}{W_i} = \frac{47.38 \times 10^6}{4430460} = 10.69 \text{ MPa}$$

$$\sigma_{pc} = \frac{N_p}{A_0} + \frac{N_p e_0}{W_0} = \frac{222320}{60338} + \frac{222320 \times 125.87}{4430460} = 10.0 \text{ MPa}$$

Tensile stress is allowed while cracking is not allowed in this beam, so

$$\kappa_1 \sigma_{pc} + \kappa_2 f_{ctk} = 1.0 \times \sigma_{pc} + 0.7 \times f_{ctk} = 10.0 + 0.7 \times 3.10 = 12.17 \text{ MPa}$$

$$\sigma_{ct} = 10.69 \text{ MPa} < \kappa_1 \sigma_{pc} + \kappa_2 f_{ctk} = 12.17 \text{ MPa}$$

Hence, the crack resistance of the normal section in the service period satisfies the requirement prescribed in the code.

7.7 Calculation and verification of fatigue stress

For the prestressed concrete structures under repeated loading, fatigue should be a concern in the design. Generally, sufficient fatigue resistance can be achieved by limiting the values of maximum stress or stress amplitude, namely fatigue stress, of concrete, prestressing tendons, and nonprestressing steels. The following introduces the equivalent equal-amplitude repeated stress method to calculate and verify fatigue stress provided in the Q/CR 9300-2018 code.

In the calculation of normal fatigue stress of prestressed concrete railway beams, the following basic assumptions are adopted:

(1) Plane sections remain plane in the service period.

(2) The distribution of normal stress in the compressive zone is a triangle.

(3) Properties of the converted section are used in the calculation.

For the members with a prestress level of $\lambda > 1.0$, $1 > \lambda \geqslant 0.7$ while cracking is not allowed, $\alpha_{EP} = E_p/E_c$ and $\alpha_{ES} = E_s/E_c$ are adopted for the ratio of elastic modulus; for the members with a prestress level of $1 > \lambda \geqslant 0.7$ while cracking is allowed, $\alpha_{EP} = \alpha_{ES} = 10$ is adopted.

The stress of concrete can be calculated by

$$\sigma_{cf} = \sigma_{cp} + \frac{M_D \cdot y_0}{I_0} + (1 + \mu)_s \frac{M_L \cdot y_0}{I_0} = \sigma_{cp} + \sigma_{cD} + \sigma_{cL} \tag{7.105}$$

The stress amplitude of prestressing tendons can be calculated by

$$\Delta \sigma_p = \alpha_{EP} \sigma_{cL,p} \tag{7.106}$$

The stress amplitude of reinforcing steels (outermost layer) is given by

$$\Delta \sigma_s = \alpha_{ES} \sigma_{cL,s} \tag{7.107}$$

where σ_{cf} = concrete stress produced by the characteristic permanent actions and characteristic trainload (includes impact force and centrifugal force) in the calculating fiber; when the calculating section bears repeated positive and negative moments, M_L

should be replaced by $M_{L,max}$ and $M_{L,min}$, respectively, to obtain the maximum and minimum stresses.

$\Delta\sigma_p$ = stress amplitude of prestressing tendons produced by the characteristic trainload (includes impact force and centrifugal force).

$\Delta\sigma_s$ = stress amplitude of reinforcing steels produced by the characteristic trainload (includes impact force and centrifugal force).

σ_{cp} = concrete stress produced by the effective prestressing force in the calculating fiber.

σ_{cD} = concrete stress produced by the characteristic permanent actions in the calculating fiber.

σ_{cL} = concrete stress produced by the characteristic trainload (includes impact force and centrifugal force) in the calculating fiber.

M_D = bending moment produced by the characteristic permanent actions.

M_L = bending moment produced by the characteristic trainload (includes impact force and centrifugal force) in the calculating fiber.

$(1 + \mu)_s$ = enlarged coefficient considering the impact due to train load running.

$\sigma_{cL,p}$ = concrete stress produced by the characteristic trainload (includes impact force and centrifugal force) in the centroid of prestressing tendons.

$\sigma_{cL,s}$ = concrete stress produced by the characteristic trainload (includes impact force and centrifugal force) in the outer reinforcing steels in the tension zone.

y_0 = distance from calculating point to the centroid of the converted section.

I_0 = moment of inertia of the converted section (concrete in the tension zone is neglected).

For the prestressing tendons and reinforcing steels in the compression zone, the concrete with a stress amplitude less than $0.5\,f_{cd}$ in the compression zone, and the concrete with maximum stress less than $0.5 f_{td}$, fatigue verification is not required.

In the fatigue limit state, the maximum concrete stress and stress amplitudes of prestressing tendons and reinforcing steels in a prestressed concrete flexural member in which tensile stress of concrete is not allowed should satisfy the limitations prescribed in the code:

Concrete

$$\sigma_{cf} \leqslant 0.5 f_{cd} \tag{7.108}$$

Prestressing tendons

$$\Delta\sigma_p \leqslant \Delta f_{pfd} \tag{7.109}$$

Reinforcing steels

$$\Delta\sigma_s \leqslant \Delta f_{sfd} \tag{7.110}$$

Table 7.6 Basic fatigue strength of reinforcing steels (MPa).

Structure detail	$\Delta f'_{sfd}$
Base metal	145
Flash butt welding connection	130
Thread rolling bar connection	98
Arc welding connection	60

where f_{cd} = characteristic strength of concrete in compression.

Δf_{pfd} = permissible fatigue strength range of prestressing tendons, given in Table 2.25.

Δf_{sfd} = permissible fatigue strength range of reinforcing steels, 130 MPa for HPB300 base material, which can be derived from Eq. (7.111) for HRB400 and HRB500 steels.

The fatigue strength of HRB400 and HRB500 steels is given by

$$\Delta f_{sfd} = \gamma_1 \cdot \gamma_2 \cdot \gamma_3 \cdot \Delta f'_{sfd} \qquad (7.111)$$

where $\Delta f'_{sfd}$ = permissible basic fatigue strength range of HRB400 and HRB500 steels, given in Table 7.6.

γ_1 = influence coefficient of stress ratio, for base metal and flash butt welding connection it can be checked in Table 7.7, for thread rolling bar connection and arc welding connection it is taken to be equal to 1.0.

γ_2 = influence coefficient of steel bar diameter, given in Table 7.8.

γ_3 = strength grade coefficient of reinforcing steels, given in Table 7.9.

When the tensile stress of concrete is allowed while cracking is not allowed, Eqs. (7.109) and (110) become

$$1.5\Delta\sigma_p \leqslant \Delta f_{pfd} \qquad (7.112)$$

Table 7.7 Influence coefficient of stress ratio γ_1.

Stress ratio ρ	0	0.1	0.2	0.3	0.4	0.5	0.6	0.7	0.8	0.9
γ_1	1.000	0.926	0.891	0.851	0.783	0.703	0.606	0.486	0.343	0.177

Note: The stress ratio ρ is the ratio of the minimum stress to the maximum stress.

Table 7.8 Influence coefficient of steel bar diameter γ_2.

Structure detail	Diameter $d < 20$mm	Diameter $d \geqslant 20$ mm
Base metal	1	1
Flash butt welding connection	1	0.72
Thread rolling bar connection	0.55	1
Arc welding connection	1	1

Table 7.9 Influencing coefficient of reinforcing steel strength grade γ_3.

Structure detail	HRB400	HRB500
Base metal	1.0	1.04
Flash butt welding connection	1.0	1.1
Thread rolling bar connection	1.0	1.2
Arc welding connection	1.0	1.0

$$1.5\Delta\sigma_s \leqslant \Delta f_{sfd} \tag{7.113}$$

When the cracking is allowed, the stress amplitudes of prestressing tendons and reinforcing steels can be obtained by

$$\Delta\sigma_p = \Delta\sigma_{p2} + \Delta\sigma_{p3} - \sigma_{pg} \tag{7.114}$$

$$\Delta\sigma_s = \Delta\sigma_{s3} - \sigma_{sg} \tag{7.115}$$

where $\Delta\sigma_{p2}$ = stress change in prestressing tendons from the state under permanent actions and prestressing force to the decompression state, given by Eq. (7.75), $\Delta\sigma_{p2} = \sigma_{p2}$.

$\Delta\sigma_{p3}$ = stress change in prestressing tendons from the decompression state to the cracked state produced by the characteristic permanent actions and characteristic trainload (includes impact force and centrifugal force), which can be derived from Eq. (7.84), $\Delta\sigma_{p3} = \sigma_{p3}$.

σ_{pg} = stress in prestressing tendons produced by the characteristic permanent actions.

$\Delta\sigma_{s3}$ = stress change in reinforcing steels from the decompression state to the cracked state produced by the characteristic permanent actions and characteristic trainload (includes impact force and centrifugal force), which can be derived from Eq. (7.85), $\Delta\sigma_{s3} = \sigma_{s3}$.

σ_{sg} = stress in reinforcing steels produced by the characteristic permanent actions.

7.8 Stress analysis in the anchorage zone of pretensioned members

There is no anchorage at the ends of the prestressing tendons in a pretensioned member, where the tension in the prestressing tendons is transmitted to the concrete by their bond force to generate compression in concrete. Obviously, the stress at the end of the prestressing tendon is zero, and the stress reaches the effective stress (σ_{pe}) after a certain distance from the end, as shown in Fig. 7.12C. In the serviceability limit state, the length from the end to the effective stress is called the stress–transmission length of prestressing tendons, l_{tr}.

When the jacking force is released, the diameter of the prestressing tendons at the member end recovers to the original value, and the prestressing tendons retract into

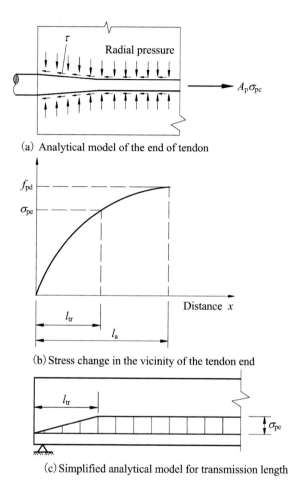

(a) Analytical model of the end of tendon

(b) Stress change in the vicinity of the tendon end

(c) Simplified analytical model for transmission length

Figure 7.12 Schematic diagram of force transmission in a pretensioned member.

the members resulting in part of the bond force close to the member end being invalid. Hence, a gradual change of tendon diameter forms in the vicinity of the member end, acting as an anchor wedge, as shown in Fig. 7.12A, then the radial pressure is generated on the prestressing tendons by the surrounded concrete, and the friction between the tendons and the concrete plays the main role in transmission force. It can be seen that the stress distribution in the stress-transmission length in a pretensioned member is complex, as shown in Fig. 7.12B.

To simplify the analysis, it is commonly assumed that the change of tendon stress in the transmission length is linear, as shown in Fig. 7.11C, so the transmission length can be derived approximately by

$$l_{tr} = \alpha \frac{\sigma_{pe}}{f_{ctd}} d \qquad (7.116)$$

where l_{tr} = stress-transmission length.

σ_{pe} = effective stress in prestressing tendons.

α = coefficient on the shape of prestressing tendons, generally 0.13–0.18 is taken.

d = nominal diameter of prestressing tendons, equivalent diameter $\sqrt{n}d$ is adopted for a bundle of tendons, where n is the total number of single tendons or wires and d is the nominal diameter of a strand or wire.

f_{ctd} = design value of tensile strength of concrete.

The coefficients on the shape of commonly used prestressing tendons, wires, and strands, in the pretensioned members, are compiled in Table 7.10.

The JTG 3362-2018 code provides the transmission lengths of commonly used prestressing tendons in the pretensioned members when $\sigma_{pe} = 1000$ MPa according to Eq. (7.114), as compiled in Table 7.11. When the effective prestress in the prestressing tendon is different from this value, the transmission length should be increased or decreased in proportion according to the table values.

For railway bridges, the Q/CR 9300-2018 code prescribes that the transmission length of prestressing strands in the pretensioned members can be taken as equal to $80d$ in most cases.

In the ultimate limit state, the maximum stress in the prestressing tendons reaches the design value of tensile strength. In order to ensure that the tendon is not pulled out, there must be sufficient anchorage length. The length from the zero stress point (end of the tendon) to the point corresponding to the design value of tensile strength is called the

Table 7.10 Coefficient on the shape of commonly used prestressing tendons.

	The coefficient on the shape of the prestressing tendon α	
Type of prestressing tendon	Highway members (JTG 3362-2018)	Civil and industrial members (GB 50010-2010)
Helical ribs wire	0.14	0.13
2 or 3-Wire strand	0.15	0.16
7-Wire strand	0.16	0.17

Table 7.11 Stress-transmission length of the prestressing tendon.

	Concrete grade					
Type of prestressing tendon	C30	C35	C40	C45	C50	≥C55
7-Wire strand	80d	73d	67d	64d	60d	58d
Helical ribs wire	70d	64d	58d	56d	53d	51d

anchorage length of the prestressing tendon, as shown in Fig. 7.10B, which can be derived approximately by

$$l_a = \alpha \frac{f_{pd}}{f_{ctd}} d \qquad (7.117)$$

where l_a = anchorage length of the prestressing tendon.

f_{pd} = design value of tensile strength of prestressing tendons.

The anchorage length can be calculated according to Eq. (7.115), as compiled in Table 7.12 (for railway bridges) and Table 7.13 (for highway bridges).

7.9 Stress analysis in the anchorage zone of post-tensioned structures

In the bonded post-tensioned flexural structures, the prestressing tendons can be anchored to the end section of the structure (Fig. 7.13A), or the top of the top plate and bottom plate of the box section (Fig. 7.13B), or the inner side of the webs of the box girder. The volume of concrete under and around the anchorage device bears local compression due to the high prestressing concentrated force (P_d) which is called the anchorage zone. In Fig. 7.13A, the anchorage zone is sometimes called the end block, and it is named the triangular tooth block in Fig. 7.13B. To understand the mechanism of force transfer under local compression, a part of the column with end anchorage is taken for analysis, as shown in Fig. 7.14.

It is assumed that the prestressing concentrated force is applied on the end section (AA') with a small square in the middle. According to St. Venant's principle, the stress

Table 7.12 Anchorage length of the prestressing tendon (Q/CR 9300-2018).

Type of prestressing tendon	Concrete grade			
	C40	C45	C50	≥C65
Strand	125d	115d	110d	110d

Table 7.13 Anchorage length of the prestressing tendon (JTG 3362-2018).

Type of prestressing tendon	Concrete grade					
	C40	C45	C50	C55	C60	≥C65
7-Wire strand, $f_{pd} = 1260$ MPa	130d	125d	120d	115d	110d	105d
Helical ribs wire, $f_{pd} = 1200$ MPa	95d	90d	85d	83d	80d	80d

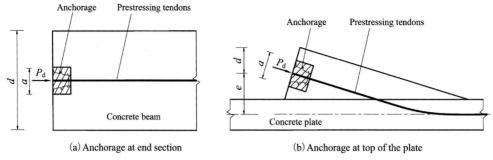

(a) Anchorage at end section   (b) Anchorage at top of the plate

Figure 7.13 Anchorage zone types.

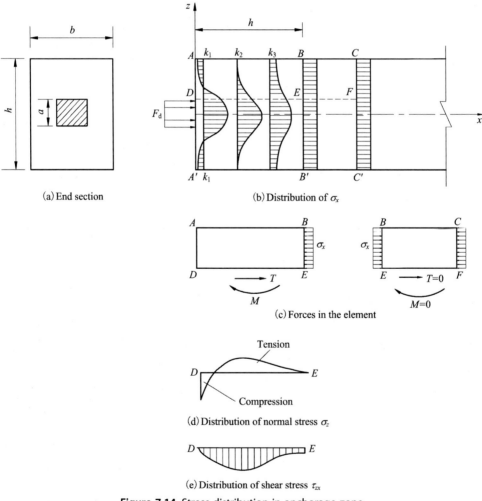

(a) End section   (b) Distribution of σ_x

(c) Forces in the element

(d) Distribution of normal stress σ_z

(e) Distribution of shear stress τ_{zx}

Figure 7.14 Stress distribution in anchorage zone.

in the cross-section at distance h approximately away from the end section becomes uniform (BB'), as shown in Fig. 7.14B. In the length AB, three typical sections $K1 - K3$ are chosen to compare the distribution of longitudinal stress. In section $K3$ close to BB', all longitudinal stresses across the section are compressive and the stress in the middle region is large, while it is small on both sides. In section $K2$, the stress on both sides is zero and the stress in the middle region becomes larger. In section $K1$ close to local compression, the stress in the middle region is very large, consequently, longitudinal tensile stress on the side regions will be generated.

Take an element $ABED$ along the outer side edge DF of the local compression, as shown in Fig. 7.14C. To counterbalance the longitudinal compression in section BE, shear force along the DE plane will be generated, resulting in a shear force on the DE plane, as shown in Fig. 7.14E. Establishing the moment equilibrium at the reference point E, the bending moment acting on section DE can be derived as

$$M_{DE} = \frac{(h-a)^2 F_l}{8h} \qquad (7.118)$$

where M_{DE} = bending moment acting on section DE due to local compression.

h = depth of the cross-section.

a = width of the square acted on by the prestressing concentrated force.

F_l = local compression due to the prestressing force.

The M_{DE} should be in a clockwise direction, so the concrete close to end section AA' bears compression and that close to section BB' bears tension, as shown in Fig. 7.14D. Researches show that the maximum normal tensile stress of concrete in section DE occurs in the region of 0.15–0.3 h approximately (Fig. 7.15), which may result in

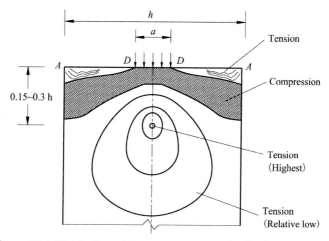

Figure 7.15 Distribution of transverse stress under local compression.

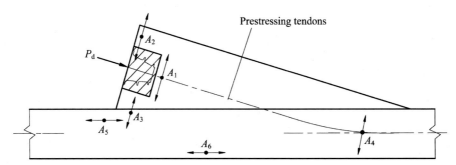

Figure 7.16 Tension regions in the triangular tooth block anchorage zone.

developing bursting cracks. Hence, the transverse reinforcement in this region perpendicular to section *DE* should be arranged.

In the transverse direction, the concrete in region *AD* of section *K*1 close to end section *AA'* bears tension to counterbalance the compression in the middle region, as shown in Fig. 7.15, which may lead to splitting cracking. Therefore, the transverse reinforcement around the anchorage device is also required.

For the anchorages that are arranged at the top of the top plate or bottom plate of the box section (Fig. 7.13B), the above analysis shows that at least five regions in the vicinity of anchorage and prestressing tendons are tensioned, *A*1 − *A*5, as shown in Fig. 7.16, which may result in burst cracking, splitting cracking, or spalling cracking. Under the action of compression in the tooth block due to the prestressing concentrated force, tension will be generated in the opposite edge of the concrete plate, *A*6, as shown in Fig. 7.16. Therefore, the reinforcing steels must be arranged in the regions *A*1 − *A*6 to resist the tension forces.

Suggested readings

[1] GB 50010: 2015. Code for design of concrete structures. Beijing: China Construction Industry Press; 2015.
[2] Q/CR 9300: 2018. Code for design on railway bridge and culvert. Beijing: China Railway Publishing House; 2018.
[3] JTG 3362: 2018. Specifications for design of highway reinforced concrete and prestressed concrete bridges and culverts. Beijing: People's Communications Press; 2018.
[4] Naaman Antoine E. Prestressed concrete structure analysis and design. 2nd ed. New York: McGraw Hill; 2004.
[5] Nawy Edward G. Prestressed concrete: a fundamental approach. 5th ed. Upper Saddle River: Prentice-Hall, Inc; 2006.
[6] Hu Di. The steel restraint influence coefficient method to analyze time-dependent effect in prestressed concrete bridges. Eng Mech 2006;23(6):120−6.
[7] Hu Di. Basic principles of prestressed concrete structure design. 2nd ed. Beijing: China Railway Publishing House; 2019.
[8] Hu Di. Theory of Creep effect on concrete structures. Beijing: Science Press; 2015.

CHAPTER 8

Calculation and control of deformations and cracks

Contents

Analysis and Design of Prestressed Concrete
ISBN 978-0-12-824425-8, https://doi.org/10.1016/B978-0-12-824425-8.00008-7

8.1 The significance of deformation and crack control

Prestressed concrete is widely used to fabricate medium- and long-span highway and railway beams. The beam bends when trains or vehicles run over it, so the deflection of the beam should be limited to a reasonable value to allow passengers to remain comfortable. For railway bridges, especially for high-speed railway bridges, the excessive deflection caused by the trains reduces the smoothness of the rails, which not only decreases the comfort level of passengers but also increases the train impact on the track system and bridge, even affecting the safety of passengers and the bridge. Therefore, the deflection of the bridge beams caused by trains must be controlled to a limited value stipulated in the design codes to guarantee the required smoothness of the rails.

As discussed in Chapter 5, concrete creep and shrinkage and tendon relaxation reduce the eccentric prestressing force in a prestressed concrete beam, resulting in a time-dependent downward deflection. On the other hand, the beam generates time-dependent camber due to the concrete creep under the action of the dead load and the eccentric prestressing force. Consequently, the prestressed concrete beams' deflection in the service period changes with time, and the direction (downward or upward) of the deflection depends on many factors, such as the characteristics of structural system, concrete creep, proportion of the dead load in the design load, and the eccentric prestressing force, etc. Observation shows that the long-term camber generally occurs in a simply supported beam, while long-term downward deflection possibly occurs in a continuous beam with long spans. For a simply supported beam in a high-speed railway with a ballastless track system used widely in China, the excessive time–dependent camber not only decreases the smoothness of the rails but can cause difficulty in amending the rail level to the required position by adjusting the rail fasteners, because the adjustable gap by the rail fasteners is very limited. Therefore, the long-term camber of high-speed railway beams must be controlled with strict precision. For instance, the maximum incremental camber of the simply supported beams with a span of 32 m in a high-speed railway must be less than 10 mm since the track system has been constructed.

When trains are running over a bridge, the end of the beam rotates around the bearing, as shown in Fig. 8.1. For a high-speed railway beam with a ballastless track system, the upward vertical displacement of the rails due to beam end rotation leads to compression in the rail fasteners above the beam end region and tension in the adjacent fasteners. If the repeated compression and tension are large, the performance of the fasteners is affected seriously, even resulting in fatigue failure. The change in the vertical displacement depends on the rotation angle and the distance between the center of the bearing and the beam end (also called the overhanging length of beam end), so the end rotation angle caused by the trains is also required to be strictly limited.

As for the prestressed concrete beams in the buildings, the deflection also needs to be controlled. Excessive deflection of the beam affects the installation and normal operation

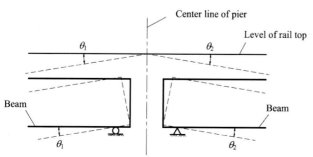

Figure 8.1 Schematic diagram of rotation of the beam end.

of the machinery or equipment on the floor, and may give rise to cracking in the connected nonbearing wall, damage to the attached decoration and cause misalignment of the adjacent doors and windows. Excessive deflection of the roof beam brings large deflection of the panels, and can even cause cracking, resulting in water accumulation in the concave area and leakage in the roof.

For a partially prestressed concrete member, cracking is allowed in some cases. Research and investigations have shown that the corrosion of reinforcement is related to the crack width—the larger the crack width the more susceptible the reinforcements are to corrosion. The diameter of the wires in prestressing strands and wire bundles is small, so their performance is sensitive to corrosion in a state of high tensile stress. In the case of a small rust pit, the stress concentration may occur in the steel wire, which is easy to break. Overall, cracks with excessive width not only reduce the durability of the structures but discount the sectional stiffness, even resulting in serviceability failure. Hence, the crack width must be controlled for the prestressed concrete structures.

In general, the deflection, rotation angle of beam end, and crack width are important indexes in serviceability limit state, which should be controlled in design.

8.2 Calculation and control of deflection and rotation angle

8.2.1 Flexural behavior and assumptions in deflection calculations

The relation between the deflection and the bending moment at midspan in a flexural member from stretching tendons to failure is illustrated in Fig. 7.2, which can be simplified to Fig. 8.2. The beam exhibits a camber at transfer, so point ② in Fig. 8.2 represents the camber caused by the prestressing force and the self-weight of the beam. As external load increases, the section reaches the decompression state, and then the critical cracking state

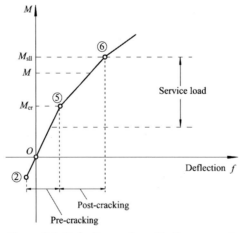

Figure 8.2 Deflection varies with the moment.

(point ⑤ in Fig. 8.2). To simplify this analysis, it is assumed that the concrete beam performance is elastic between points ② and ⑤, although the portion of the concrete in the vicinity of the tension edge has taken up plastic deformation, as shown in Fig. 7.7B. If cracking is allowed, the maximum bending moment will be greater than M_{cr} (M_{cr} is the cracking moment), while it is less than M_{sll} (M_{sll} is the service limit load, point ⑥ in Fig. 7.2). Since the crack width is controlled in an allowable value specified in the codes between points ⑤ and ⑥, the deflection of the beam varies linearly approximately with the moment.

The following assumptions are often adopted in calculations of the deflection:

(1) The deformation of the section conforms to the plane section assumption.

(2) The elastic formulas can be used to calculate the deflections both in the uncracked and cracked members in the service period, and the principle of superposition is applied.

(3) The concrete gross cross-sectional area is used before cracking ($M \leq M_{cr}$), while the tensile concrete is disregarded after cracking ($M > M_{cr}$).

(4) Deflection due to shear deformation is omitted.

(5) The value of concrete modulus specified in the codes is adopted.

The deflection generated at applying loads is called the short-term deflection, while the deflection considering the effect of concrete creep and shrinkage in the service period is called the long-term deflection.

8.2.2 Short-term deflection of the flexural members

Let point B correspond to the bending moment at service loads, point B may be located below point ⑤ or between points ⑤ and ⑥, as shown in Fig. 8.2. To derive the typical formulas of deflection in the service period, let point B be between points ⑤ and ⑥ in which cracking is allowed, as shown in Fig. 8.3. There are two approaches to calculate the deflection corresponding to point B, one is $f_B = f_{O \to A} + f_{A \to B}$ where the gross

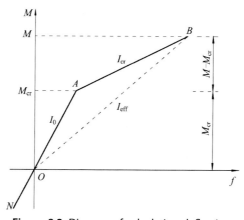

Figure 8.3 Diagram of calculating deflection.

cross-sectional moment of inertia I_g is used in the calculation of $f_{O \to A}$ while the cracked sectional moment of inertia I_{cr} is used in the calculation of $f_{A \to B}$. The other is $f_B = f_{O \to B}$ where the effective moment of inertia I_{eff} is used whether the section is cracked or not. The former approach is called the bilinear calculation method and the latter is called the single-line calculation method or effective-moment-of-inertia method.

8.2.2.1 Bilinear method

At the time of applying bending moment $M(x)$, according to Fig. 8.3, the short-term deflection of the flexural member can be calculated as follows:

When $M(x) \leq M_{cr}$

$$f_s = \int_0^L \frac{\overline{M}(x) \cdot M(x)}{B_1} dx \tag{8.1}$$

When $M(x) > M_{cr}$

$$f_s = \int_0^L \overline{M}(x) \left[\frac{M_{cr}(x)}{B_1} + \frac{M(x) - M_{cr}(x)}{B_2} \right] dx \tag{8.2}$$

where $M(x)$ = design bending moment along the length of the member.
$\overline{M}(x)$ = bending moment due to unit force acting on the considered position.
$M_{cr}(x)$ = cracking moment.
B_1 = flexural rigidity of the uncracked section, $B_1 = E_c I_c$.
B_2 = flexural rigidity of the cracked section, $B_2 = E_c I_{cr}$.
E_c = modulus of elasticity of concrete.
I_c = moment of inertia of the uncracked section.
I_{cr} = moment of inertia of the cracked section.
L = effective span of the member.
Eqs. (8.1) and (8.2) can be written uniformly as

$$f_s = \int_0^L \frac{\overline{M}(x) \cdot M(x)}{B_s} dx \tag{8.3}$$

where B_s = short-term flexural rigidity of the member.
For a prismatic beam, Eq. (8.3) can be rewritten as

$$f_s = \alpha_f \frac{M \cdot L^2}{B_s} \tag{8.4}$$

where α_f = factor of deflection, related to the load types and support conditions. For instance, $\alpha_f = 5/48$ is taken for the deflection at midspan of a simply supported beam subjected to uniformly distributed load, and $\alpha_f = 1/4$ is taken for the deflection at end of a cantilever beam subjected to uniformly distributed load.

M = bending moment in a specific section, generally it refers to the midspan section for a simply supported or continuous beam, and the section close to the rigid support for a cantilever beam.

The bilinear method, based on Eqs. (8.1)−(8.4), is adopted to calculate the short-term deflection of the prestressed concrete flexural members recommended in the *Code for Design on Railway Bridge and Culvert* (Q/CR 9300-2018), *Specifications for Design of Highway Reinforced Concrete and Prestressed Concrete Bridges, and Culverts* (JTG 3362-2018), and *Codes for design of concrete structures* (GB 50010-2010).

8.2.2.1.1 Calculation of short-term deflection for railway bridges

According to the Q/CR 9300-2018 code, the short-term deflection of a prestressed concrete railway beam is calculated by Eqs. (8.3) and (8.4), in which the short-term flexural rigidity is specified as

$$B_s = \beta_p \beta_1 \beta_{cr} E_c I_0 \tag{8.5}$$

where E_c = modulus of elasticity of concrete.

I_0 = moment of inertia of section converted to concrete.

β_1 = stiffness reduction factor considering the effect of fatigue, as compiled in Table 8.1, $\beta_1 = 1$ is taken for calculation of the deflection caused by the prestressing force.

Table 8.1 Values of the stiffness reduction factor.

Section types		
Reduction factor	Cracking is not allowed	Cracking is allowed
β_p	$\left\| \frac{1+\lambda}{2} \right.$	$\left\| \frac{1+\lambda}{2} \right.$
β_1	$\left\| \frac{\lambda - 0.5}{0.95\lambda - 0.45} \right.$	$\left\| \frac{\lambda - 0.5}{0.95\lambda - 0.45} \right.$
β_{cr}	1	$\left\| \frac{\beta_2 M}{\beta_2 M_{cr} + \beta_p (M - M_{cr})} E_c I_0 \right.$
Notes	λ = Prestress level β_2 = Reduction factor considering the influence of reinforcement ratio, $\beta_2 = 0.1 + 2\alpha_{EP}\rho \leq 0.50$ ρ = Reinforcement ratio, $\rho = \frac{A_p + A_s}{bh_0}$ α_{EP} = Ratio of the elastic modulus of the prestressing tendon to concrete, $\alpha_{EP} = 10$ for concrete C40 − C60 M_{cr} = Cracking moment, $M_{cr} = (\sigma_{pc} + \gamma f_{tk}) W_0$ M = Total bending moment b = Width of rectangular section or the web width of the T and I sections h_0 = Effective depth of the section		

β_p = stiffness reduction factor considering the effect of prestress level, as compiled in Table 8.1, $\beta_p = 1$ is taken for calculation of the deflection caused by the prestressing force.

β_{cr} = stiffness reduction factor considering the effect of cracking, as compiled in Table 8.1.

8.2.2.1.2 Calculation of short-term deflection for highway bridges

The JTG 3362-2018 code prescribes that the short-term deflection of a prestressed concrete highway beam should be calculated by Eqs. (8.1) and (8.2) and the short-term flexural rigidity is specified as

$$B_1 = 0.95 E_c I_0 \tag{8.6a}$$

$$B_2 = E_c I_{cr} \tag{8.6b}$$

where B_1 = flexural rigidity of the uncracked section.
B_2 = flexural rigidity of the cracked section.
I_0 = moment of inertia of the converted section.
I_{cr} = moment of inertia of the cracked section.

8.2.2.1.3 Calculation of short-term deflection for building beams

The GB 50010-2010 code prescribes that the short-term deflection of a prestressed concrete beam in the buildings is calculated from Eqs. (8.1) and (8.2) and the flexural rigidity is specified as follows.

When beam cracking is not allowed:

$$B_s = 0.85 E_c I_0 \tag{8.7}$$

When beam cracking is allowed:

$$B_s = \frac{0.85 E_c I_0}{k_{cr} + w(1 - k_{cr})} \tag{8.8}$$

$$k_{cr} = \frac{M_{cr}}{M_k} \tag{8.9}$$

$$M_{cr} = \left(\sigma_{pc} + \gamma f_{ctk}\right) W_0 \tag{8.10}$$

$$w = \left(1.0 + \frac{0.21}{\alpha_E \rho}\right)\left(1 + 0.45\lambda_f\right) - 0.7 \tag{8.11}$$

$$\lambda_f = \frac{\left(b_f - b\right) h_f}{h_0} \tag{8.12}$$

where M_{cr} = cracking moment.

M_k = bending moment caused by the characteristic combination of actions.

$k_{cr} = \frac{M_{cr}}{M_k}$, $k_{cr} = 1.0$ is taken if $k_{cr} > 1.0$.

W_0 = section modulus of converted section.

f_{ctk} = characteristic tensile strength of concrete.

σ_{pc} = compressive stress of concrete in the edge fiber at the tension zone caused by the prestressing force.

b_f = width of the flange in the tension zone.

h_f = depth of the flange in the tension zone.

ρ = ratio of tensile reinforcement, $\rho = \frac{\alpha_1 A_p + A_s}{bh_0}$ for flexural members, where $\alpha_1 = 1.0$ is for bonded post-tensioned tendons and $\alpha_1 = 0.3$ is for unbonded post-tensioned tendons.

γ = correction coefficient considering the influence of plasticity of concrete, given by Eq. (7.66).

α_E = ratio of the elastic modulus of the reinforcement to concrete.

b = width of the rectangular section or the web width of the T- or I-section.

h_0 = effective depth of the section.

For a cracked prestressed concrete beam with a constant section, the reduction coefficient of flexural rigidity in Eq. (8.8) can be derived from Fig. 8.4. Let B correspond to the bending moment due to the service loads, and point C corresponds to the typical value of k_{cr} with 0.4 ($M_{0.4} < M_{sll}$) introduced as a reference point.

According to Eq. (8.4), the different deflections caused by the different moments are obtained as

$$f_{cr} = \alpha_f \frac{M_{cr} \cdot L^2}{\beta_{cr} E_c I_0} \tag{8.13}$$

$$f = \alpha_f \frac{M_k \cdot L^2}{\beta E_c I_0} \tag{8.14}$$

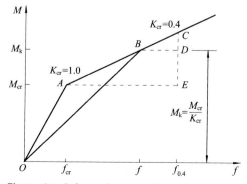

Figure 8.4 Relation between K_{cr} and deflection.

$$f_{0.4} = \alpha_f \frac{M_{0.4} \cdot L^2}{\beta_{0.4} E_c I_0} \tag{8.15}$$

where $M_{0.4}$ = bending moment corresponding to $K_{cr} = 0.4$ (point C in Fig. 8.4).

β = reduction coefficient of flexural rigidity corresponding to M_k.

β_{cr} = reduction coefficient of flexural rigidity corresponding to M_{cr}.

$\beta_{0.4}$ = reduction coefficient of flexural rigidity corresponding to $M_{0.4}$.

The following geometric relation can be attained from Fig. 8.4:

$$\frac{f_{0.4} - f}{f_{0.4} - f_{cr}} = \frac{CD}{CE} = \frac{k_{cr} - 0.4}{1 - 0.4}$$

Substituting Eqs. (8.13)−(8.15) into the above equation yields

$$\frac{1}{\beta} = \frac{1}{\beta_{0.4}} + \frac{k_{cr} - 0.4}{0.6}\left(\frac{1}{\beta_{cr}} - \frac{1}{\beta_{0.4}}\right)$$

Experimental results show that $\beta_{cr} \approx 0.85$ and $\frac{1}{\beta_{0.4}} \approx \left(0.8 + \frac{0.15}{\alpha_E\rho}\right)(1 + 0.45\lambda_f)$,

hence, the reduction coefficient of flexural rigidity in Eq. (8.8) is derived as

$$\beta = \frac{0.85}{k_{cr} + \left[\left(1.0 + \frac{0.21}{\alpha_E\rho}\right)(1 + 0.45\lambda_f) - 0.7\right](1 - k_{cr})} \tag{8.16}$$

8.2.2.2 Single-line method

Another approach to obtain the deflection corresponding to point B in Fig. 8.3 can be achieved by establishing the relation between the deflection and the bending moment approximately based on the dotted line OB, in which the single equivalent moment of inertia is used. If the bending moment due to service loads exceeds the cracking moment, this equivalent moment of inertia should consider the comprehensive contribution of the moment of inertia in the cracked and uncracked sections. The single-line method, also known as the effective-moment-of-inertia method, was developed by Branzon. The effective moment of inertia is given by

$$I_{eff} = \left(\frac{M_{cr}}{M_a}\right)^3 I_g + \left[1 - \left(\frac{M_{cr}}{M_a}\right)^3\right]I_{cr} \leq I_g \tag{8.17}$$

where I_{eff} = effective moment of inertia.

I_g = moment of inertia of the gross section.

I_{cr} = moment of inertia of the cracked section.

M_{cr} = cracking moment.

M_a = bending moment caused by the combination of actions specified in the codes.

Thus, the deflection can be calculated by

$$f_s = \int_0^L \frac{\overline{M(x)} \cdot M(x)}{E_c I_{eff}} \, dx \qquad (8.18)$$

or

$$f_s = \alpha_f \frac{M \cdot L^2}{E_c I_{eff}} \qquad (8.19)$$

8.2.3 Long-term deflection of the flexural members

8.2.3.1 General concerns on long-term deflection calculation

Since a prestressed concrete flexural member has been constructed, the change to the vertical camber or deflection with time is inevitably due to the following factors:

(1) The eccentric prestressing force reduces gradually due to prestress losses, resulting in change of deflection with time.

(2) The curvature of the flexural section under the eccentric prestressing force and the dead load increases with time due to concrete creep.

(3) The free deformation of concrete is restrained by the prestressing tendons and reinforcing steels, leading to change of curvature with time.

Therefore, the deflection (or camber) of a prestressed concrete flexural member is a time-dependent variable from stretching tendons to the entire service period, so the time-dependent analysis should be used to calculate the deflection. Considering that the coupling of the above factors affects the deflection and the time-dependent analysis is very complicated and lengthy, which must be completed by the computer, the simplified approaches are commonly provided in the codes.

Take a simply supported plain concrete beam as an example (Fig. 8.5) to illustrate the effect of creep on the long-term sectional curvature. Assuming that the beam does not crack under the action of the self-weight, the neutral axis keeps in the same position and the modulus of elasticity of concrete is constant.

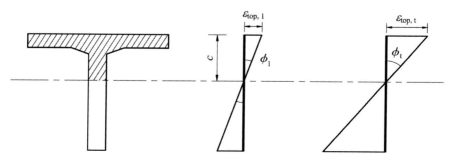

Figure 8.5 Variation of the sectional curvature.

Let the initial elastic curvature of the midspan section at applying self-weight at a time t_1 be $\phi_g(t_1)$, the curvature at time t ($t > t_1$) can be obtained as

$$\phi_g(t, t_1) = \frac{\varepsilon_{top,t}}{c} = \frac{\varepsilon_{top,1}[1 + \varphi(t, t_1)]}{c} = \phi_g(t_1)[1 + \varphi(t, t_1)] \tag{8.20}$$

where $\varepsilon_{top,1}$ = elastic strain of concrete in the top fiber caused by the self-weight at time t_1.

$\varepsilon_{top,t}$ = concrete strain in the top fiber at time t ($t > t_1$).

c = depth of compression zone of concrete.

$\phi_g(t_1)$ = elastic curvature of the midspan section caused by the self-weight at time t_1.

$\phi_g(t, t_1)$ = curvature of the midspan section at time t ($t > t_1$) when applying load at time t_1.

$\varphi(t, t_1)$ = creep coefficient of concrete at time t when applying load at time t_1.

Eq. (8.20) demonstrates that the sectional curvature increases with time due to creep, known as creep curvature. Consequently, the deflection of the flexural member alters with time.

For a prestressed concrete beam, the calculation of the time-dependent sectional curvature is far more complicated than Eq. (8.20), since not only the development of the concrete creep is constrained by the reinforcement, but also the stress redistribution in the section caused by the reinforcement constraint affects the time-dependent curvature. The degree of reinforcement constraint on the creep effect depends on the reinforcement ratio, the position of the reinforcement, modulus ratio of reinforcement to concrete, and the value of creep, etc. To simplify the analysis, a constraint reduction coefficient can be introduced to consider the comprehensive effect of reinforcement constraint, hence, the time-dependent curvature of a prestressed concrete beam caused by the self-weight is attained as

$$\phi_g(t) = \phi_g(t_1)[1 + \gamma_s(t, t_1)\varphi(t, t_1)] \tag{8.21}$$

where $\gamma_s(t, t_1)$ = constraint reduction coefficient for creep curvature by the reinforcement at the time duration $t - t_1$.

Actually, whether it's a pretensioned or post-tensioned flexural member, the beam bends at transfer, a synthesis result of the camber caused by the prestressing force and the deflection caused by the self-weight. Therefore, the calculation of time-dependent sectional curvature caused by the prestressing force and the self-weight should be taken into account simultaneously. The sectional curvature at transfer at time t_1 is given by

$$\phi_{g,p}(t_1) = \phi_g(t_1) + \phi_p(t_1) = \frac{M_g(x)}{E_c I_c} + \frac{A_p \sigma_{pe}(x, t_1)e_p + M_{p,s}(x)}{E_c I_c} \tag{8.22}$$

where $\phi_p(t_1)$ = elastic curvature of midspan section at time t_1.

$M_g(x)$ = bending moment caused by the self-weight.

$M_{p,s}(x)$ = secondary bending moment caused by the prestressing force in the post-tensioned statically indeterminate structures.

A_p = cross-section area of the prestressing tendons.

$\sigma_{pe}(x, t_1)$ = effective stress in prestressing tendons at time t_1.

E_c = modulus of elasticity of concrete.

I_c = moment of inertia.

e_p = eccentricity of the prestressing tendons.

I_c is obtained from the converted section for pretensioned structures while it is obtained from the net section for post-tensioned structures.

As discussed in Section 5.3, the creep and shrinkage of concrete and relaxation of tendon interact with the time-dependent effect on prestressed concrete structures, all of which cause prestress losses resulting in a reduction of the eccentric effective prestressing force, which alters the sectional curvature. Except that concrete creep introduces a change of curvature as shown in Eq. (8.20), the shrinkage alone will generate nonuniform longitudinal shortening when the centroid of all reinforcements (prestressing tendons and reinforcing steels) does not coincide with the centroid of the concrete section which also leads to a change of curvature. Considering a structure with variable sections along the span arranged with curved tendons, referring to Eq. (5.62), the change of curvature at any section under the prestressing force and the self-weight can be derived as

$$\Delta\phi(x, t, t_1) = \gamma_{r,sp}(x, t, t_1)\phi_{g,p}(x, t_1)\varphi(t, t_1) \tag{8.23}$$

where $\gamma_{r,sp}(x, t, t_1)$ = constraint reduction coefficient for creep curvature by reinforcement, estimated by Eq. (8.24).

$$\gamma_{r,sp}(x, t, t_1) = 1$$
$$- \kappa\left\{\varepsilon_{c,sp}(x, t_1)\varphi(t, t_1) + \varepsilon_{sh}(t, t_1) - \rho\alpha_{EP}\mu_p\lambda(t, t_1)[1 + \chi(t, t_1)\varphi(t, t_1)]\frac{\sigma_r(t, t_1)}{E_p}\right\} \tag{8.24}$$

$$\kappa = \frac{e_{ps}}{\phi_{g,p}(t_1)\varphi(t, t_1)r^2} \cdot \frac{\left(\alpha_{Ep}\mu_p + \alpha_{ES}\mu_s\right)[1 + \chi(t, t_1)\varphi(t, t_1)]}{1 + \rho\left(\alpha_{Ep}\mu_p + \alpha_{ES}\mu_s\right)[1 + \chi(t, t_1)\varphi(t, t_1)]} \tag{8.25}$$

where $\varepsilon_{c,sp}(x, t_1)$ = concrete strain at the centroid of all reinforcements caused by the prestressing force and the self-weight at time of loading t_1.

$\varphi(t, t_1)$ = creep coefficient during the time period $t - t_1$.

$\varepsilon_{sh}(t, t_1)$ = shrinkage strain during the time period $t - t_1$.

$\chi(t, t_1)$ = aging coefficient of concrete at the time of loading t_1 to the calculating time t, varies from 0.6 to 0.9, generally, 0.82 can be taken.

A_c = area of the net section.

I_c = moment of inertia of the net section.

$r = \sqrt{\frac{I_c}{A_c}}$ = radius of gyration of the net section.

e_{ps} = distance from the centroid of all reinforcements to the centroid of the concrete section.

μ_p = prestressing tendon ratio, $\mu_p = \frac{A_p}{A_c}$.

μ_s = reinforcing steel ratio, $\mu_s = \frac{A_s}{A_c}$.

$\alpha_{ES} = \frac{E_s}{E_c}$, ratio of the modulus of elasticity between reinforcing steel and concrete.

$\alpha_{EP} = \frac{E_p}{E_c}$, ratio of the modulus of elasticity between prestressing tendon and concrete.

Considering $EI\phi(x) = M(x)$ and the obtained Eqs. (8.22) and (8.23), the time-dependent camber or deflection of a prestressed concrete beam under the prestressing force and the self-weight can be obtained as

$$f_{g,p}(t) = \int_0^L \frac{\overline{M}(x) \cdot M_{g,p}(x, t)}{E_c I(x)} dx = \int_0^L \overline{M}(x) \phi(x, t) dx$$

$$= \int_0^L \overline{M}(x) \left[\phi_{g,p}(t_1) + \Delta\phi(x, t, t_1) \right] dx \tag{8.26}$$

$$= f_{g,p}(t_1) + \int_0^L \overline{M}(x) \cdot \left[\gamma_{r,sp}(x, t, t_1) \phi_{g,p}(t_1) \varphi(t, t_1) \right] dx$$

where $f_{g,p}(t)$ = camber or deflection of the beam at time $t(t > t_1)$ when the prestressing force and the self-weight are applied at time t_1.

$f_{g,p}(t_1)$ = initial camber or deflection at time t_1, $f_{g,p}(t_1) = \int_0^L \overline{M}(x) \cdot \phi_{g,p}(x, t_1) dx$, $\phi_{g,p}(x, t_1)$ is given by Eq. (8.22).

$\overline{M}(x)$ = bending moment produced by the unit force acting on the position considered.

L = effective span of the prestressed concrete structure.

When the second-stage dead load, such as pavement system in the highway bridges and rail system in the railway bridges, is applied subsequently at a time $t_2(t_2 > t_1)$, it contributes to the time-dependent effect with the prestressing force and the self-weight. To simplify the analysis, the deflection change due to the second-stage dead load can be calculated by referring to Eq. (8.26), hence,

$$f(t) = f_{g,p}(t_1) + f_d(t_2) + \int_0^L \overline{M}(x) \cdot \phi_{g,p}(t_1)\gamma_{r,sp}(x,t,t_1)\varphi(x,t_1)dx$$

$$+ \int_0^L \overline{M}(x) \cdot \Delta\phi_d(x,t,t_2)dx \tag{8.27}$$

where $f(t)$ = total camber or deflection of the beam under the prestressing force and all dead loads.

$f_d(t_2)$ = deflection caused by the second-stage dead load at time t_2, $f_d(t_2) = \int_0^L \frac{\overline{M}(x)M_d(x)}{E_cI_0}dx$.

$\Delta\phi_d(x,t,t_2)$ = change of sectional curvature caused by the second-stage dead load.
I_0 = moment of inertia of the section converted to concrete.

For further simplified calculation, the second-stage dead load is treated to be applied simultaneously with the self-weight, so that its effect can be included in Eqs. (8.22), (8.23), and (8.26), in this case, Eqs. (8.23), (8.24), and (8.26) are rewritten as

$$f(t) = f_{p,d}(t_1) + \int_0^L \overline{M}(x) \cdot \Delta\phi(x,t,t_1)dx \tag{8.28}$$

$$\Delta\phi(x,t,t_1) = \gamma_{r,sp}(x,t,t_1)\phi_{p,d}(x,t_1)\varphi(t,t_1) \tag{8.29}$$

$\gamma_{r,sp}(x,t,t_1) = 1$

$$-\kappa\left\{\varepsilon_c(x,t_1)\varphi(t,t_1) + \varepsilon_{sh}(t,t_1) - \rho\alpha_{EP}\mu_p\lambda(t,t_1)[1 + \chi(t,t_1)\varphi(t,t_1)]\frac{\sigma_r(t,t_1)}{E_p}\right\}$$

$$\tag{8.30}$$

where $f(t)$ = total camber or deflection of the beam.

$f_{p,d}(t_1)$ = camber or deflection caused by the prestressing force and all dead loads at time t_1.

$\phi_{p,d}(x,t_1)$ = sectional curvature caused by the prestressing force and all dead loads at time t_1.

$\varepsilon_c(x,t_1)$ = concrete strain at the centroid of all reinforcements caused by the prestressing force and all dead loads at time t_1.

Eq. (8.26) or (8.28) shows that the long-term vertical deformation due to concrete creep and shrinkage and tendon relaxation could be camber or deflection, depending on the change of sectional curvature related to the moment ratio due to the prestressing force and all dead loads, the magnitude of prestress losses and the characteristics of the structural system, etc. Generally, the time-dependent effect leads to camber in the simply supported prestressed concrete railway or highway beams, while there is deflection in the continuous prestressed concrete beams with long spans.

The approach to predict the long-term deflection based on Eq. (8.26) or (8.28) is called the time-dependent coefficient-modified method. In addition, the long-term deflection can be estimated by modifying the short-term flexural rigidity as shown in Eq. (8.3) or (8.4), which is called the stiffness-modified method.

8.2.3.2 Time-dependent coefficient-modified method

For arbitrary forms of prestressed concrete flexural structures under the prestressing force and the dead loads, the long-term deflection can be predicted by Eq. (8.26) or (8.28), while it is difficult to achieve the theoretical solution since the change of sectional curvature due to concrete creep and shrinkage and tendon relaxation is difficult to be expressed with an equation that can be integrated. Hence, the curvatures of three or five typical sections are adopted to obtain the approximate formula. When three curvatures, two curvatures at the support section and one curvature at the midspan section, in a beam are used and all dead loads are assumed to be applied simultaneously, the time-dependent deflection at midspan can be estimated by

$$f_M(t) = f_M(t_1) + \frac{L^2}{96}[\Delta\phi_A(t, t_1) + 10\Delta\phi_M(t, t_1) + \Delta\phi_B(t, t_1)] \qquad (8.31)$$

where $f_M(t)$ = time-dependent deflection at midspan.

$f_M(t_1)$ = initial deflection at midspan at loading.

$\Delta\phi_M(t, t_1)$ = change of sectional curvature at the midspan section during the time period $t - t_1$.

$\Delta\phi_A(t, t_1)$ = change of sectional curvature at the support section during the time period $t - t_1$.

$\Delta\phi_B(t, t_1)$ = change of sectional curvature at the support section during the time period $t - t_1$.

L = effective span of the beam.

In Eq. (8.31), $\Delta\phi_A(t, t_1)$, $\Delta\phi_M(t, t_1)$, and $\Delta\phi_B(t, t_1)$ can be calculated by Eq. (8.29) along with Eqs. (8.25) and (8.30). For the general prestressed concrete structures under normal ambient environment with an average relative humidity of 60%–80%, Eq. (8.29) can be simplified into

$$\Delta\phi(x, t, t_1) = \phi_{p,d}(x, t_1)\varphi(t, t_1)$$

$$- \frac{8\mu}{1 + 18\mu} \cdot \frac{e_{ps}}{r^2}\left[\varepsilon_c(x, t_1)\varphi(t, t_1) + \varepsilon_{sh}(t, t_1) - 12\rho\mu_p\frac{\sigma_r(t, t_1)}{E_p}\right] \qquad (8.32)$$

where $\varepsilon_c(x, t_1)$ = concrete strain at the centroid of all reinforcements caused by prestressing force and the dead loads at time t_1.

μ = reinforcement ratio, $\mu = \mu_p + \mu_s$.

For preliminary design purposes, a comprehensive influence coefficient can be used to estimate the long-term deflection for prestressed concrete highway beams recommended in the JTG 3362-2018 code, as

$$f_l = \eta_{\theta,s} f_s \tag{8.33}$$

where f_l = long-term deflection of the beam.

f_s = deflection produced by the frequent combination of action, calculated by Eqs. (8.2) and (8.6).

$\eta_{\theta,s}$ = incremental coefficient of long-term deflection, 1.6 is taken for C40 and lower-grade concrete, 1.45−1.35 is taken for C40−C80 concrete.

In the calculation of long-term camber caused by the effective prestressing force in the service period, $\eta_{\theta,p} = 2.0$ is generally taken and the converted section stiffness $B_s = E_c I_0$ is used.

The ACI 318 code provides a formula proposed by Branson to estimate the long-term deflection of a reinforced concrete flexural member:

$$f_l = f_e(t_0)(1 + \lambda_s \zeta) \tag{8.34}$$

$$\lambda_s = \frac{1}{1 + 50\rho'} \tag{8.35}$$

where f_l = long-term deflection.

$f_e(t_0)$ = elastic deflection at the time of applying loads.

λ_s = constraint reduction coefficient by reinforcement.

ζ = coefficient considering the effect of concrete creep and shrinkage, 1.0, 1.2, 1.4, and 2.0 correspond to durations of 3 months, 6 months, 1 year, and more than 5 years.

ρ' = compressive reinforcement ratio.

If λ_s is adjusted by considering the effect of tendon ratio, the prestressing force, and the prestress losses, Eq. (8.35) can still be used to estimate the long-term deflection of a simply supported prestressed concrete flexural member.

8.2.3.3 Stiffness-modified method

The comprehensive effect due to concrete creep and shrinkage and tendon relaxation along with the restraint by the reinforcement on the long-term deflection of the prestressed concrete beams can be reflected by the change of the flexural rigidity, so the long-term deflection can be predicted by

$$f_l = \int_0^L \frac{\overline{M}(x)M(x)}{\beta_\theta B_s} dx \tag{8.36}$$

where f_l = long-term deflection.

B_s = flexural rigidity at the time of applying loads.

β_θ = reduction factor of flexural rigidity considering the effect of long-term loading due to creep and shrinkage of concrete, relaxation of tendon, and fatigue, etc.

The GB50010-2010 code prescribes that the reduction factor of flexural rigidity can be given by

$$\beta_\theta = \frac{M_k}{M_q(\theta - 1) + M_k} \tag{8.37}$$

where M_k = bending moment caused by the characteristic combination of actions.

M_q = bending moment produced by the quasi-permanent combination of actions.

θ = incremental influence coefficient by the long-term loading, generally $\theta = 2.0$ is taken.

8.2.4 Calculation of rotation angle at the beam end

The beam end rotates when trains are running over it. The trainload mode stipulated in the codes can be decomposed into a uniformly distributed load and several concentrated loads, so the rotation angle at the beam end caused by the trains can be calculated separately. For a simply supported beam with a uniform section subjected to uniformly distributed load (Fig. 8.6), the rotation angle at the beam end can be obtained as

$$\theta_A = -\theta_B = \frac{qL^3}{24E_cI_0} \tag{8.38}$$

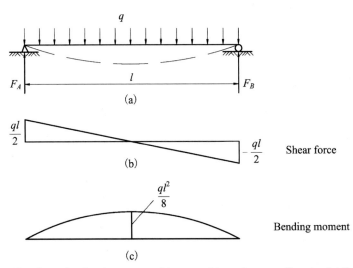

Figure 8.6 Bending of a simply supported beam subjected to a uniformly distributed load.

Under the concentrated load (Fig. 8.7), the rotation angle at the beam end is obtained as

$$\theta_A = \frac{Fab(L+b)}{6E_cI_0L} \tag{8.39a}$$

$$\theta_B = -\frac{Fab(L+a)}{6E_cI_0L} \tag{8.39b}$$

In a continuous beam, based on the compatibility equation of the section bending angle, the approximate calculation formulas of beam end angle under train load also can be obtained.

8.2.5 Control of deformation and set of camber during construction

8.2.5.1 Limits for highway beam deflection

The JTG 3362-2018 prescribes that the maximum deflection of the prestressed concrete highway beams produced by the frequent combination of actions should satisfy:

(1) Maximum deflection of the simply supported or continuous beam should not exceed 1/600 of the effective span.

(2) Maximum deflection at the end of a cantilever should not exceed 1/300 of the effective length of the cantilever.

8.2.5.2 Limits for railway beam deflection and end angle

The TB10002-2017 code prescribes that the maximum deflection of the railway beams subjected to the static train loads (excluding impact effect) should not exceed the limited values compiled in Table 8.2.

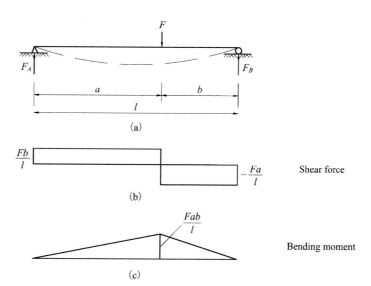

Figure 8.7 Simply supported beam subjected to a concentrated load.

Table 8.2 Limits for railway beam maximum deflection.

Type of railway	Design speed/km·h^{-1}	$L \leq 40$ m	40 m $< L \leq 80$ m	$L > 80$ m
High-speed railway	350	$L/1600$	$L/1900$	$L/1500$
	300	$L/1500$	$L/1600$	$L/1100$
	250	$L/1400$	$L/1400$	$L/1000$
Intercity railway	200	$L/1750$	$L/1600$	$L/1200$
	160	$L/1600$	$L/1350$	$L/1100$
	120	$L/1350$	$L/1100$	$L/1100$
Mixed passenger and freight railway	200	$L/1200$	$L/1000$	$L/900$
	160	$L/1000$	$L/900$	$L/800$
Heavy-haul railway	≤ 120	$L/900$	$L/800$	$L/700$

Note: (1) L is the effective span. (2) The data in the table apply to three or more simple supported beams in the double-line railway. (3) 1.1 times of the data in the table is taken for one continuous beam constituting of three or more spans; 1.4 times of the data is taken for one continuous beam constituting of two spans, and for two simple supported beams in the double−line railway. (4)0.6 times of the data is taken for the simply supported and continuous beams in the single−line railway.

In a railway with speed more than 200 km/h, TB10002-2007 prescribes that the incremental long-term camber or deflection Δf_{cr} due to concrete creep and shrinkage and tendon relaxation since the track system has been constructed should meet the following requirements:

(1) For prestressed concrete beams with a ballast track system:

$$\Delta f_{cr} \leq 20 \text{ mm}$$

(2) For prestressed concrete beams with a ballastless track system:

$\Delta f_{cr} \leq 10$ mm when $L \leq 50$ m

$\Delta f_{cr} \leq L/5000$ and $\Delta f_{cr} \leq 10$ mm when $L > 50$ m

When the trains are running over the bridge, the beams bend and the beam end rotates around the bearing, as shown in Fig. 8.8. For the prestressed concrete beams in the intercity railway and mixed passenger and freight railways with a speed of less than 200 km/h, the rotation angle at the beam end caused by static train loads should meet the following requirements:

(1) Rotation angle at beam end of side beam adjacent to abutment: $\theta \leq 3.0 \times 10^{-3}$ rad

(2) Rotation angle at beam end of two adjacent beams: $\theta_1 + \theta_2 \leq 6 \times 10^{-3}$ rad

For the prestressed concrete beams in the intercity railway, the maximum rotation angle at the beam end under the static train loads should meet the requirements stipulated by the TB10002-2007 code compiled in Table 8.3.

Figure 8.8 Schematic diagram of rotation at the beam end.

Table 8.3 Limits for an intercity railway beam end's maximum rotation angle.

Type of track	Position	Limited value (rad)	Note
Ballast track system	Between abutment and beam	$\leq 2.0 \times 10^{-3}$	
	Between adjacent beams	$\theta_1 + \theta_2 \leq 6.0 \times 10^{-3}$	
Ballastless track system	Between abutment and beam	$\theta \leq 2.1 \times 10^{-3}$ $\theta \leq 1.5 \times 10^{-3}$ $\theta \leq 1.0 \times 10^{-3}$	$L_c \leq 0.3\text{m}$ $0.3\text{m} < L_c \leq 0.55\text{m}$ $0.55\text{m} < L_c \leq 0.75\text{m}$
	Between adjacent beams	$\theta_1 + \theta_2 \leq 4.2 \times 10^{-3}$ $\theta_1 + \theta_2 \leq 3.0 \times 10^{-3}$ $\theta_1 + \theta_2 \leq 2.0 \times 10^{-3}$	$L_c \leq 0.3\text{m}$ $0.3\text{m} < L_c \leq 0.55\text{m}$ $0.55\text{m} < L_c \leq 0.75\text{m}$

Note: L_c is the distance between the center of the bearing and the beam end.

For the prestressed concrete beams in a high-speed railway, the maximum rotation angle at the beam end under the static train loads should meet the requirements stipulated by the *Code for design of high-speed railway bridge* (TB10621-2014) compiled in Table 8.4.

8.2.5.3 Limits for building beam deflection

The maximum deflection of the prestressed concrete beams in buildings caused by the characteristic combination of actions considering the long-term effect should not exceed the limits specified in the GB 50010-2010 code, which can be expressed as:

$$f_l - \eta_\theta f_p \leq [f] \tag{8.40}$$

where f_l = beam deflection caused by the characteristic combination of actions.

f_p = short-term camber caused by the prestressing force.

η_θ = incremental coefficient of long-term camber, 2.0 can be taken in most cases.

$[f]$ = limit for beam deflection specified in the code, compiled in Table 8.5.

8.2.5.4 Set of camber for prestressed concrete beams during construction

The camber of prestressed concrete flexural members caused by the eccentric prestressing forces will gradually increase due to the creep of concrete. When the incremental long-term camber is greater than the downward deflection caused by the design load, it is not necessary to set the camber during construction, while the adverse effect of excessive camber on the structure should be of concern. For high-speed railway beams with a ballastless track system, the excessive camber decreases the smoothness of the rails, and brings

Table 8.4 Limits for a high-speed railway beam end's maximum rotation angle.

	Position	Limited value (rad)	Note
Ballast track system	Between abutment and beam	$\theta \leq 2.1 \times 10^{-3}$	
	Between adjacent beams	$\theta_1 + \theta_2 \leq 4.0 \times 10^{-3}$	
Ballastless track system	Between abutment and beam	$\theta \leq 1.5 \times 10^{-3}$ $\theta \leq 1.0 \times 10^{-3}$	$L_c \leq 0.55$m 0.55m $< L_c \leq 0.75$m
	Between adjacent beams	$\theta_1 + \theta_2 \leq 3.0 \times 10^{-3}$ $\theta_1 + \theta_2 \leq 2.0 \times 10^{-3}$	$L_c \leq 0.55$m 0.55m $< L_c \leq 0.75$m

Table 8.5 Limits for a building beam's maximum deflection.

Type of beam	Limited value of deflection
Crane beam: Manual crane	$L_0/500$
Electric crane	$L_0/600$
Roof, floor, and stair components:	
When $L_0 < 7$m	$L_0/200(L_0/250)$
When 7m $\leq L_0 \leq 9$m	$L_0/250(L_0/300)$
When $L_0 > 9$m	$L_0/300(L_0/400)$

Note: (1) L_0 is the effective span. (2) The values in brackets in the table apply to the components with higher requirements for deflection. (3) For cantilevers, the effective span is taken as two times the actual cantilever length.

difficulty in adjusting the rail fasteners since the adjustable gap by the rail fasteners is limited; for highway beams, excessive camber may give rise to cracking in the deck pavement. Otherwise, it is generally necessary to set the camber for the beams during construction to make the beam at an agreeable elevation at service loads.

For prestressed concrete highway beams, JTG 3362-2018 stipulates that the camber should be set during construction if the incremental long-term camber is less than the deflection caused by the frequent combination of actions, and the set value should be equal to the difference between the above two terms.

For prestressed concrete railway beams with a ballast track system, TB10002-2007 stipulates that it is not necessary to set the camber during construction if the deflection caused by dead load and static live load does not exceed 15 mm or 1/1600th of the effective span, the effect of this deflection on the track system could be eliminated by adjusting the thickness of the ballast. Otherwise, camber set is required and the set value should be equal to the sum of the deflection caused by the dead load and half static live load. For prestressed concrete railway beams with a ballastless track system, it is necessary to set the camber during construction if the deflection caused by the dead load and static live load exceeds 5 mm.

8.3 Example 8.1

A simply supported beam in a mixed passenger and freight railway with a design speed of 160 km/h is made of C50 concrete, stress-relieved strands $3\phi^S15.2$ ($3-7\phi5, f_{ptk} = 1860$ MPa), and $6\Phi20$ (HRB400) reinforcing steel. The effective span of the beam is 16 m, the dimension and arrangement of reinforcement at the midspan section are shown in Fig. 8.9. At midspan, the bending moment due to dead load is $M_g = 1440$ kN·m, the bending moment due to static live load is $M_l = 1749$ kN·m, the maximum design moment due to dead and live loads is $M = 3849.0$ kN·m, the cracking moment is $M_{cr} = 3200.0$ kN·m. The prestress level $\lambda = 0.82$ is adopted.

Calculate and verify the beam deflection according to the TB10002-2017 code:

(1) Deflection caused by the dead load.

(2) Deflection caused by the dead load and live load.

(3) Verify the deflection caused by the static live load.

8.4 Solution

8.4.1 Parameters

$E_s = 2.0 \times 10^5$ MPa, $E_p = 1.95 \times 10^5$ MPa, $E_c = 3.55 \times 10^4$ MPa.

$f_{ptk} = 1860$ MPa, $f_{pd} = 1200$ MPa, $f_{sd} = 320$ MPa,

$A_p = 3 \times 980 = 2940$ mm^2, $A_s = 6 \times \frac{\pi}{4} \times 20^2 = 1885$ mm^2

$a_p = 120$ mm, $a_s = 55$ mm, $I_0 = 0.1270395$ m^4

$a = \frac{f_{pd}A_p a_p + f_{sd}A_s a_s}{f_{pd}A_p + f_{sd}A_s} = \frac{1200 \times 2940 \times 120 + 320 \times 1885 \times 55}{1200 \times 2940 + 320 \times 1885} = 110.5$ mm

$h_0 = h - a = 1200 - 110.5 = 1089.5$ mm.

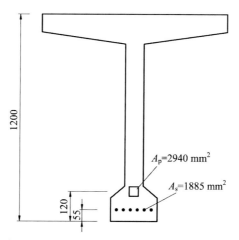

Figure 8.9 Schematic diagram of reinforcement at the midspan section.

8.4.2 Flexural rigidity of the uncracked section

$$\beta_p = \frac{1+\lambda}{2} = \frac{1+0.82}{2} = 0.910$$

$$\beta_1 = \frac{\lambda - 0.5}{0.95\lambda - 0.45} = \frac{0.82 - 0.5}{0.95 \times 0.82 - 0.45} = 0.973$$

The flexural rigidity of the uncracked section is obtained from Eq. (8.5):

$$B_s = \beta_p \beta_1 E_c I_0 = 0.910 \times 0.973 \times 3.55 \times 10^4 \times 0.1270395 = 3.993 \times 10^3 \text{ MN·m}^2$$

8.4.3 Flexural rigidity of the cracked section

$$\rho = \frac{A_p + A_s}{bh_0} = \frac{2940 + 1885}{260880} = 0.0185$$

$$\alpha_{EP} = \frac{E_p}{E_c} = \frac{1.95 \times 10^5}{3.55 \times 10^4} = 5.493$$

$\beta_2 = 0.1 + 2\alpha_{EP}\rho = 0.1 + 2 \times 5.493 \times 0.0185 = 0.303 < 0.5$, So. $\beta_2 = 0.303$
Substituting $M = 3849.0$ kN·m, $M_{cr} = 3200.0$ kN·m into Eq. (8.5) and Table 8.1, the flexural rigidity of the cracked section is obtained:

$$B_{cr} = \beta_1 \frac{\beta_p \beta_2 M}{\beta_2 M_f + \beta_p (M - M_{cr})} E_c I_0$$

$$= 0.973 \times \frac{0.910 \times 0.303 \times 3849}{0.303 \times 3200 + 0.910 \times (3849 - 3200)} \times 3.55 \times 10^4$$

$$\times 0.1270395 = 2984.9 \text{ MN·m}^2$$

8.4.4 Calculation of deflection

Since $M_g = 1440$ kN·m $< M_{cr} = 3200.0$ kN·m, the flexural rigidity of the uncracked section is adopted, the deflection at the midspan caused by the dead load is obtained:

$$f_g = \frac{5M_g L^2}{48B_s} = \frac{5 \times 1740 \times 16^2 \times 10^3}{48 \times 3.993 \times 10^6} = 0.0162 \text{ m} = 0.96 \text{ cm}$$

Since $M = 3849.0$ kN·m $> M_{cr} = 3200.0$ kN·m, the flexural rigidity of the cracked section is adopted, the deflection at the midspan caused by the dead and live loads is obtained:

$$f_d = \frac{5M_d L^2}{48B_{cr}} = \frac{5 \times 3849 \times 16^2 \times 10^3}{48 \times 2984.9 \times 10^6} = 0.0344 \text{ m} = 3.44 \text{ cm}$$

8.4.5 Verify the deflection caused by the static live load

The deflection at the midspan caused by the static live load is obtained:

$$f_l = \frac{5M_l L^2}{48B_{cr}} = \frac{5 \times 1749 \times 16^2 \times 10^3}{48 \times 2984.9 \times 10^6} = 0.0156 \text{ m} = 1.56 \text{ cm}$$

In a mixed passenger and freight railway with a design speed of 160 km/h, the allowable deflection due to the static live loads is

$$[f_l] = \frac{L}{1000} = \frac{1600}{1000} = 1.6 \text{ cm}$$

Since $f_l = 1.56$ cm $< [f_l] = 1.6$ cm, the maximum deflection at the midspan of the railway beam subjected to the static train load satisfies the requirement stipulated in the B10002-2017 code.

8.5 Example 8.2

A post-tensioned simply supported box girder (TQ20051-2322-II) in a double–line high-speed railway with a ballastless track system is made of C50 concrete, bonded strands $\phi^S 15.2$ ($7\phi 5$, $f_{ptk} = 1860$ MPa) and HRB335 reinforcing steels. The total length and the effective span of the girder are 32.6 and 31.5 m, respectively. The box girder has a constant depth of 3.05 m, the top and bottom plates and webs become thicker in the vicinity of the girder ends. The dimension of the box girder at the midspan section is shown in Fig. 8.10, where the area of the concrete section is 8.95×10^6 mm^2, and the areas of the strands and reinforcing steels are 37380 and

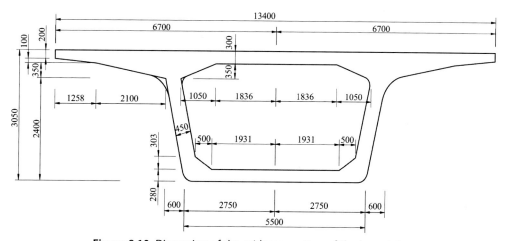

Figure 8.10 Dimension of the midspan section of the box girder.

36983.7 mm^2, respectively. The distances from the centroids of gravity of concrete section, strands, and reinforcing steels to the bottom edge are 1981.2 mm, 359.61 mm, and 1966.01 mm respectively. The moment of inertia of the concrete section with respect to the centroid is $1.10 \times 10^{13} \text{ mm}^4$. The weight of the track system, accessory equipment, and sideways above the girder is 184 kN/m. Calculate and verify the rotation angles of the girder end during train running according to the *Code for design of high-speed railway bridge* (TB10621-2014).

The ZK live load model (Fig. 8.11) is used as the trainloads in the high-speed railway.

8.6 Solution

8.6.1 Parameters

$A_c = 8.95 \times 10^6 \text{ mm}^2, A_p = 37380 \text{ mm}^2, A_s = 36983.7 \text{ mm}^2, I_c = 1.10 \times 10^{13} \text{ mm}^4$
$h = 3050 \text{ mm}, y_c = 1981.2 \text{ mm}, a_p = 359.61 \text{ mm}, a_s = 1966.01 \text{ mm}.$
$E_c = 3.55 \times 10^4 \text{ MPa}, E_p = 1.95 \times 10^5 \text{ MPa}, E_s = 2.0 \times 10^5 \text{ MPa}.$
$\alpha_{EP} = \frac{E_p}{E_c} = 5.49, \alpha_{ES} = \frac{E_s}{E_c} = 5.63$

8.6.2 Geometrical properties

The area of the midspan section converted to concrete:

$$A_0 = A_c + (\alpha_{EP} - 1)A_p + (\alpha_{ES} - 1)A_s = 8.95 \times 10^6 + (5.49 - 1)$$
$$\times 37380 + (5.63 - 1) \times 36983.7 = 9.29 \times 10^6 \text{ mm}^2$$

The distance from the centroid of the converted section to the bottom edge of the girder:

$$d_0 = \frac{S_0}{A_0} = \frac{A_c \times y_c + (\alpha_{EP} - 1)A_p \times \left(\frac{h}{2} - a_p\right) + (\alpha_{ES} - 1)A_s \times \left(\frac{h}{2} - a_s\right)}{9.29 \times 10^6}$$

$$= \frac{8.95 \times 10^6 \times 1981.2 + (5.49 - 1) \times 37380 \times \left(\frac{3050}{2} - 359.61\right) + (5.63 - 1) \times 36983.7 \times \left(\frac{3050}{2} - 1966.01\right)}{9.29 \times 10^6}$$

$$= 1921.62 \text{ mm}$$

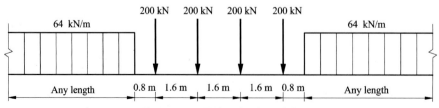

Figure 8.11 ZK live load model.

The distance from the centroid of the concrete section to the centroid of the converted section:

$$d_1 = y_c - d_0 = 1981.2 - 1921.62 = 59.58 \text{ mm}$$

The distance from the centroid of the prestressing tendons to the centroid of the converted section:

$$e_{01p} = d_0 - a_p = 1921.62 - 359.61 = 1562.01 \text{ mm}$$

The distance from the centroid of the reinforcing steels to the centroid of the converted section:

$$e_{01s} = d_0 - a_s = 1921.62 - 1966.01 = -44.39 \text{ mm}$$

The moment of inertia of the converted section:

$$I_0 = I_c + A_c d_1^2 + (\alpha_{EP} - 1)A_p e_{01p}^2 + (\alpha_{ES} - 1)A_s e_{01s}^2$$

$$= 1.10 \times 10^{13} + 8.95 \times 10^6 \times 59.58^2 + (5.49 - 1) \times 37380 \times 1562.01^2$$

$$+ (5.63 - 1) \times 36983.7 \times (-44.39)^2$$

$$= 114.41 \times 10^{11} \text{mm}^4$$

8.6.3 Train loads

To use Eqs. (8.38) and (8.39), the train loads exerted on the girder in the single-line high-speed railway shown in Fig. 8.11 can be converted to the equivalent loads, as shown in Fig. 8.12.

8.6.4 Rotation angle at the girder end

The rotation angles at the girder ends caused by the equivalent loads of ZK are calculated separately as follows:

$$\theta_{A,q} = \frac{qL^3}{24E_c I_0} = \frac{64 \times 31500^3}{24 \times 3.55 \times 10^4 \times 114.41 \times 10^{11}} = 2.05 \times 10^{-4} \text{ rad}$$

$$\theta_{B,q} = -\frac{qL^3}{24E_c I_0} = -\frac{64 \times 31500^3}{24 \times 3.55 \times 10^4 \times 114.41 \times 10^{11}} = -2.05 \times 10^{-4} \text{ rad}$$

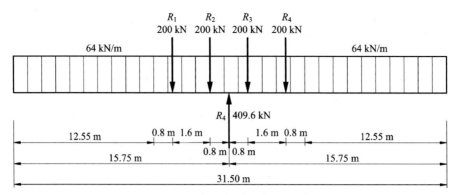

Figure 8.12 Equivalent loads of ZK on the girder.

$$\theta_{A,R1} = \frac{Fab(L+b)}{6EIL} = \frac{2 \times 10^5 \times 13350 \times 18150 \times (31500 + 18150)}{6 \times 3.55 \times 10^4 \times 114.41 \times 10^{11} \times 31500}$$
$$= 0.31 \times 10^{-4} \text{ rad}$$

$$\theta_{B,R1} = -\frac{Fab(L+a)}{6EIL} = -\frac{2 \times 10^5 \times 13350 \times 18150 \times (31500 + 13350)}{6 \times 3.55 \times 10^4 \times 114.41 \times 10^{11} \times 31500}$$
$$= -0.28 \times 10^{-4} \text{ rad}$$

$$\theta_{A,R2} = \frac{Fab(L+b)}{6EIL} = \frac{2 \times 10^5 \times 14950 \times 16550 \times (31500 + 16550)}{6 \times 3.55 \times 10^4 \times 114.41 \times 10^{11} \times 31500}$$
$$= 0.31 \times 10^{-4} \text{ rad}$$

$$\theta_{B,R2} = -\frac{Fab(L+a)}{6EIL} = -\frac{2 \times 10^5 \times 14950 \times 16550 \times (31500 + 14950)}{6 \times 3.55 \times 10^4 \times 114.41 \times 10^{11} \times 31500}$$
$$= -0.30 \times 10^{-4} \text{ rad}$$

$$\theta_{A,R3} = -\theta_{B,R2}$$

$$\theta_{B,R3} = -\theta_{A,R2}$$

$$\theta_{A,R2} = -\theta_{B,R1}$$

$$\theta_{B,R4} = -\theta_{A,R1}$$

$$\theta_{A,R5} = -\frac{Fab(L+b)}{6EIL} = -\frac{409600 \times 15750 \times 15750 \times (31500 + 15750)}{6 \times 3.55 \times 10^4 \times 114.41 \times 10^{11} \times 31500}$$
$$= -0.63 \times 10^{-4} \text{ rad}$$

$$\theta_{B,R5} = \frac{Fab(L+a)}{6EIL} = \frac{409600 \times 15750 \times 15750 \times (31500+15750)}{6 \times 3.55 \times 10^4 \times 114.41 \times 10^{11} \times 31500}$$

$$= 0.63 \times 10^{-4}$$

Hence, the rotation angles at the girder ends during train running in the single-line railway are:

$$\theta_A = \theta_{A,q} + \theta_{A,R1} + \theta_{A,R2} + \theta_{A,R3} + \theta_{A,R4} + \theta_{A,R5}$$

$$= 2.05 \times 10^{-4} + 0.31 \times 10^{-4} + 0.31 \times 10^{-4} + 0.30 \times 10^{-4} + 0.28 \times 10^{-4}$$

$$-0.63 \times 10^{-4} = 2.62 \times 10^{-4} \text{ rad}$$

$$\theta_B = \theta_{B,q} + \theta_{B,R1} + \theta_{B,R2} + \theta_{B,R3} + \theta_{B,R4} + \theta_{B,R5}$$

$$= -2.05 \times 10^{-4} - 0.28 \times 10^{-4} - 0.30 \times 10^{-4} - 0.31 \times 10^{-4} - 0.31 \times 10^{-4}$$

$$+0.63 \times 10^{-4} = -2.62 \times 10^{-4} \text{ rad}$$

The rotation angles at the girder ends during train running in the double-line railway are:

$$\theta_A = -\theta_B = 5.24 \times 10^{-4} \text{ rad}$$

At each end, the distance between the center of bearing to the girder end (overhanging length of beam end) is:

$$L_c = \frac{L_0 - L_e}{2} = \frac{32.6 - 31.5}{2} = 0.32 \text{ m} < 0.55 \text{ m}$$

According to Table 8.5, $[\theta] = 1.5 \times 10^{-3}$ rad and $\theta_1 + \theta_2 < [\theta] = 3.0 \times 10^{-3}$ rad. The rotation angle of the girder end when the girder is adjacent to an abutment is:

$$\theta_A = -\theta_B = 5.24 \times 10^{-4} \text{ rad} < [\theta] = 1.5 \times 10^{-3} \text{ rad}$$

The rotation angle of the girder end of two adjacent girders is:

$$\theta_A + |\theta_B| = 1.04 \times 10^{-3} \text{ rad} < [\theta] = 3.0 \times 10^{-3} \text{ rad}$$

Hence, the rotation angles at the girder ends meet the requirements stipulated in the TB10621-2014 code.

8.7 Example 8.3

A partially prestressed concrete simply supported highway beam (Class B) with an effective span of 28.66 m is made of C50 concrete and bonded strands. The moments of inertia of the converted section and cracked section are $I_0 = 321.5 \times 10^9 \text{ mm}^4$ and

$I_{cr} = 232.4 \times 10^9$ mm^4, respectively, and the cracking moment is $M_{cr} = 1875$ kN·m. Eq. (8.4) is used to calculate the midspan deflection caused by the bending moment $M_s = 2631.45$ kN·m due to the frequent combination of actions and $\alpha_f = 5/48$. The bending moment along the entire effective span due to the effective prestressing force is $M_{pe} = 1738.74$ kN·m. For the incremental long-term coefficient, $\eta_{\theta,s} = 1.43$ is taken for service loads and $\eta_{\theta,p} = 2$ is taken for the effective prestressing force.

Complete the following tasks according to the JTG 3362-2018 code:
(1) Calculate the short-term deflection due to the frequent combination of actions.
(2) Calculate the camber produced by the effective prestressing force.
(3) Set camber for the beam during construction according to the code.

8.8 Solution

8.8.1 Parameters

$E_c = 3.45 \times 10^4$ MPa, $I_0 = 321.5 \times 10^9$ mm^4, $I_{cr} = 232.4 \times 10^9$ mm^4
$M_{pe} = 1738.74$ kN·m, $M_s = 2631.45$ kN·m.
$\eta_{\theta,s} = 1.43$, $\eta_{\theta,p} = 2$

8.8.2 The deflection due to the frequent combination of actions

Since $M_s = 2631.45 > M_{cr} = 1875$ kN·m.

The short-term deflection due to the frequent combination of actions is given by

$$f_s = \alpha \left[\frac{M_{cr}}{B_1} + \frac{M_s - M_{cr}}{B_2} \right] = \alpha \left[\frac{M_{cr}}{0.95 E_c I_0} + \frac{M_s - M_{cr}}{E_c I_{cr}} \right]$$

$$= \frac{5}{48} \times \frac{1875 \times 10^6 \times 28660^2}{0.95 \times 3.45 \times 10^4 \times 321.5 \times 10^9} + \frac{5}{48} \times \frac{(2631.45 - 1875) \times 10^6 \times 28660^2}{3.45 \times 10^4 \times 232.4 \times 10^9}$$

$$= +21.36 \text{ mm } (\downarrow)$$

The long-term deflection due to the frequent combination of actions is obtained as

$$f_{s,l} = \eta_{\theta,s} f_s = 1.43 \times 21.36 = 30.54 \text{ mm } (\downarrow)$$

8.8.3 The camber produced by the effective prestressing force

The short-term camber produced by the effective prestressing force is given by

$$f_p = \int_0^L \frac{M_p \overline{M}_x}{E_c I_0} dx = -\frac{M_{pe} L^2}{8 \times E_c I_0} = -\frac{1738.74 \times 10^6 \times 28660^2}{8 \times 0.95 \times 3.45 \times 10^4 \times 282.847 \times 10^9}$$

$$= -23.18 \text{ mm } (\uparrow)$$

The long-term camber due to the effective prestressing force is obtained as

$$f_{p,l} = \eta_{\theta,e} f_p = 2 \times (-23.18) = -46.36 \text{ mm } (\uparrow)$$

8.8.4 Camber set during construction

The midspan deflection of the beam under the effective prestressing force and the frequent combination of actions considering the long-term effect is given by

$$\Delta f_l = f_{s,l} + f_{p,l} = 30.54 - 46.36 = -15.82 \text{ mm } (\uparrow)$$

Therefore, camber set during construction is not required.

8.9 Example 8.4

A prestressed concrete beam in a building with an effective span of 10 m is made of C40 concrete, bonded strands $\phi^S 15.2$ ($7\phi 5$, $f_{ptk} = 1860$ MPa), and HRB335 reinforcing steel. The beam is designed as Class III in crack control. The depth and width of the rectangular section are 620 and 300 mm, respectively. The service load is treated as a uniformly distributed load, and the bending moment produced by the characteristic combination of actions and quasi-permanent combination of actions are $M_K = 350.2$ kN·m and $M_q = 210.5$ kN·m, respectively. The decompression moment and cracking moment are $M_0 = 276.8$ kN·m and $M_{cr} = 296.7$ kN·m, respectively.

Other data are given as follows: $a_p = 100$ mm, $a_s = 40$ mm, $A_s = 452.3$ mm^2, $A_p = 789.6$ mm^2, $f_{pd} = 1320$ MPa, $f_{sd} = 300$ MPa, $E_s = 2.0 \times 10^5$ MPa, $E_p = 1.95 \times 10^5$ MPa, $E_c = 3.25 \times 10^4$ MPa.

Verify the deflection of the beam according to the GB 50010-2010 code.

8.10 Solution

8.10.1 Sectional properties

$$a = \frac{f_{py} A_p a_p + f_y A_s a_s}{f_{py} A_p + f_y A_s} = \frac{1320 \times 789.6 \times 100 + 300 \times 452.3 \times 40}{1320 \times 789.6 + 300 \times 452.3} = 93 \text{ mm}$$

$$h_0 = h - a = 620 - 93 = 527 \text{ mm}$$

The properties of the converted section are obtained as

$$A_0 = 0.1935699 \text{ mm}^2, \quad S_0 = 0.05812563 \text{ m}^3, \quad y = 0.3002824 \text{ m},$$
$$I_0 = 0.006413607 \text{ m}^4$$

8.10.2 Short-term flexural rigidity

Flexural rigidity of the uncracked section:

$$B_s = 0.85 E_c I_0 = 0.85 \times 3.25 \times 10^4 \times 0.006413607 = 177.176 \text{ MN} \cdot \text{m}^2$$

Flexural rigidity of the cracked section:

$$B_s = \frac{0.85 E_c I_0}{K_{cr} + (1 - K_{cr})\omega}$$

where $K_{cr} = \frac{M_a}{M_K} = \frac{296.7}{350.2} = 0.847$

$$\omega = \left(1.0 + \frac{0.21}{\alpha_E \rho}\right)(1 + 0.45\gamma_f) - 0.7$$

$$\alpha_{EP} = E_p / E_c = 2.0 \times 10^5 / 3.25 \times 10^4 = 6.154$$

$$\rho_{ps} = \frac{A_p + A_s}{bh_0} = \frac{452.3 + 789.6}{300 \times 527} = 0.00786$$

$\gamma_f = 0$ for rectangular section, so

$$\omega = \left(1.0 + \frac{0.21}{\alpha_E \rho_{ps}}\right)(1 + 0.45\gamma_f) - 0.7 = \left(1.0 + \frac{0.21}{6.154 \times 0.00786}\right)$$
$$\times (1 + 0) - 0.7 = 4.641$$

$$B_s = \frac{0.85 E_c I_0}{K_{cr} + (1 - K_{cr})\omega} = \frac{0.85 \times 3.25 \times 10^4 \times 0.006413607}{0.847 + (1 - 0.847) \times 4.641} = 113.789 \text{ MN} \cdot \text{m}^2$$

8.10.3 Long-term flexural rigidity

Eq. (8.37) gives the reduction factor of flexural rigidity:

$$\beta_\theta = \frac{M_k}{M_q(\theta - 1) + M_k} = \frac{350.2}{210.5 \times (2 - 1) + 350.2} = 0.625$$

Hence, the long-term flexural rigidity is obtained as

$$B = \beta_\theta B_s = 0.625 \times 113.789 = 71.11 \text{ MN} \cdot \text{m}^2$$

8.10.4 Verification of deflection

The long-term deflection at the midspan caused by the characteristic combination of actions is

$$f_l = \frac{5}{48} \frac{M_K L^2}{B} = \frac{5}{48} \times \frac{350.2 \times 10^2}{71.069 \times 10^3} = 0.0513 \text{ m}$$

The long-term camber at the midspan caused by the effective prestressing force is

$$f_{p,l} = \eta_\theta f_p = 2 \times \frac{M_0 L^2}{8 E_c I_0} = 2 \times \frac{276.8 \times 10^2}{83.25 \times 10^7 \times 0.006413607} = 0.0332 \text{ m}$$

Hence,

$$f_l - \eta_\theta f_p = 0.0513 - 0.0332 = 0.0181 \text{ m} = 1.81 \text{ cm} < [f] = \frac{L}{300} = \frac{1000}{300}$$

$$= 3.33 \text{ cm}$$

Therefore the long-term deflection of the beam meets the requirement specified in the code.

8.11 Calculation and control of cracking

8.11.1 General concerns on cracking in the prestressed concrete structures

Cracking occurs in concrete structures when the concrete tensile stress exceeds its tensile strength, which may be caused either by the loads or by nonload actions, such as temperature and shrinkage of concrete, or both. In a flexural member, the load-type cracks are typically distinguished into flexure cracks in the tension zone caused by the excessive normal tensile stress of concrete, oblique cracks (web-shear cracks and shear-flexure cracks) in the webs caused by the excessive principal tensile stress of concrete, and spalling or splitting cracks in the anchorage zone caused by the excessive spalling or splitting force. Researches show that the developing mechanism and propagation mode for oblique cracks and spalling cracks are complicated, they are difficult to describe using analytical expressions. The oblique cracks and spalling cracks cannot close automatically once they occur, therefore, they should be prohibited by arranging sufficient reinforcing steels in the corresponding regions. In this section, the calculation and control of flexure cracking in partially prestressed concrete structures are discussed.

The crack width mainly depends on the deformation difference between the concrete and the reinforcement in the tension zone, which changes due to concrete creep and shrinkage and tendon relaxation under the service loads and the prestressing force, so the crack width is a time-dependent variable. The short-term cracks refer to those that occur at the time of cracking, while the long-term cracks reflect the time-dependent effect.

In design, it is always assumed that the flexural member cracks when the live loads are applied (in the case of a combination of dead load and live load prescribed in the code). For a real prestressed concrete member, the interval between the completion of construction and the beginning of applying live loads lasts several weeks or several months, which changes the concrete and reinforcement stresses due to the time-dependent effect that influence the cracking moment, which accordingly affects the short-term crack width when live loads are applied. Once cracking occurs, the crack width alters with time due to concrete creep and shrinkage. Therefore, both short-term and long-term cracks are related to the time-dependent effect. To simplify the analysis, an incremental coefficient that is greater than 1.0 is commonly introduced in the crack width formula to demonstrate the long-term effect, such as the formulas provided in the Q/CR 9300-2018, JTG 3362-2018, and GB 50010-2010 codes.

8.11.2 Cracking behavior and cracking analysis

In a prestressed concrete flexural member, the cracking triggers at the extreme tension fiber once the bending moment exceeds the cracking moment, as shown in Fig. 8.13. The concrete at the tension zone between two adjacent cracks shrinks at the onset of cracking, consequently, the stress in the concrete reduces and the decreased stress is less than the tensile strength. If the external load keeps constant, the distance between two adjacent cracks can be taken to be constant, which is generally defined as the stabilized minimum cracking space, although the creep and shrinkage of concrete possibly change this value. When the external load increases to the point that the maximum tensile stress in the concrete reaches the tensile strength again, a new crack forms between two adjacent cracks.

In the analysis of cracking, it is assumed that the strain in reinforcement and concrete is proportional to the distance from the neutral axis both in the cracked and uncracked sections.

The crack width can be treated as the deformation difference between the reinforcement and the concrete between two adjacent cracks, where there is

$$w = \int_0^{l_{cr}} [\varepsilon_{sx} - \varepsilon_{cx}] dx = \int_0^{l_{cr}} \left[\frac{\sigma_{sx}}{E_s} - \frac{\sigma_{cx}}{E_c} \right] dx \tag{8.41}$$

where ε_{sx} = tensile strain of reinforcement at any section.

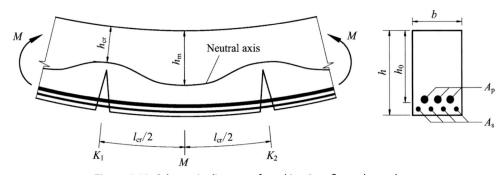

Figure 8.13 Schematic diagram of cracking in a flexural member.

ε_{cx} = strain of concrete in the effective area surrounding the tensile reinforcement.

σ_{sx} = tensile stress of reinforcement at any section.

σ_{cx} = stress of concrete in the effective area surrounding the tensile reinforcement.

E_c = modulus of elasticity of concrete.

E_s = modulus of elasticity of reinforcing steel.

In Fig. 8.13, the tensile stress in reinforcement reaches its maximum at a crack, while the minimum is in the middle of two adjacent cracks, meaning that the stress in the reinforcement can be approximately expressed as

$$\sigma_{sx} = \sigma_{s,cr} - B_0\left(1 - \cos\frac{2\pi x}{l_{cr}}\right) \tag{8.42}$$

where $\sigma_{s,cr}$ = tensile stress of reinforcement in the crack.

l_{cr} = crack spacing, equal to twice the length over which a slip between the concrete and reinforcement occurs.

B_0 = coefficient related to the bond stress between the concrete and reinforcement, $B_0 = \frac{\zeta f_{ctk}}{\rho_{s,ef}}$, $\zeta \le 0.5$.

f_{ctk} = characteristic tensile strength of concrete.

$\rho_{s,ef}$ = effective ratio of reinforcement $(A_{ps}/A_{c,ef})$.

$A_{c,ef}$ = effective area of concrete in tension.

A_{ps} = cross-sectional area of the prestressing and reinforcing steels at the tension zone, $A_{ps} = A_p + A_s$.

Considering that the tensile stress in concrete is relatively small compared to the tensile stress in reinforcing steels, the concrete stress σ_{cx} in Eq. (8.41) can be disregarded. Substituting Eq. (8.42) into Eq. (8.41) yields

$$w = \frac{\sigma_{s,cr}}{E_s}\left[1 - \frac{\zeta f_{ctk}}{\rho_{s,ef}\sigma_{s,cr}}\right]l_{cr} \tag{8.43}$$

Eq. (8.43) shows that the crack width mainly depends on the tensile stress in reinforcement at a crack, crack spacing, ratio of reinforcement, and the distance from the center of all tensile reinforcements to the tensile edge which relates to the concrete cover.

When Eq. (8.43) is used to calculate the crack width in a prestressed concrete flexural member at the applied loads, the nominal tensile stress $\sigma_{s,cr}$ of the reinforcing steels should be the stress change beyond the decompression state, because the strain (stress) in reinforcement in Eq. (8.41) should be the increment compared to the original section where the concrete strain is zero. Also, $\sigma_{s,cr}$ should be replaced by the mean stress between two adjacent cracks, so an adjusting coefficient needs to be introduced to demonstrate the effect of nonuniformity in the distribution of stress when $\sigma_{s,cr}$ is still adopted.

When the concrete cover does not exceed 65 mm, statistical analysis of experimental data on crack spacing in various types of reinforced and prestressed concrete members shows that the mean crack spacing can be expressed as

$$l_{m,cr} = \alpha_1 \left(\alpha_2 c_c + \alpha_3 \frac{d}{\rho_{s,ef}} \right) \tag{8.44}$$

where $l_{m,cr}$ = mean crack spacing.

$\alpha_1 - \alpha_3$ = parameters related to the types of members, the surface shape of reinforcement, and the characteristic of loads.

c_c = concrete cover.

d = diameter of reinforcement, the equivalent diameter of all reinforcements d_{eq} is adopted when various reinforcements with different diameters are arranged.

Generally, the equivalent diameter of all reinforcements can be derived from

$$d_{eq} = \frac{\sum n_i d_i^2}{\sum n_i \nu_i d_i} \tag{8.45}$$

where d_{eq} = equivalent diameter of all reinforcements.

n_i = the number of bars (or wires) with a diameter d_i.

ν_i = coefficient of relative bond property of reinforcement.

When the effect of bond property of reinforcement is neglected, Eq. (8.45) turns into

$$d_{eq} = \frac{\sum n_i d_i^2}{\sum n_i d_i} \tag{8.46}$$

or can be rewritten as

$$d_{eq} = \frac{4(A_s + A_p)}{u} \tag{8.47}$$

The equivalent stress of tensile reinforcements at a crack in Eq. (8.43) can be derived from Eq. (7.80), or calculated approximately by

$$\sigma_{s,cr} = \frac{M \pm M_{p2} - N_{p0}(z - e_{p,s})}{(\alpha_1 A_p + A_s)z} \tag{8.48}$$

$$e = e_{p,s} + \frac{M \pm M_{p2}}{N_{p0}} \tag{8.49}$$

$$e_{p,s} = y_{ps} - e_{p0} \tag{8.50}$$

$$z = \left[0.87 - 0.12(1 - \gamma_f') \left(\frac{0}{e} \right)^2 \right] h_0 \leq 0.87 h_0 \tag{8.51}$$

where M = bending moment due to the short-term combination of actions.

M_{p2} = secondary bending moment due to stretching the prestressing tendons in the post-tensioned statically indeterminate structures.

N_{p0} = generalized prestressing force when normal stress of concrete equals zero.

α_1 = reduction factor of unbonded prestressing tendons, 0.3 is taken generally, 1.0 is taken for bonded prestressing tendons.

z = distance between the action points of the resultant of the longitudinal prestressing and reinforcing steels at the tension zone and the resultant of compression.

$e_{p,s}$ = distance between N_{p0} and the resultant of the longitudinal prestressing and reinforcing steels in the tension zone.

y_{ps} = eccentricity of N_{p0}.

e_{p0} = eccentricity of prestressing and reinforcing steels in the tension zone.

γ_f' = ratio of the section area of the compression (flange) to the effective section area of the web, given by

$$\gamma_f' = \frac{\left(b_f' - b\right)h_f'}{bh_0} \tag{8.52}$$

where b_f' = width of compression (flange).

8.11.3 Calculation of crack width in the codes

The mechanism of the formation and process of propagation and the time-dependent effect on the cracking in the prestressed concrete structures are very complicated, so the calculation of crack width in the service period is far more complex than the above cracking analysis. The present design codes provide approximate calculation formulas for crack width.

8.11.3.1 Calculation of crack width for railway beams

For cracking-allowed prestressed concrete railway beams with rectangular, T-, I-, and box sections, the "characteristic crack width" (the guarantee rate corresponding to this crack width below the characteristic value is 95%) may be obtained by

$$w_{fk} = \alpha_2 \alpha_3 \left(2.4c_c + v\frac{d_{eq}}{\rho_{s,ef}} \right) \frac{\Delta \sigma_{s2}}{E_s} \tag{8.53}$$

$$\rho_{s,ef} = \frac{A_s + A_p}{A_{c,ef}} \tag{8.54}$$

where w_{fk} = characteristic crack width.

c_c = concrete cover, the shortest distance from the surface of prestressing tendons and reinforcing steels to the tension zone edge.

d_{eq} = nominal diameter of the reinforcement, given by Eq. (8.46).

$\rho_{s,ef}$ = effective ratio of longitudinal reinforcement.

ν = bond coefficient of reinforcement; 0.02 is taken for ribbed reinforcement and 0.04 is taken for steel wire or strand; for post-tensioning tendons inside the grouted pipe, 0.04 is taken for deformed bars and 0.06 is taken for steel wires and strands; when two kinds of steel bars are used together, the weighted average value can be used.

α_2 = increasing coefficient of the characteristic crack width compared with the mean crack width, generally, 1.8 can be taken.

α_3 = increasing coefficient due to fatigue at the service loads, generally 1.5 can be taken.

$\Delta\sigma_s$ = incremental stress of reinforcing steels at the tension zone calculated based on the cracked section after decompression; in the post-tensioned statically indeterminate beams the secondary bending moment due to stretching tendons should be taken into account, given by Eq. (8.48).

A_p = area of prestressing tendons.

A_s = area of reinforcing steels.

u = circumference of all reinforcements.

$A_{c,ef}$ = effective area of concrete in tension, which can be obtained from Fig. 8.14, mm^2.

8.11.3.2 Calculation of crack width for highway members

For cracking-allowed prestressed concrete highway beams with rectangular, T-, I-, and box sections, the formula of maximum crack width in the JTG 3362-2018 code obtained by the statistical method from a large number of experimental data is expressed as

$$w_{max} = C_1 C_2 C_3 \left(\frac{c_c + d}{0.30 + 1.4\rho_{s,ef}} \right) \frac{\sigma_{ss}}{E_s} \tag{8.55}$$

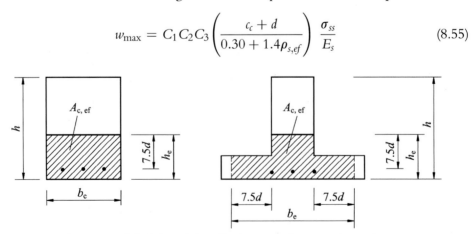

Figure 8.14 Calculation of the effective area of concrete in tension.

where $w_{\max}$ = maximum crack width.

C_1 = coefficient related to the surface shape of reinforcement, 1.4 is taken for smooth reinforcement, 1.0 is taken for ribbed reinforcement, and 1.15 is taken for epoxy resin-coated ribbed reinforcement.

C_2 = long-term coefficient related to action, $C_2 = 1 + 0.5\frac{M_l}{M_s}$, where M_l and M_s are the design bending moments calculated by the quasi-permanent combination of actions and frequent combination of actions, respectively.

C_3 = coefficient related to the type of member, 1.0 is taken for flexural members and 1.2 is taken for axial tension members.

σ_{ss} = equivalent stress of longitudinal tensile reinforcement in the cracked section caused by the frequent combination of actions, which can be calculated by Eqs. (8.48)−(8.51).

c_c = concrete cover, 50 mm is taken if $c_c > 50$ mm.

$\rho_{s,ef}$ = effective ratio of longitudinal reinforcement, 0.1 is taken if $\rho_{te} > 0.1$ and 0.01 is taken if $\rho_{te} \leq 0.01$.

$A_{c,ef}$ = effective area of concrete in tension.

When various reinforcements with different diameters are arranged in the tension zone, the equivalent diameter can be obtained by Eq. (8.45). d_i is the diameter of reinforcement, and its values are taken as follows: $d_i = d$ for a single bar (or wire), $d_{se} = \sqrt{n}d$ for a bundle of bars with the same diameter d (n is the number), and $d_{pe} = \sqrt{n}d_p$ for a bundle of wires or strands with nominal diameter d_p of a single wire or strand.

In the JTG 3362-2018 code, the effective area of concrete in tension in Eq. (8.55) is defined as follows: the total cross-sectional area is taken for axial tension member, $2a_s b$ is taken for flexural, eccentric tension, and compression members, where a_s is the distance from the centroid of all reinforcements to the tension zone edge, b is the width of rectangular section or the effective width of flange of reverse T-, I-, and box sections.

8.11.3.3 Calculation of crack width for members in the buildings

For cracking-allowed prestressed concrete axial tension, flexural and eccentric compression members used in the industrial and civil buildings, the GB 50010-2010 code prescribes that the maximum crack width should be calculated by

$$w_{\max} = \alpha_{cr}\psi\frac{\sigma_{ss}}{E_s}\left(1.9c_c + 0.08\frac{d_{eq}}{\rho_{s,ef}}\right) \tag{8.56}$$

where α_{cr} = coefficient related to the type of member, 1.5 is taken for flexural and eccentric compression members, 2.2 is taken for axial tension members.

ψ = coefficient of nonuniformity distribution of the tensile stress in the reinforcement, 0.2 is taken if $\psi < 0.2$, 1.0 is taken if $\psi > 1.0$, and 1.0 is taken for the members directly subjected to repeated loading.

σ_{ss} = equivalent stress of longitudinal tensile reinforcement in the cracked section caused by the characteristic combination of actions, can be calculated by Eqs. (8.48)–(8.51).

d_{eq} = equivalent diameter of reinforcement, obtained from Eq. (8.45), where the coefficient values of the relative bond property of reinforcement ν_i are listed in Table 8.6.

The coefficient of nonuniformity distribution of the tensile stress in the reinforcement between two adjacent cracks in a flexural member can be derived from

$$\psi = 1.1 - 0.65 \frac{f_{ctk}}{\rho_{s,ef}\sigma_{ss}} \tag{8.57}$$

where f_{ctk} = characteristic tensile strength of concrete.

8.11.4 Crack control in the prestressed concrete members

In general, crack control of a prestressed concrete structure can be divided into two categories, ensuring the crack resistance for those in which cracking is not allowed and limiting the crack width for those in which cracking is allowed. The crack resistance can be achieved by limiting the normal stress and principal stresses of concrete for normal sections and oblique sections, respectively, as discussed in Section 7.4. For members in which cracking is allowed three approaches can be adopted to control cracking: by limiting the crack width directly, by limiting the nominal tensile stress of concrete or the tensile stress of reinforcement at the tension zone, and by limiting the spacing of reinforcement. The Q/CR 9300-2018, JTG 3362-2018, GB 50010-2010, and *fip* Model Code 2010 codes prescribe that crack control can be accomplished by limiting the crack width at service loads. In the ACI 318 code, crack control is achieved by limiting the spacing of reinforcement and the stress change of bonded prestressing tendons beyond the decompression stress.

8.11.4.1 General expression for crack control by limiting crack width

The crack width in a prestressed concrete member should satisfy

$$w_{\max} \leq [w] \tag{8.58}$$

Table 8.6 Coefficient values of the relative bond property of reinforcement.

Type of steel	Reinforcing steels		Pretensioning steels			Post-tensioning steels		
	Smooth bar	Ribbed bar	Ribbed bar	Ribbed wire	Indented wire and strand	Ribbed bar	Strand	Smooth wire
ν_i	0.7	1.0	1.0	0.8	0.6	0.8	0.5	0.4

Note: the relative bond characteristic coefficients in the table should be discounted by 0.8 for epoxy resin–coated ribbed bars.

or

$$w_{fk} \leq [w] \tag{8.59}$$

where w_{max} = maximum crack width.

w_{fk} = characteristic crack width.

$[w]$ = allowable crack width stipulated in the codes.

8.11.4.2 Allowable crack width in the railway members

For the prestressed concrete flexural members in which cracking is allowed, the Q/CR 9300-2018 code prescribes that the tensile stress in the extreme fiber of concrete in the precompressed tension zone under permanent actions should be greater than 0.1 MPa, and the characteristic crack width under design loads should be in accordance with the following stipulations:

(1) Under the combination of characteristic values of permanent actions and frequent values of basic variable actions, the characteristic crack width should not exceed $[w] = 0.10$ mm.

(2) Under the combination of characteristic values of permanent actions, frequent values of basic variable actions, and other quasi–permanent values of variable actions, the characteristic crack width should not exceed $[w] = 0.15$ mm.

8.11.4.3 Allowable crack width in highway members

For the prestressed concrete members allowing cracking (Class B), the maximum crack width calculated by the formulas provided by the JTG 3362-2018 code should not exceed the values listed in Table 8.7.

The JTG 3362-2018 code also prescribes that cracking is not allowed in II-D and II-E of the destruction environment by freezing and thawing, III-D, III-E, and III-F of marine chloride environment, IV-D and IV-E of deicing salt and other chloride environments, V-F of salt crystallization environment, VI-D and VI-E of chemical corrosion environment, and VII-D of abrasive environment.

8.11.4.4 Allowable crack width in building members

Under the characteristic combination of their actions, considering the long-term effect, the maximum crack width in building members calculated by Eq. (8.56) should not exceed the allowable values specified in the GB 50010-2010 code, as shown in Table 8.8.

8.12 Example 8.5

A post-tensioned simply supported beam in a mixed passenger and freight railway with a design speed of 160 km/h is made of C50 concrete, four groups of bonded strands $9\phi^S 15.2$ (9-7ϕ5, $f_{ptk} = 1860$ MPa) and 7Φ 28 (HRB400) reinforcing steels. The

Table 8.7 Allowable maximum crack width.

Type of environment	Grade of environment	[w]/mm
Carbonation environment	I-A	0.20
	I-B	0.15
	I-C	0.10
Destruction environment by freezing and thawing	II-C	0.10
Marine chloride environment	III-C	
Deicing salt and other chloride environments	IV-C	
Salt crystallization environment	V-E	
Chemical corrosion environment	VI-C	
Abrasive environment	VII-C	

Table 8.8 Allowable maximum crack width.

Description	[w]/mm
Indoor dry environment	0.20
Nonaggressive hydrostatic environment	
Indoor wet environment	0.10
Open-air environment in nonsevere cold areas	

effective span of the beam is 32 m. The dimension and arrangement of reinforcements at the midspan section are shown in Fig. 8.15. The incremental stress of reinforcing steel calculated based on cracking section after decompression is 30.2 MPa at service loads, and the allowable crack width prescribed in the Q/CR 9300-2018 code is 0.10 mm. Calculate and verify the crack width at the midspan section.

8.13 Solution

8.13.1 Parameters

$E_s = 2.0 \times 10^5$ MPa, $E_p = 1.95 \times 10^5$ MPa, $E_c = 3.55 \times 10^4$ MPa

$f_{pd} = 1200$ MPa, $f_{sd} = 320$ MPa

$A_p = 4 \times 1.26 \times 10^3 = 5.04 \times 10^3$ mm^2, $A_s = 4.310 \times 10^3$ mm^2

$a_p = 120$ mm, $a_s = 60$ mm

$$a_{ps} = \frac{A_s f_{sd} a_s + A_p f_{pd} a_p}{A_s f_{sd} + A_p f_{pd}} = \frac{4.310 \times 10^3 \times 320 \times 60 + 5.04 \times 10^3 \times 1200 \times 120}{4.310 \times 10^3 \times 320 + 5.04 \times 10^3 \times 1200} = 108.9 \text{ mm}.$$

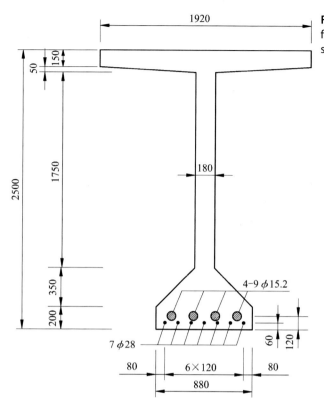

Figure 8.15 Dimension and reinforcement arrangement at the midspan section.

8.13.2 Calculation and verification of crack width

$$u_p = 4 \times \sqrt{9} \times \pi \times 15.2 = 573 \text{ mm}$$
$$u_s = 7 \times \pi \times 28 = 616 \text{ mm}$$
$$u = u_p + u_s = 573 + 616 = 1189 \text{ mm}$$
$$\alpha_2 = 1.8, \alpha_3 = 1.5$$
$$c_c = 60 - 14 = 46 \text{ mm} > 40 \text{ mm, so } c_c = 40 \text{ mm is adopted}$$
$$\nu_s = 0.04, \nu_p = 0.06$$

$$\nu = \frac{\nu_s \times u_s + \nu_p \times u_p}{u} = \frac{0.04 \times 616 + 0.06 \times 573}{1189} = 0.050$$

$$d_{eq} = \frac{4(A_s + A_p)}{u} = \frac{4(5.04 \times 10^3 + 4.310 \times 10^3)}{1189} = 31.5 \text{ mm}$$

The height of the effective area of concrete in tension can be obtained from Fig. 8.14, as

$$h_e = a_{ps} + 7.5 d_{eq} = 108.9 + 7.5 \times 31.5 = 345.2 \text{ mm}$$

In the trapezoid at the bottom of the midspan section, $h_1 = 200$ mm, $h_2 = 350$ mm, $b_e = 800$, and $b_t = 180$ mm.

Considering $h_e = 345.2$ mm$< h_1 + h_2 = 200 + 350 = 550$ mm, the width of the effective area of concrete in tension is obtained as

$$b_{e,top} = b_t + \frac{h_1 + h_2 - h_e}{h_2}(b_e - b_t) = 180 + \frac{200 + 350 - 345.2}{350} \times (880 - 180)$$

$$= 589.6 \text{ mm}$$

The effective area of concrete in tension can be calculated from Fig. 8.15, as

$$A_{c,ef} = b_e \times h_1 + \frac{1}{2}(h_e - h_1) \times (b_e + b_{e,top}) = 880 \times 200 + \frac{1}{2}(345.2 - 200)$$

$$\times (880 + 589.6) = 282693.0 \text{ mm}^2$$

$$\rho_{s,ef} = \frac{A_s + A_p}{A_{c,ef}} = \frac{(4.310 + 5.040) \times 10^3}{282693.0} = 0.032$$

$$w_{fk} = \alpha_2\alpha_3\left(2.4c_c + \nu\frac{d_{eq}}{\rho_{s,ef}}\right)\frac{\Delta\sigma_{s2}}{E_s} = 1.8 \times 1.5 \times \left(2.4 \times 40 + 0.05 \times \frac{31.5}{0.032}\right)$$

$$\times \frac{30.2}{1.95 \times 10^5} = 0.06 \text{ mm} < [w] = 0.10 \text{ mm}$$

Thus, the characteristic crack width at the midspan section of the railway beam at service loads meets the requirement prescribed in the code.

8.14 Example 8.6

A partially prestressed concrete simply supported highway beam (Class B) with an effective span of 24 m is made of C40 concrete, bonded strands 3 $\phi^s 12.7$ (3-7ϕ4, $f_{ptk} = 1860$ MPa), and 5Φ 16 (HRB400) reinforcing steels. The dimension and arrangement of reinforcements at the midspan section are shown in Fig. 8.16. The bending moments produced by the quasi-permanent combination of actions and the frequent combination of actions are $M_l = 4213.50$ kN·m and $M_s = 4712.40$ kN·m, respectively. In the state of decompression in the section, the generalized prestressing force is $N_{po} = 3112.20$ kN. The allowable crack width prescribed in the code is 0.15 mm. Verify the crack width at the midspan section according to the JTG 3362-2018 code.

8.15 Solution

8.15.1 Parameters

$E_s = 2.0 \times 10^5$ MPa, $E_p = 1.95 \times 10^5$ MPa, $E_c = 3.55 \times 10^4$ MPa.

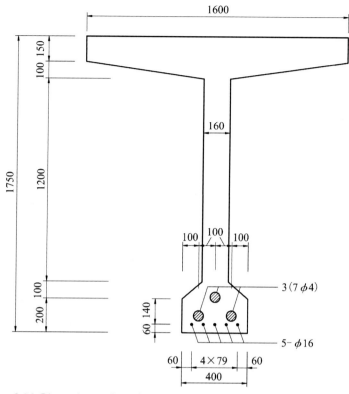

Figure 8.16 Dimension and reinforcement arrangement at the midspan section.

$$f_{pd} = 1260 \text{ MPa}, f_{sd} = 330 \text{ MPa}$$

$$A_p = 3 \times 690.9 = 2072.7 \text{ mm}^2, A_s = 1005 \text{ mm}^2$$

$$a_p = \frac{1 \times 210 + 2 \times 100}{1 + 2} = 136.7 \text{ mm}, a_s = 60 \text{ mm}$$

$$a_{ps} = \frac{A_s f_{sd} a_s + A_p f_{pd} a_p}{A_s f_{sd} + A_p f_{pd}} = \frac{1005 \times 330 \times 60 + 2072.7 \times 1260 \times 136.7}{1005 \times 330 + 2072.7 \times 1260} = 144 \text{ mm}$$

$$h_0 = h - a_{ps} = 1750 - 144 = 1606 \text{ mm}$$

8.15.2 Calculation of equivalent stress in tensile reinforcement

This is a simply supported highway beam and there is no reinforcement at compression zone, so $M_{p2} = 0$, $e_{p,s} = 0$

$$e = \frac{M_s}{N_{po}} = \frac{4712.4}{3112.2} = 1.514 \text{ m}$$

$$\gamma_f' = \frac{\left(b_f' - b\right)h_f'}{bh_0} = \frac{(1600 - 160) \times (150 + 100/2)}{160 \times 1606} = 1.12$$

$$Z = \left[0.87 - 0.12\left(1 - \gamma'_f\right)\left(\frac{h_0}{e}\right)^2\right]h_0 = \left[0.87 - 0.12 \times (1 - 1.12) \times \left(\frac{1606}{1514}\right)^2\right]$$
$$\times 1606 = 1371.2 \text{ mm} < 0.87h_0 = 0.87 \times 1606 = 1397.2 \text{ mm}$$

$$\sigma_{ss} = \frac{M_s \pm M_{p2} - N_{po}\left(Z - e_{p,s}\right)}{\left(A_p + A_s\right)Z} = \frac{4712.4 \times 10^6 - 3112.2 \times 10^3 \times 1371.2}{(2072.7 + 1005) \times 1371.2}$$

$$= 105.6 \text{ MPa}$$

8.15.3 Calculation and verification of crack width

$$C_1 = 1.0 \quad C_2 = 1 + 0.5\frac{M_l}{M_s} = 1 + 0.5 \times \frac{4213.50}{4712.40} = 1.447, \quad C_3 = 1.0$$

$$d_{pe} = \sqrt{n}d = \sqrt{7} \times 12.7 = 33.6 \text{ mm}$$

$$d = \frac{3 \times 33.6^2 + 5 \times 16^2}{3 \times 33.6 + 5 \times 16} = 25.81 \text{ mm}$$

$$c_c = 60 - 8 = 54 \text{ mm} > 50 \text{ mm, so } c = 50 \text{ mm}$$

$$\rho_{s,ef} = \frac{A_s + A_p}{A_{c,ef}} = \frac{A_s + A_p}{2ab} = \frac{2072.9 + 1005}{2 \times 129 \times 400} = 0.03$$

$$w_{max} = C_1 C_2 C_3 \left(\frac{c_c + d}{0.30 + 1.4\rho_{s,ef}}\right)\frac{\sigma_{ss}}{E_s}$$

$$= 1.0 \times 1.447 \times 1.0 \times \frac{105.6}{2.00 \times 10^5} \times \left(\frac{50 + 25.81}{0.3 + 1.4 \times 0.03}\right)$$

$$= 0.169 \text{ mm} > [w] = 0.15 \text{ mm}$$

Thus, the maximum crack width at the midspan section of the highway beam in the service period does not meet the requirement prescribed in the code.

8.16 Example 8.7

A partially post-tensioned simply supported beam in a building with an effective span of 10 m is made of C40 concrete, bonded strands $\phi^s 12.7$ $(2 - 7\phi4, f_{ptk} = 1860 \text{ MPa})$, and $4\Phi 12$ (HRB335) reinforcing steels. The level of crack control of the beam is Class III. The dimension and arrangement of reinforcements at the midspan section are shown in Fig. 8.17. The bending moments produced by the characteristic combination of actions is $M_k = 350.20 \text{ kN·m}$. In the state of decompression in the section, the generalized prestressing force is $N_{po} = 720.50 \text{ kN}$. The allowable crack width prescribed in the code is 0.20 mm. Verify the crack width at the midspan section according to the GB 50010-2010 code.

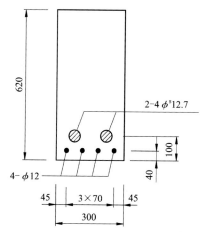

Figure 8.17 Dimension and reinforcement arrangement at the midspan section.

8.17 Solution

8.17.1 Parameters

$f_{pd} = 1320$ MPa, $f_{sd} = 300$ MPa, $f_{ctk} = 2.39$ MPa

$A_p = 98.7 \times 4 \times 2 = 789.6$ mm^2, $A_s = 452.3$ mm^2

$a_p = 100$ mm, $a_s = 40$ mm

$$a = \frac{A_s f_{sd} a_s + A_p f_{pd} a_p}{A_s f_{sd} + A_p f_{pd}} = \frac{1320 \times 789.6 \times 100 + 300 \times 452.3 \times 40}{1320 \times 789.6 + 300 \times 452.3} = 93.1 \text{ mm}$$

$h_0 = 620 - 93.1 = 526.9$ mm

8.17.2 Calculation of equivalent stress in tensile reinforcement

This is a simply supported highway beam and there is no reinforcement at the compression zone, so $M_{p2} = 0$, $e_{p,s} = 0$.

For rectangular section, $b'_f = b$, so $\gamma'_f = 0$.

$$e = e_p + \frac{M_K \pm M_2}{N_{po}} = \frac{M_K}{N_{po}} = \frac{350.2}{720.5} = 0.486 \text{ m} = 486 \text{ mm}$$

$$Z = \left[0.87 - 0.12\left(1 - \gamma'_f\right)\left(\frac{h_0}{e}\right)^2\right]h_0 = \left[0.87 - 0.12\left(\frac{526.9}{486}\right)^2\right] \times 526.9$$

$$= 384.1 \text{ mm} < 0.87 h_0 = 0.87 \times 526.9 = 458.4 \text{ mm}$$

$$\sigma_{ss} = \frac{M_K - N_{po}Z}{\left(A_p + A_s\right)Z} = \frac{350.2 - 720.5 \times 384.1 \times 10^{-3}}{(789.6 + 452.3) \times 10^{-6} \times 384.1 \times 10^{-3}}$$

$$= 1.540 \times 10^{-4} \times 10^9 \text{kN/mm}^2 = 154.0 \text{ MPa}$$

8.17.3 Calculation and verification of crack width

$\alpha_{cr} = 2.1$

$c_c = 34$ mm.

$\nu_p = 0.5$ (post-tensioned strand)

$\nu_s = 1.0$ (ribbed reinforcing steel)

$$d_{eq} = \frac{\sum n_i d_i^2}{\sum n_i \nu_i d_i} = \frac{2 \times 4 \times 12.7^2 + 4 \times 12^2}{2 \times 4 \times 0.5 \times 12.7 + 4 \times 1.0 \times 12} = 18.9 \text{ mm}$$

$$\rho_{s,ef} = \frac{A_s + A_p}{A_{c,ef}} = \frac{A_s + A_p}{0.5bh} = \frac{789.6 + 452.3}{0.5 \times 300 \times 620} = 0.013$$

$$\psi = 1.1 - 0.65 \frac{f_{tk}}{\rho_{s,ef}\sigma_{ss}} = 1.1 - 0.65 \times \frac{2.39}{0.013 \times 154} = 0.324$$

$$w_{\max} = \alpha_{cr}\psi \frac{\sigma_{ss}}{E_s}\left(1.9c_c + 0.08\frac{d_{eq}}{\rho_{s,ef}}\right) = 2.1 \times 0.324 \times \frac{1.540}{1.95 \times 10^5}$$

$$\times \left(1.9 \times 34 + 0.08 \times \frac{18.9}{0.0134}\right) = 0.097 \text{ mm} < [w] = 0.2 \text{ mm}$$

Thus, the maximum crack width at the midspan section of the beam in the service period meets the requirement prescribed in the code.

Suggested readings

[1]   Hu Di. The steel restraint influence coefficient method to analyze time-dependent effect in prestressed concrete bridges. Eng Mech 2006;23(6):120−6.
[2]   Branson DE. Deformation of concrete structures. New York: McGraw Hill; 1977.
[3]   Gilbert RI. Deflection calculation for reinforced concrete structures-Why we sometimes get wrong. ACI Struct J 1990;96(6):1027−32.
[4]   Ghali A, Azarnejad A. Deflection prediction of members of any concrete strength. ACI Struct J 1999; 96(5):807−16.
[5]   Hu Di. Basic principles of prestressed concrete structure design. 2nd ed. Beijing: China Railway Publishing House; 2019.
[6]   Hu Di. Theory of creep effect on concrete structures. Beijing: Science Press; 2015.
[7]   Lu Q, Hu Di. Stress of reinforcement in reinforced concrete flexural members with controlled crack width. Industrial Construction 2008;38(4):46−9.
[8]   GB 50010: 2015. Code for design of concrete structures. Beijing: China Construction Industry Press; 2015.
[9]   Q/CR 9300: 2018. Code for design on railway bridge and culvert. Beijing: China Railway Publishing House; 2018.
[10] JTG 3362: 2018. Specifications for design of highway reinforced concrete and prestressed concrete bridges and culverts. Beijing: People's Communications Press; 2018.
[11] Visintin P, Sturm AB, Oehlers DJ. Long- and short-term serviceability behavior of reinforced concrete: mechanics models for deflections and crack widths. Struct Concr 2018;19:489−507.

CHAPTER 9

Ultimate bearing capacity of flexural members

Contents

Analysis and Design of Prestressed Concrete
ISBN 978-0-12-824425-8, https://doi.org/10.1016/B978-0-12-824425-8.00009-9

9.1 General concepts of ultimate bearing capacity of flexural members

The calculation and verification of the bearing capacity of the prestressed concrete members are needed to guarantee structural safety under the construction loads and service loads. The ultimate bearing capacity of a section is also called the ultimate strength, or the strength for short. For the most broadly used flexural members, bending moment and shear force are inevitably generated in the sections, so the flexural strength in the normal sections, the shear strength and flexural strength in the oblique sections should be verified, expressed as

$$M_d \leq M_u \tag{9.1}$$

$$V_d \leq V_u \tag{9.2}$$

where M_d = design value of bending moment in the section under consideration.
V_d = design value of shear force in the oblique section under consideration.
M_u = flexural bearing capacity of the section, also called flexural strength.
V_u = shear bearing capacity of the oblique section, also called shear strength.

9.2 Flexural bearing capacity of the normal section

9.2.1 Failure patterns and assumptions in the calculation of flexural bearing capacity

The critical neutral axis depth, x_b, in a section of the prestressed concrete flexural member is defined as the neutral axis depth of the balanced section at flexural failure, corresponding to the state that the concrete in the compression zone is crushed once all reinforcements (including reinforcing steels and prestressing tendons) in the tension zone reach their yield strengths. Considering the yield strength of reinforcement corresponds to a certain amount of tensile strain, there will be a clear warning of impending failure of the member in the form of large deflections and a large number of visible cracks, so the failure pattern in the balanced section is a type of plastic failure.

The actual neutral axis depth, x_a, in a section can be obtained directly from the equilibrium equation at failure. If $x_a < x_b$, the maximum stress of concrete in the compression zone is less than the compressive strength at the moment all tensile reinforcements reach their yield strengths, the section is under-reinforced. On the other hand, if $x_a > x_b$, the concrete in the compression zone will rupture while the tensile reinforcements will not reach their yield strengths when the section is over-reinforced. For over-reinforced sections, large quantities of reinforcement are employed and the member fails abruptly with negligible deflection and very few cracks, which is a typical brittle failure pattern. Hence, over-reinforcement must be avoided in design.

In the calculation of flexural bearing capacity of the normal section in the prestressed concrete members, the following assumptions are adopted:

(1) The cross-section remains plane immediately before failure.

(2) The concrete in the tension zone is disregarded for sectional bearing capacity.

(3) For bonded prestressed concrete members, the reinforcements and the surrounded concrete have the same deformation.

(4) The stress in longitudinal reinforcement equals the product of strain and modulus of elasticity, and it should satisfy the following requirements:

$$-f'_{sd} \leq \sigma_{si} \leq f_{sd} \tag{9.3}$$

$$-\left(f'_{pd} - \sigma'_{p0i}\right) \leq \sigma_{pi} \leq f_{pd} \tag{9.4}$$

where σ_{si} = stress in longitudinal reinforcing steels in the ith layer, "+" for tension and "−" for compression.

σ_{pi} = stress in longitudinal prestressing tendons in the ith layer.

σ'_{p0i} = stress in longitudinal prestressing tendons in the ith layer in the compression zone when the concrete stress at this level is zero.

f_{sd} = design tensile strength of reinforcing steel.

f'_{sd} = design strength of reinforcing steel in compression.

f_{pd} = design tensile strength of prestressing tendon.

f'_{pd} = design strength of prestressing tendon in compression.

The *Codes for Design of Concrete Structures* (GB 50010-2010) specifies the relations between the stress and strain of concrete for short-term uniaxial compression:

When $\varepsilon_c \leq \varepsilon_0$

$$\sigma_c = f_{cd}\left[1 - \left(1 - \frac{\varepsilon_c}{\varepsilon_0}\right)^n\right] \tag{9.5}$$

When $\varepsilon_0 < \varepsilon_c \leq \varepsilon_{cu}$

$$\sigma_c = f_{cd} \tag{9.6}$$

$$n = 2 - \frac{1}{60}\left(f_{cu,k} - 50\right) \tag{9.7}$$

$$\varepsilon_0 = 0.002 + 0.5\left(f_{cu,k} - 50\right) \times 10^{-5} \tag{9.8}$$

$$\varepsilon_{cu} = 0.0033 - \left(f_{cu,k} - 50\right) \times 10^{-5} \tag{9.9}$$

where σ_c = compressive stress in concrete corresponding to compressive strain ε_c.

f_{cd} = design strength of concrete in compression.

ε_0 = concrete strain corresponding to f_{cd}, 0.002 is taken if ε_0 is less than 0.002.

ε_{cu} = ultimate strain of concrete in compression, 0.033 is taken if ε_{cu} is less than 0.033 for uneven compression, ε_0 is taken for axial compression.

$f_{cu,k}$ = characteristic value of cubic compressive strength of concrete.

n = exponent, 2.0 is taken if n is great than 2.

The ultimate compressive strain of concrete varies from 0.002 to 0.008, generally, 0.002–0.035 is taken in the codes. The *Specifications for design of highway reinforced concrete and prestressed concrete bridges and culverts* (JTG 3362-2018) stipulates that 0.0033 is taken for concrete with grade C50 or below, 0.003 is taken for concrete with grade C80, the interval values can be determined by the linear interpolation method. If necessary, the ultimate strain of the extreme compression fiber of a concrete core confined by closed ties may be increased.

In the limit state, the stress in concrete close to the extreme fiber in the compression zone reaches the maximum value while it is zero in the neutral axis, the stress distribution between the extreme fiber and the neutral axis is very complicated, as shown in Fig. 9.1B. To simplify the analysis, an equivalent stress block is commonly adopted to simulate the concrete stress distribution in the compression zone by which they have the same magnitude and action point of the resultant of compression in concrete.

From Fig. 9.1B, the resultant of compression in concrete and its action point can be attained, respectively, as

$$D_c = \int_0^{x_a} \sigma_c(\varepsilon_{c,y}) b(y) dy \tag{9.10a}$$

$$y_D = \frac{\int_0^{x_a} \sigma_c(\varepsilon_{c,y}) b(y)(x_a - y) dy}{\int_0^{x_a} \sigma_c(\varepsilon_{c,y}) b(y) dy} \tag{9.10b}$$

where x_a = actual neutral axis depth

$\sigma_c(\varepsilon_{c,y})$ = concrete stress corresponding to concrete strain $\varepsilon_{c,y}$ at level y.

$\varepsilon_{c,y}$ = concrete strain at level y, $\varepsilon_{c,y} = \frac{y}{x_a}\varepsilon_{cu}$.

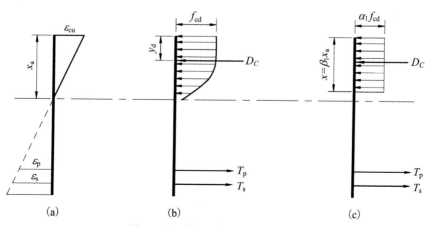

Figure 9.1 Equivalent stress block.

y = distance from the extreme compression fiber of concrete to the level under consideration.

ε_{cu} = ultimate strain of concrete in compression.

$b(y)$ = width of the section at level y.

y_D = distance from the point of the resultant in the compression zone to the extreme compression fiber of concrete.

In Fig. 9.1C, the resultant of compression in concrete and its action point are given by

$$D_c = \alpha_1 f_{cd} \int_0^{\beta_1 x_a} b(y)dy \tag{9.10c}$$

$$y_D = \beta_1 \frac{\int_0^{\beta_1 x_a}(x_a - y)b(y)dy}{\int_0^{\beta_1 x_a} b(y)dy} \tag{9.10d}$$

where f_{cd} = design strength of concrete in compression.

α_1 = ratio of the width of stress block to the maximum width of actual stress diagram.

β_1 = ratio of the depth of stress block to the actual neutral axis depth.

α_1 and β_1 can be solved using Eq. (9.10), then the stress in concrete in the compression zone and the depth of the stress block are obtained, as

$$\sigma_c = \alpha_1 f_{cd} \tag{9.11}$$

$$x = \beta_1 x_a \tag{9.12}$$

where σ_c = concrete stress in the equivalent stress block.

x = depth of equivalent stress block.

Different codes specify different values for α_1 and β_1, generally 0.65–1.0 and 0.74–8.0 for α_1 and β_1, respectively. In Chinese design codes, $\alpha_1 = 1.0$ is taken for railway and highway beams when the design strength of concrete in compression is taken from the corresponding codes. For industrial and civil buildings, the GB 50010-2010 code specifies that 1.0 is taken for C50 and lower grade concrete and 0.94 for C80 concrete, respectively, and the values for intermediate strength between C50 and C80 can be obtained by the linear interpolation method. The values of β_1 should be taken from Table 9.1.

9.2.2 Relative boundary depth of the compression zone

The critical neutral axis depth can be derived from the strain relationship between concrete and reinforcement in the balanced section at failure. It is assumed that the maximum

Table 9.1 Values of β_1.

Grade of concrete	≤ C50	C55	C60	C65	C70	C75	C80
β_1	0.80	0.79	0.78	0.77	0.76	0.75	0.74

compressive strain in the extreme fiber of concrete reaches ε_{cu} at crushing, and the maximum tensile strain in reinforcement in the extreme fiber corresponds to the yield strength. For prestressing steel wires and strands, $f_{0.2}$ is adopted as the "nominal yield stress," since they have no obvious yield point. In the codes, the design tensile strength of prestressing tendons, f_{pd}, is replaced to calculate the yield strain. In Fig. 9.2, three lines represent the strain distributions corresponding to three failure patterns.

Considering a prestressed concrete balanced section with prestressing tendons in the tension zone only, as shown in Fig. 9.2, by which the critical neutral axis depth is obtained as

$$\frac{x_a}{h_0} = \frac{\varepsilon_{cu}}{\left(\Delta\varepsilon + \varepsilon_{p,po}\right) + \varepsilon_{cu}} \tag{9.13}$$

where x_a = depth of actual neutral axis.

h_0 = effective depth of the section, the distance from the extreme compressive fiber of concrete to the point of the resultant of tensile force in the tension zone.

ε_{cu} = ultimate compressive strain of concrete.

$\varepsilon_{p,po}$ = strain change in the extreme fiber of prestressing tendons from the state of decompression section to failure.

$\Delta\varepsilon$ = additional strain adjusting for steel wires and strands, generally 0.002 is taken.

To establish the strain relationship between concrete and reinforcement in the section, only the strain change in the extreme fiber of tensile reinforcement from the

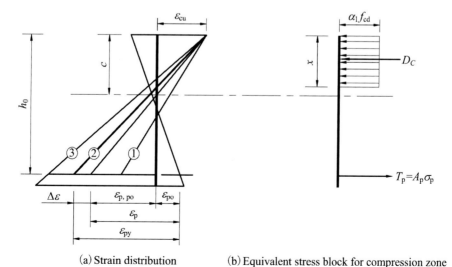

(a) Strain distribution   (b) Equivalent stress block for compression zone

Figure 9.2 Stress and strain in a prestressed concrete flexural section: ①, failure of the over-reinforced section; ②, failure of the balanced section; ③, failure of the under-reinforced section.

decompression section where the strain in concrete and prestressing steels is zero to the critical section at failure can be taken into account, hence:

$$\varepsilon_{p,po} = \varepsilon_{pd} - \varepsilon_{po} \tag{9.14}$$

where ε_{pd} = strain in prestressing tendons at failure, $\varepsilon_{pd} = f_{pd}/E_p$.

ε_{po} = strain in prestressing tendons at decompression state that the concrete strain at the centroid of prestressing tendons is zero.

f_{pd} = design tensile strength of prestressing tendons.

As discussed above, the over-reinforced sections can be avoided by limiting the depth of the neutral axis at failure. In Eq. (9.13), the total strain in the extreme fiber of prestressing tendons at failure (ε_{py}) will increase as the depth of neutral axis (x_a) decreases in a given section, meaning that the prestressing tendons yield easily in the case of a low depth of the neutral axis. Namely, $x_a \leq x_b$ could be the premise for prestressed concrete flexural members to exhibit plastic behavior at failure. For convenience, the relative neutral axis depth ratio, defined as the ratio of the neutral axis depth to the effective depth of the section, is introduced to replace the neutral axis depth in design and analysis, expressed as

$$\xi = \frac{x}{h_0} \tag{9.15}$$

where ξ = relative neutral axis depth ratio.

x = depth of equivalent rectangular, $x = \beta_1 x_a$.

h_0 = effective depth of the section.

Accordingly, the relative boundary neutral axis depth ratio or the relative boundary ratio of the compression zone depth is defined as

$$\xi_b = \frac{x_b}{h_0} \tag{9.16}$$

where ξ_b = relative boundary ratio of the compression zone depth.

x_b = critical depth of the neutral axis, also called the boundary depth of the neutral axis.

9.2.2.1 ξ_b for the prestressing tendons with an obvious yield point

For the prestressing tendons with an obvious yield point, such as high-strength hot-rolled threaded steel bars, the strain in the tendon corresponds to the design tensile strength that can be obtained by $\varepsilon_{pd} = f_{pd}/E_p$ directly, hence, $\Delta\varepsilon = 0$ in Eq. (9.13). Substituting $x = \beta_1 x_a$ into Eqs. (9.13) and (9.15) yields

$$\xi_b = \frac{x_b}{h_0} = \frac{\beta_1 \varepsilon_{cu}}{\varepsilon_{cu} + \varepsilon_{p,po}} = \frac{\beta_1}{1 + \dfrac{\varepsilon_{p,po}}{\varepsilon_{cu}}} = \frac{\beta_1}{1 + \dfrac{f_{pd} - \sigma_{po}}{E_p \varepsilon_{cu}}} \tag{9.17}$$

where E_p = modulus of elasticity of the prestressing tendon.

9.2.2.2 ξ_b for the prestressing tendons without an obvious yield point

The prestressing tendons without an obvious yield point such as steel wires and strands, $\Delta\varepsilon = 0.002$ should be taken into account in the total strain in tendons at failure, as shown in Fig. 9.3, namely:

$$\varepsilon_{py} - \varepsilon_{po} = \Delta\varepsilon + \varepsilon_{p,po} = 0.002 + \varepsilon_{p,po} = 0.002 + \varepsilon_{pd} - \varepsilon_{po} \qquad (9.18)$$

where ε_{py} = total strain in the extreme fiber of the prestressing tendons at failure.

f_{pd} = design tensile strength of the prestressing tendons without an obvious yield point.

ε_{pd} = elastic strain in the prestressing tendons corresponding to the design tensile strength, $\varepsilon_{pd} = \frac{f_{pd}}{E_p}$

Substituting $x = \beta_1 x_a$ and Eq. (9.18) into Eqs. (9.13) and (9.15) yields

$$\xi_b = \frac{\beta_1 \varepsilon_{cu}}{\varepsilon_{cu} + 0.002 + \varepsilon_{pd} - \varepsilon_{po}} = \frac{\beta_1}{1 + \dfrac{0.002}{\varepsilon_{cu}} + \dfrac{f_{pd} - \sigma_{po}}{E_p \varepsilon_{cu}}} \qquad (9.19)$$

Combining with the internal force equilibrium in cross-section, the maximum reinforcement ratio can be deduced from the relative boundary ratio of the compression zone depth. For the flexural member with a rectangular section with the prestressing tendons in the tension zone only, the maximum reinforcement ratio can be obtained by

$$\rho_{\max} = \frac{A_p}{bh} \approx \frac{A_p}{bh_0} = \frac{\xi_b f_{ck}}{f_{pd}} \qquad (9.20)$$

where f_{ck} = characteristic strength of concrete in compression.

Hence, the failure patterns of a flexural section can be described as follows:

Failure of an under-reinforced section, $\varepsilon_p > \varepsilon_{pd}$, $x < x_b = \xi_b h_0$, $\xi < \xi_b$, $\rho < \rho_{\max}$.

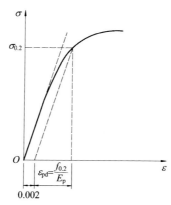

Figure 9.3 Stress versus strain in steel wires and strands.

Failure of a balanced section (boundary failure), $\varepsilon_p = \varepsilon_{pd}, x = x_b = \xi_b h_0, \xi = \xi_b$
$\rho = \rho_{max}$.

Failure of an over-reinforced section, $\varepsilon_p < \varepsilon_{pd}, x > x_b = \xi_b h_0, \xi > \xi_b, \rho > \rho_{max}$.

9.2.2.3 ξ_b specified in the codes

The relative boundary ratio of the compression zone depth, ξ_b, can be obtained from Eqs. (9.17) and (9.19). The values of ξ_b for railway and highway bridges are compiled in Tables 9.2 and 9.3, respectively. When different types of reinforcement are employed in the tension zone, the smallest ξ_b should be adopted.

9.2.3 Flexural bearing capacity of the normal section

9.2.3.1 Flexural bearing capacity of the rectangular section

A typical prestressed concrete rectangular section subjected to the prestressing force and external bending moment is shown in Fig. 9.4. Considering the most general situation, the prestressing tendons and reinforcing steels are arranged both in the tension zone and compression zone.

Table 9.2 Values of relative boundary ratio of the compression zone depth (ξ_b) for railway bridges.

Reinforcement	Concrete	
	C50 and below	C55 and C60
HPB300	0.59	0.57
HRB400	0.54	0.52
HRB500	0.50	0.48
Steel wires and strands	0.40	0.38
Prestressing threaded steel bar	0.40	0.38

Table 9.3 Values of relative boundary ratio of the compression zone depth (ξ_b) for highway bridges.

Reinforcement	Concrete			
	C50 and below	C55, C60	C65	C75, C80
HPB300	0.58	0.56	0.54	—
HRB400, HRBF400, RRB400	0.53	0.51	0.49	
HRB500	0.49	0.47	0.46	—
Steel wires and strands	0.40	0.38	0.36	0.35
Prestressing threaded steel bar	0.40	0.38	0.36	—

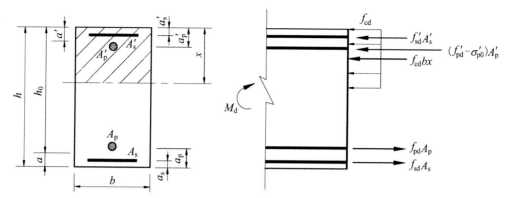

Figure 9.4 Typical prestressed concrete rectangular section subjected to bending moment.

At flexural failure, it is assumed that all reinforcements in the tension zone reach their design strengths. Establishing the moment equilibrium at the reference point of the resultant of prestressing tendons and reinforcing steels in the tension zone, the flexural bearing capacity is derived as follows:

$$M_u = f_{cd}bx\left(h_0 - \frac{x}{2}\right) + f'_{sd}A'_s\left(h_0 - d'_s\right) + \sigma'_p A'_p\left(h_0 - d'_p\right)$$

or

$$M_u = f_{cd}bx\left(h_0 - \frac{x}{2}\right) + f'_{sd}A'_s\left(h_0 - d'_s\right) + \left(f'_{pd} - \sigma'_{p0}\right)A'_p\left(h_0 - d'_p\right) \tag{9.21}$$

where the depth of the neutral axis can be obtained from

$$f_{sd}A_s + f_{pd}A_p = f_{cd}bx + f'_{sd}A'_s + \sigma'_p A'_p \tag{9.22}$$

And the depth of the neutral axis should satisfy

$$x \leq \xi_b h_0 \tag{9.23}$$

To ensure that the strain of the prestressing tendons and reinforcing steels in the compression zone reaches a certain value at failure, the depth of the neutral axis should also satisfy

$$x \geqslant 2a' \tag{9.24}$$

If only reinforcing steels are arranged in the compression zone, or σ'_p is tensile (negative) when both the prestressing tendons and reinforcing steels are arranged in the compression zone, the depth of the neutral axis should also satisfy

$$x \geqslant 2a'_s \tag{9.25}$$

where M_u = flexural bearing capacity of the section under consideration.

f_{cd} = design strength of concrete in compression.

f_{pd} = design tensile strength of the prestressing tendon.

f'_{pd} = design strength of the prestressing tendon in compression.

f_{sd} = design tensile strength of the reinforcing steel.

f'_{sd} = design strength of the reinforcing steel in compression.

σ'_p = stress in the prestressing tendons in the compression zone, $\sigma'_p = f'_{pd} - \sigma'_{p0}\sigma'_{p0}$ = stress in the prestressing tendons in the compression zone corresponding to zero concrete strain at the same level.

A_p = cross-sectional area of the prestressing tendons in the tension zone.

A'_p = cross-sectional area of the prestressing tendons in the compression zone.

A_s = cross-sectional area of the reinforcing steels in the tension zone.

A'_s = cross-sectional area of the reinforcing steels in the compression zone.

b = width of the rectangular section.

h_0 = effective depth of the section, $h_0 = h - a$.

a = distance from the centroid of all reinforcements in the tension zone to the same side edge, $a = \frac{a_s A_s + a_p A_p}{A_s + A_p}$.

d' = distance from the centroid of all reinforcements in the compression zone to the same side edge, $d' = \frac{A'_s d'_s + A'_p d'_p}{A'_s + A'_p}$.

a_p = distance from the centroid of the prestressing tendons in the tension zone to the same side edge.

d'_p = distance from the centroid of the prestressing tendons in the compression zone to the same side edge.

a_s = distance from the centroid of the reinforcing steels in the tension zone to the same side edge.

d'_s = distance from the centroid of the reinforcing steels in the compression zone to the same side edge.

The following explains why $\sigma'_p = f'_{pd} - \sigma'_{p0}$ is adopted as the stress in compressive prestressing tendons in Eq. (9.21) when the section fails.

Set $\Delta\varepsilon'_p$ as the strain change in the prestressing tendons in the compression zone from the decompression state (the strain of concrete in the whole section is zero) to the flexural failure state. At failure, the maximum compressive strain in the extreme fiber of concrete reaches ε_{cu}, as shown in Fig. 9.5. If $x \geqslant 2d'$, $\Delta\varepsilon'_p$ possibly reaches two-thirds of ε_{cu}, therefore, $\Delta\varepsilon'_p \approx 0.002$ when $\varepsilon_{cu} = 0.003$ is adopted, and the corresponding stress change in tendons from the decompression state to the failure state is approximately given by

$$\Delta\sigma'_p = E_p\Delta\varepsilon'_p \approx E_p \times 0.002 \approx 400 \text{ MPa}$$

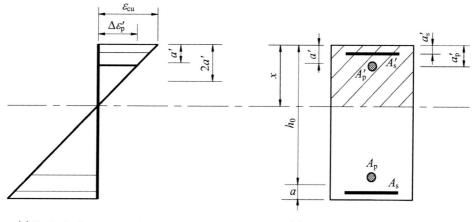

(a) Strain in the compression zone   (b) Reinforcements in the section

Figure 9.5 Strain relation in the compression zone.

Referring to Tables 2.20 and 2.22, $f'_{pd} = 390 - 410$ MPa, thereby

$$\Delta\sigma'_p \approx f'_{pd} \tag{9.26}$$

In the compression zone, a positive sign is taken for compressive stress in concrete, hence, the prestressing tendons that provide compressive bearing capacity in the compression zone are also assigned a positive sign for compression. Considering the stress of prestressing tendons corresponding to the state of the decompression section, σ'_{p0} (tensile stress), the stress in prestressing tendons in the compression zone at failure is obtained as

$$\sigma'_p = -\sigma'_{p0} + \Delta\sigma'_p = f'_{pd} - \sigma'_{p0} \tag{9.27}$$

9.2.3.2 Flexural bearing capacity of the T-section

A typical prestressed concrete T-section subjected to the prestressing force and external bending moment is shown in Fig. 9.6.

When a flexural T-section fails, the neutral axis may fall within the flange or within the web, which leads to different formulas in calculating the flexural bearing capacity. Therefore, the location of the neutral axis should be determined first.

If the neutral axis falls within the flange, the following equation should be satisfied

$$f_{sd}A_s + f_{pd}A_p \leq f_{cd}b'_f h'_f + f'_{sd}A'_s + \left(f'_{pd} - \sigma'_{p0}\right)A'_P \tag{9.28}$$

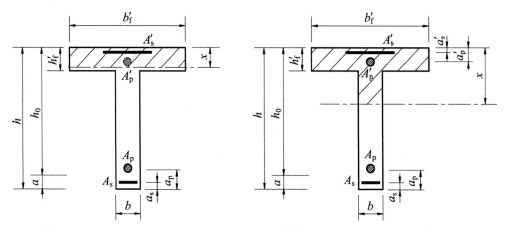

(a) Neutral axis falls within the flange   (a) Neutral axis falls within the web

Figure 9.6 Typical prestressed concrete T-section.

And the flexural bearing capacity of the T-section can be obtained as in the rectangular section with a width of flange, which gives

$$M_u = f_{cd}b'_f x\left(h_0 - \frac{x}{2}\right) + f'_{sd}A'_s\left(h_0 - d'_s\right) + \sigma'_p A'_p\left(h_0 - d'_p\right) \tag{9.29}$$

where the depth of the neutral axis is given by

$$f_{sd}A_s + f_{pd}A_p = f_{cd}b'_f x + f'_{sd}A'_s + \sigma'_p A'_p \tag{9.30}$$

If the neutral axis falls outside the flange, the neutral axis within the web can be obtained by

$$f_{sd}A_s + f_{pd}A_p = f_{cd}\left[bx + \left(b'_f - b\right)h'_f\right] + f'_{sd}A'_s + \left(f'_{pd} - \sigma'_{p0}\right)A'_p \tag{9.31}$$

And the corresponding flexural bearing capacity is given by

$$\begin{aligned} M_u = f_{cd}&\left[bx\left(h_0 - \frac{x}{2}\right) + \left(b'_f - b\right)h'_f\left(h_0 - \frac{h'_f}{2}\right)\right] \\ &+ f'_{sd}A'_s\left(h_0 - d'_s\right) + \left(f'_{pd} - \sigma'_{p0}\right)A'_p\left(h_0 - d'_p\right) \end{aligned} \tag{9.32}$$

where h'_f = depth of the flange in a T-section.

b'_f = width of the flange in a T-section.

Similarly, the depth of the neutral axis should satisfy Eqs. (9.23)–(9.25).

The established Eqs. (9.28)–(9.32) can be used to calculate the flexural bearing capacity of the box or I-section.

9.2.3.3 Discussion on restriction conditions for formulas

Only when Eqs. (9.23)−(9.25) and (9.27) are satisfied, can Eqs. (9.21), (9.29), and (9.32) be used to calculate the flexural bearing capacity.

Actually, $x \leq \zeta_b h_0$ in Eq. (9.23) limits the depth of the neutral axis to avoid brittle failure due to over-reinforcement, and $x \geq 2d'$ in Eq. (9.24) ensures that the incremental strain in compressive prestressing tendons from decompression state to failure state can reach ε'_{pd} ($\varepsilon'_{pd} \approx 0.002$) so that Eq. (9.27) can be used effectively.

Fig. 9.7 shows the strains of tendons in the compression zone in different sections, where line ② corresponds to the boundary condition in the balanced section.

According to the assumption of plane section and the compatibility condition of deformation, the following relation is attained:

$$\frac{x}{d'} = \beta_1 \frac{c}{d'} = \frac{\beta_1 \varepsilon_{cu}}{\varepsilon_{cu} - \varepsilon_{cp}}$$

where ε_{cp} = strain change in tendons in the compression zone from decompression state to the failure state.

To guarantee the effective usage of Eq. (9.27), $\varepsilon_{cp} = \varepsilon'_{pd} = \frac{f'_{pd}}{E_p}$ is required, hence

$$\frac{x}{d'} = \frac{\beta_1 \varepsilon_{cu}}{\varepsilon_{cu} - \dfrac{f'_{pd}}{E_p}} \tag{9.33}$$

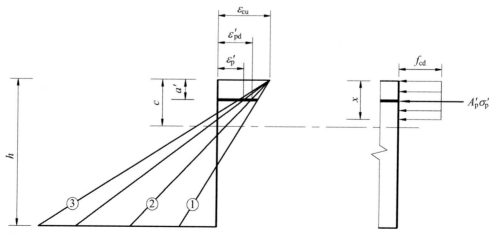

Figure 9.7 Strains of prestressing tendons in the compression zone: ①, $\varepsilon'_p > \varepsilon'_{pd}$; ②, $\varepsilon'_p = \varepsilon'_{pd}$; ③, $\varepsilon'_p < \varepsilon'_{pd}$.

Referring to Tables 2.20 and 2.22, the design values of compressive strength of prestressing steels is within 380–410 MPa, so

$$\frac{f'_{pd}}{E_p} \approx \frac{400}{200 \times 10^3} = 0.002$$

If $\varepsilon_{cu} = 0.003, \frac{f'_{pd}}{E_p} \approx 0.002,$ and $\beta_1 = 0.74 - 0.80$ are substituted into Eq. (9.33), $\frac{x}{d} \approx 2.0$ is derived. Thus, if $x \geq 2d'$, Eq. (9.27) can be satisfied.

Similarly, if $x \geq 2d'_s$, the reinforcing steels in the compression zone can reach the design value of compressive strength when the section fails.

If $x < 2.0d'$, showing that the neutral axis is close to the tendons in the compression zone and $\varepsilon'_p < \varepsilon'_{pd}$, in this case, the stress in tendons can be obtained from Fig. 9.7, as

$$\sigma'_p = \varepsilon_{cu} E_p \left(\frac{\beta_1 d'_p}{x} - 1 \right) + \sigma'_{p0} \tag{9.34}$$

To avoid calculating the stress of tendons in the compression zone, a simplified approach is commonly adopted as follows.

When both the prestressing tendons and reinforcing steels are arranged in the compression zone and the stress in the tendons is compressive (positive) when the section fails, it is assumed that the point of compression of concrete coincides with the point of compression of tendons and reinforcing steels, then the flexural bearing capacity can be obtained by establishing the moment equilibrium at this action point, as

$$M_u = f_{pd} A_p \left(h - a_p - d' \right) + f_{sd} A_s \left(h - a_s - d' \right) \tag{9.35}$$

When only the prestressing tendons are arranged in the compression zone, or both the tendons and reinforcing steels are arranged while the stress in tendons is compressive (positive) when the section fails, it is assumed that the point of compression in concrete coincides with the point of compression in reinforcing steels, then the flexural bearing capacity can be obtained by establishing the moment equilibrium at this action point, as

$$M_u = f_{pd} A_p \left(h - a_p - d'_s \right) + f_{sd} A_s \left(h - a_s - d'_s \right) - \left(f'_{pd} - \sigma'_{p0} \right) A'_P \left(d'_p - d'_s \right) \tag{9.36}$$

9.3 Example 9.1

A simply supported beam with an effective span of 32.0 m is used in a mixed passenger and freight railway with a design speed of 160 km/h. The dimension and reinforcement at the midspan section are shown in Fig. 9.8, where C45 concrete and five groups of

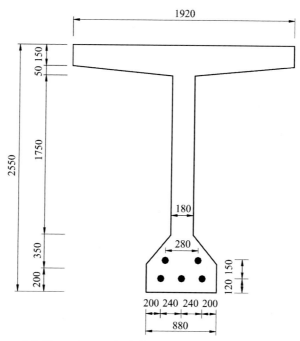

Figure 9.8 Dimension and reinforcements at the midspan section.

bonded strands $9\phi^S15.2$ (9-7ϕ5) with a characteristic tensile strength of 1860 MPa are used. The design value of the bending moment at the midspan is 9836.8 kN•m. Calculate and verify the flexural bearing capacity of the midspan section according to the *Code for Design of Railway Bridges and Culverts* (Q/CR 9300-2018).

9.4 Solution

9.4.1 Parameters

$f_{cd} = 22.5$ MPa, $f_{ptk} = 1860$MPa, $f_{pd} = 1260$ MPa, $M_d = 0.984 \times 10^4$ kN·m

$A_{p0} = 9 \times 140 = 1.26 \times 10^3$ mm^2 (each $9\phi^S15.2$)

$A_p = 5 \times 1.26 \times 10^3 = 6.3 \times 10^3$ mm^2

$a_p = \dfrac{3 \times 120 + 2 \times 270}{5} = 180$ mm

$h_0 = h - a_p = 2500 - 180 = 2320$ mm.

9.4.2 Judge the location of the neutral axis

$f_{pd}A_p = 1260 \times 6.3 \times 10^3 = 7.94 \times 10^6$ N.

$f_{cd}b'_f h'_f = 22.5 \times 1920 \times (150 + 50/2) = 7.56 \times 10^6$ N.

So

$$f_{pd}A_p = 7940 \text{ kN} > f_{cd}b_f'h_f' = 7560 \text{ kN}$$

The neutral axis falls within the web.

9.4.3 Calculate and verify the flexural bearing capacity

The following equilibrium equation can be obtained at the midspan section:

$$f_{pd}A_p = f_{cd}\left[bx + \left(b_f' - b\right)h_f'\right]$$

From which

$$x = \frac{1}{b}\left[\frac{f_{pd}A_p}{f_{cd}} - \left(b_f' - b\right)h_f'\right]$$

$$= \frac{1260 \times 6.3 \times 10^3/22.5 - (1920 - 180) \times 175}{180} = 268.3 \text{ mm}$$

Referring to Table 9.2, the value of the relative boundary ratio of the compression zone depth is 0.4, $\xi_b = 0.4$.

$$x = 268.3 \text{ mm} < \xi_b h_p = 0.4h_p = 0.4 \times 2320 = 928.0 \text{ mm}$$

Hence, Eq. (9.32) can be used to calculate the flexural bearing capacity:

$$M_u = f_{cd}\left[bx\left(h_0 - \frac{x}{2}\right) + \left(b_f' - b\right)h_f'\left(h_0 - \frac{h_f'}{2}\right)\right]$$

$$= 22.5 \times \left[180 \times 268.3 \times \left(2320 - \frac{268.3}{2}\right)\right.$$

$$\left. + (1920 - 180) \times 175 \times \left(2320 - \frac{175}{2}\right)\right]$$

$$= 1.767 \times 10^{10} \text{ N} \cdot \text{mm} = 1.767 \times 10^4 \text{kN} \cdot \text{m} > M_d = 0.984 \times 10^4 \text{ kN} \cdot \text{m}$$

Thus, the flexural bearing capacity at the midspan section meets the requirement specified in the code.

9.5 Example 9.2

The midspan section of a simply supported highway beam with an effective span of 24.2 m is shown in Fig. 9.9. C40 concrete and four groups of bonded strands $7\phi^S12.7$

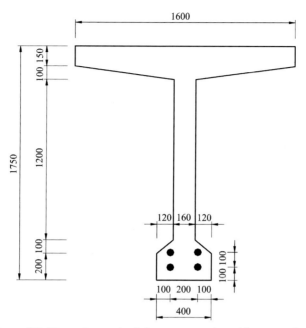

Figure 9.9 Dimension and reinforcement at the midspan section.

(7-7ϕ4) with a characteristic tensile strength of 1860 MPa are used. The design value of the bending moment at the midspan is 4853.2 kN·m. Calculate and verify the flexural bearing capacity of the midspan section according to the *Specifications for design of highway reinforced concrete and prestressed concrete bridges and culverts* (JTG 3362-2018).

9.6 Solution

9.6.1 Parameters

$f_{cd} = 18.4$ MPa, $f_{ptk} = 1860$ MPa, $f_{pd} = 1260$ MPa

$A_{p0} = 7 \times 98.7 = 690.9$ mm^2 (each $7\phi^S 12.7$)

$A_p = 4A_{p0} = 4 \times 690.9 = 2763.6$ mm^2

$a_p = \dfrac{2 \times 100 + 2 \times 210}{4} = 155$ mm

$h_0 = h - a_p = 1750 - 155 = 1595$ mm.

9.6.2 Judge the location of the neutral axis

$$f_{pd}A_p = 1260 \times 2763.6 = 3.48 \times 10^6 \text{ N}$$

$$f_{cd}b'_f h'_f = 18.4 \times 1600 \times 150 = 4.42 \times 10^6 \text{ N}$$

Therefore,

$$f_{cd}b'_f h_f = 4420 \text{ kN} > f_{pd}A_p = 3480 \text{ kN}$$

The neutral axis falls within the flange.

9.6.3 Calculate and verify the flexural bearing capacity

$$x = \frac{f_{pd}A_p}{f_{cd}b'_f} = \frac{1260 \times 2763.6}{18.4 \times 1600} = 118 \text{ mm}$$

Referring to Table 9.3, the value of relative boundary ratio of the compression zone depth is 0.4, $\xi_b = 0.4$.

$$x = 118 \text{ mm} < \xi_b h_p = 0.4 h_p = 1595 = 638 \text{ mm}$$

Hence, Eq. (9.29) can be used to calculate the flexural bearing capacity, as

$$M_u = f_{cd}bx\left(h_0 - \frac{x}{2}\right)$$

$$= 18.4 \times 1600 \times 118 \times \left(1595 - \frac{118}{2}\right) = 5.336 \times 10^9 \text{ N·mm}$$

$$= 5.336 \times 10^3 \text{ kN·m} > M_d = 1.0 \times 4.853 \times 10^3 \text{ kN·m}$$

Thus, the flexural bearing capacity of the midspan section meets the requirement specified in the code.

9.7 Example 9.3

In a frame structure of a building, a post-tensioned beam has a total length of 12 m and an effective span of 11.7 m, the dimension and arrangement of prestressing strands at the midspan section are shown in Fig. 9.10. C45 concrete and two groups of bonded strands $5\phi^S9.5$ (5−7ϕ3) with a characteristic tensile strength of 1860 MPa are used. The design value of the bending moment at the midspan is 623.8 kN·m Calculate and verify the flexural bearing capacity of the midspan section according to the GB 50010-2010 code.

9.8 Solution

9.8.1 Parameters

$f_{cd} = 23.1$ MPa, $f_{ptk} = 1860$ MPa, $f_{pd} = 1320$ MPa

$A_{p0} = 5 \times 54.8 = 274.0 \text{ mm}^2$ (each $5\phi^S9.5$)

$A_p = 2A_{p0} = 2 \times 274.0 = 548.0 \text{ mm}^2$

$a_p = 60$ mm, $h_0 = h - a_p = 1050 - 60 = 990$ mm.

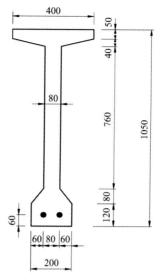

Figure 9.10 Dimension and reinforcement at the midspan section.

9.8.2 Judge the location of the neutral axis

$$f_{pd}A_p = 1320 \times 548 = 7.23 \times 10^5 \text{ N}$$

$$\alpha_1 f_{cd} b'_f h'_f = 1.0 \times 23.1 \times \left[400 \times 50 + \frac{1}{2} \times (400 + 80) \times 40 \right] = 6.84 \times 10^5 \text{ N}$$

Therefore

$$f_{pd}A_p = 723 \text{ kN} > \alpha_1 f_{cd} b'_f h'_f = 684 \text{ kN}$$

The neutral axis falls within the web.

9.8.3 Calculate and verify the flexural bearing capacity

The following equilibrium equation can be obtained at the midspan section

$$f_{pd}A_p = f_{cd} \left[bx + \left(b'_f - b \right) h'_f \right]$$

From which

$$x = \frac{1}{b}\left[\frac{f_{pd}A_p}{\alpha_1 f_{cd}} - \left(b'_f - b\right)h'_f\right]$$

$$= \frac{1}{80}\left[\frac{1320 \times 548}{1.0 \times 23.1} - (400 - 80) \times \left(50 + \frac{40}{2}\right)\right] = 111.4 \text{ mm}$$

$\xi_b = 0.4$ is adopted, thus

$$x = 111.4 \text{ mm} < \xi_b h_p = 0.4 h_p = 0.4 \times 990 = 396.0 \text{ mm}$$

Hence, Eq. (9.32) can be used to calculate the flexural bearing capacity:

$$M_u = \alpha_1 f_{cd}\left[bx\left(h_0 - \frac{x}{2}\right) + \left(b'_f - b\right)h'_f\left(h_0 - \frac{h'_f}{2}\right)\right]$$

$$= 1.0 \times 23.1 \times 80 \times 111.4 \times \left(990 - \frac{111.4}{2}\right)$$

$$+ 1.0 \times 23.1 \times (400 - 80) \times 70 \times \left(990 - \frac{70}{2}\right)$$

$$= 686.50 \times 10^6 \text{ N} \cdot \text{mm} = 686.5 \text{ kN} \cdot \text{m} > M_d = 623.8 \text{ kN} \cdot \text{m}$$

Thus, the flexural bearing capacity of the midspan section meets the requirement specified in the code.

9.9 Shear bearing capacity of the oblique section

9.9.1 Failure patterns due to shear force and the influencing factors

9.9.1.1 Failure patterns due to shear force

In a simply supported prestressed concrete flexural member subjected to service loads, three regions along the span can be distinguished approximately, as the middle region ($K2 - K2$) where the bending moment is large while the shear force is zero or relatively small, the region close to the support ($S - K1$) where the shear force is large while the bending moment is relatively small, and the transition region ($K1 - K2$), as shown in Fig. 9.11. The cracking patterns in the sections caused by the service loads relate to the magnitudes of bending moment and shear force in this section, which finally lead to different failure patterns.

At the midspan the beam exhibits normal cracking under a certain value of bending moment, the corresponding cracks starting from the bottom edge and perpendicular to the beam's axis are called flexure cracks. The flexure cracks appear in the midspan region ($K2 - K2$ in Fig. 9.11).

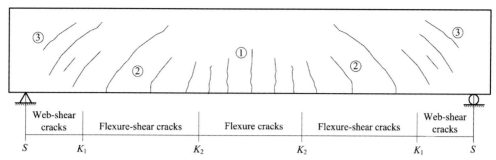

Figure 9.11 Cracking patterns in a simply supported concrete beam: ①, flexural region; ②, flexure-shear region; ③, web shear region.

Analysis of shear stress and principal stresses based on Fig. 7.6 shows that the diagonal tension along the OB direction approximately will be generated in the sections under bending moment and shear force, consequently, the direction of the cracks in the web is perpendicular to the diagonal tension or tensile principal stress, as shown in Fig. 9.12. Considering the stress in concrete in the bottom fiber is a normal tensile stress, the direction of the cracks at the bottom edge should be perpendicular to the beam's axis, therefore, the typical profile of the cracks at the transition region ($K1 - K2$ in Fig. 9.11) should be a short line perpendicular to the beam's axis starting from the bottom edge to propagate into an inclined line or curve upward, this kind of crack is called a flexure-shear crack (Fig. 9.11), which may finally lead to flexure-shear failure.

In the vicinity of the support, the shear force is large while the bending moment is relatively small (the region $S - K1$ in Fig. 9.11), so the cracks caused by shear force in the web must be inclined while they do not link to the bottom edge since the bending moment in this region cannot generate normal cracking. These cracks develop from the middle of the web, so they are called web-shear cracks, which may finally lead to web-shear failure.

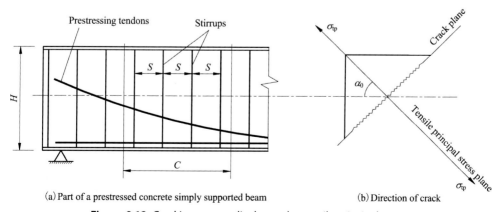

(a) Part of a prestressed concrete simply supported beam   (b) Direction of crack

Figure 9.12 Cracking perpendicular to the tensile principal stress.

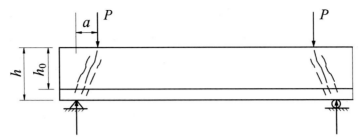

Figure 9.13 Cracking between the concentrated load and the support.

When a concentrated load is applied close to the support and its value is huge, the concrete between the support and the load works as an inclined small column, and the cracking along the column axis approximately may occur due to transverse Poisson's effect if the quantity of vertical reinforcement (stirrups and longitudinal bent up reinforcements) in the web is low, as shown in Fig. 9.13, which may quickly result in inclined compression failure. On the other hand, if the quantity of vertical reinforcement in the web in the region $K1 - K2$ (Fig. 9.11) is low, inclined tension failure may take place due to diagonal tension once the normal cracking at the bottom edge occurs.

Usually, the inclined tension failure can be prevented effectively by limiting the stirrup spacing while arranging sufficient bent-up reinforcements, and the inclined compression failure can be prevented by increasing the sectional dimensions. Hence, the shear failure due to flexure-shear or web-shear is the main concern in design.

As the flexure-shear and web-shear cracks propagate up and downward, a cracking oblique section AB along with an uncracked section BD will form before failure, as shown in Fig. 9.14. An inclined resultant $T_{s,c}$ provided by the stirrups, bent-up reinforcement and longitudinal reinforcement are acted on the section AB, a vertical shear force V_c and a compression N_c are acted on the section BD. Under shear force and compression, failure may occur due to the shear force in the oblique section AB and section BD, which is called the shear-compression failure pattern. If $T_{s,c}$ is not large enough, failure may occur due to bending from $T_{s,c}$ around the point O, which is called the flexure

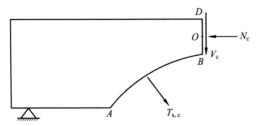

Figure 9.14 Failure patterns due to shear force.

failure pattern in the oblique section. If the bent-up reinforcement and longitudinal rein-forcement have sufficient anchor length, in most cases, the flexure failure can be prevented.

9.9.1.2 The influencing factors on shear resistance

Eqs. (7.50)–(7.57) show that the resistance ability for shear stress, principal tensile and compressive stresses of concrete in a prestressed concrete flexural member under service loads mainly relate to the prestressing, longitudinal and vertical reinforcement ratios, ratio of normal stress to shear stress and concrete strength, and these factors also affect the shear resistance by almost the same mechanism.

The prestressing

Both the longitudinal and vertical prestressing effectively reduce the tensile principal stress of concrete, as shown in Eqs. (7.50)–(7.57), hence, the longitudinal prestressing tendons, the longitudinal bent-up prestressing tendons, and the vertical prestressing ten-dons can improve the shear resistance of the oblique section. In the short- and medium-span beams, the vertical prestressing tendons are not arranged normally, in this case, the longitudinal bent-up prestressing tendons are commonly required in the regions where the shear force is large.

In a reinforced concrete flexural member, it is assumed that the shear cracks propagate at approximately 45 degrees to the beam's axis in the section in the $K1 - K2$ regions (Fig. 9.11), as $\alpha_0 = 45$ degrees shown in Fig. 9.12B. Therefore, the stirrups and vertical prestressing tendons in the length of C ($C = H$, where H is the depth of the section) contribute to the shear resistance, as shown in Fig. 9.12A. Eq. (7.50c) shows that the pre-stressing can change the value of α_0, thereby a longer inclined section of concrete and more stirrups will participate in the shear resistance when α_0 is reduced by the prestress-ing. Analysis shows that $\alpha_0 = 35$ degree $- 40$ degree is common in the region $S - K2$ (Fig. 9.11) in the prestressed concrete flexural members, and $\alpha_0 = 37.5$ degree can be adopted generally.

Ratio of normal stress to shear stress

The ratio of longitudinal normal stress to shear stress in the section under the bending moment and shear force can be expressed as

$$\frac{\sigma_{cx}}{\tau_c} = \frac{My}{I} \times \frac{Ib}{VS} = \frac{M}{V\frac{S}{by}} = \frac{M}{Vh_r} \tag{9.37}$$

where $\sigma_{cx} = $ longitudinal normal stress of concrete due to the prestressing force and the design load.

τ_c = shear stress of concrete at the level under consideration due to the prestressing force and the design load.

M = bending moment due to the prestressing force and the design load.

V = shear force in the section under consideration.

S = first static moment of the area about the neutral axis of the portion of the section outside the shear stress considered.

b = width of section at the level of shear stress considered.

y = distance from the neutral axis to the level of shear stress considered.

h_r = depth relates to the level of shear stress considered.

If h_r in Eq. (9.37) is replaced by h_0, a generalized ratio of shear span to effective depth is obtained to describe the ratio of normal stress to shear stress, as

$$m = \frac{M}{Vh_0} \tag{9.38}$$

where m = generalized shear span ratio.

h_0 = effective depth of the section.

When the section is close to the support of a simply supported beam under a concentrated compression force, as shown in Fig. 9.13, the shear force V in the section is approximately equal to the reaction on the support, in this case, Eq. (9.38) can be rewritten as

$$m = \frac{M}{Vh_0} = \frac{Va}{Vh_0} = \frac{a}{h_0} = \frac{\sigma(M)}{\tau(V)} \tag{9.39}$$

where a = distance from the support to the concentrated compression force.

In Eq. (9.39), $m = \frac{a}{h_0}$ is defined as the ratio of shear span to effective depth of the section, abbreviated as shear span ratio, which describes the ratio between the normal stress and the shear stress in the section under the concentrated load.

If the vertical prestressing is neglected, $\sigma_{cy} = 0$, substituting Eq. (9.37) into Eq. (7.50c) yields

$$\text{tg}2\alpha_0 = -\frac{1}{\xi m} \tag{9.40}$$

where ξ = coefficient relating to the sectional properties and the level considered, $\xi = \frac{h_0 b}{S} y$.

Eq. (9.40) shows that the angle between the tensile principal stress and the beam's axis changes with the shear span ratio. As m deduces, α_0 decreases, and a longer inclined section of concrete and more transverse stirrups will contribute to the shear resistance, thereby the shear resistance increases. Researches show that the shear span ratio has almost no effect on the shear resistance when $m > 3$.

Concrete strength

For the possibly occurred flexure-shear failure and web-shear failure in the oblique section under shear and compression simultaneously, the higher compressive strength of concrete will effectively improve both the compression strength and the shear resistance. Generally, the relationship between the shear resistance and the compressive strength of concrete can be expressed as a quadratic curve or approximately linear.

Ratio of longitudinal reinforcement

Longitudinal prestressing tendons and reinforcing steels can restrict the upward movement of the neutral axis at failure, consequently, the area of compressive concrete in the section increases, and the shear bearing capacity contributed by the concrete in the oblique section improves. In addition, as the longitudinal reinforcement ratio increases, the dowel action between the reinforcement and the concrete enhances, resulting in a higher ability to transmit the shear forces.

Web reinforcement

The transverse stirrups, inclined prestressing tendons, and reinforcing steels are called the web reinforcement, playing the most important role in shear resistance, as shown in Fig. 9.12. The greater the quantity of web reinforcement in the crack plane or oblique plane perpendicular to the tensile principal stress, the greater the shear bearing capacity of the oblique section is.

9.9.2 Shear bearing capacity of the oblique section
9.9.2.1 General model for shear resistance of the oblique section

A typical analytical model for shear bearing capacity based on the shear-compression failure pattern in a simply supported beam is shown in Fig. 9.15. Let C be the projection length of the oblique section on the beam's axis.

Referring to the previous discussion on the main factors contributing to the shear resistance, the general expression of the shear bearing capacity can be written as

$$V_u = V_c + V_{sv} + V_s + V_p \tag{9.41}$$

where V_u = shear bearing capacity.

V_c = shear bearing capacity provided by the concrete in the oblique section.

V_{sv} = shear bearing capacity provided by the stirrups and vertical prestressing tendons (without considering the effect of prestressing).

V_s = shear bearing capacity provided by the reinforcing steels in the oblique section.

V_p = shear bearing capacity provided by the prestressing tendons in the oblique section, including the contributions by the prestressing force, the longitudinal prestressing tendons and the bent-up prestressing tendons.

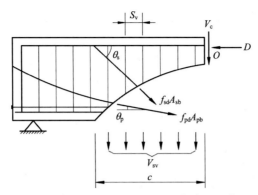

Figure 9.15 Shear bearing capacity calculation diagram.

When the inclined cracking occurs, the performance mechanism in the oblique section is very complicated and the complete theoretical formulas are difficult to derive. Thus, the presented approaches introduce numerous empirical coefficients in computations. The following general expressions are commonly adopted to calculate the partial shear resistance:

$$V_c = k_1 f_{td} bh_0 \tag{9.42}$$

$$V_{sv} = k_2 \left(\rho_{sv} f_{sd,v} + \rho_{pv} f_{pd,v} \right) bh_0 \tag{9.43}$$

$$V_s = V_{sb} = k_3 f_{sd} \sum A_{sb} \sin \theta_s \tag{9.44}$$

$$V_{pb} = k_4 f_{pd} \sum A_{pb} \sin \theta_p \tag{9.45}$$

where V_{sb} = the shear bearing capacity provided by the bent-up reinforcing steels intersecting with the oblique section.

V_{pb} = shear bearing capacity provided by the bent up prestressing tendons intersecting with the oblique section.

$k_1 - k_2$ = coefficients.

$k_3 - k_4$ = discount coefficients, mainly reflecting partial bent-up steel stresses less than the design strength at failure due to nonuniform stress in the bent-up steels, generally, 0.75−0.8 is taken.

b = width of the section at the level of the shear plane considered.

h_0 = effective depth, the distance from the point of the resultant of tensile force in the tension zone to the extreme compressive fiber of concrete.

ρ_{sv} = bent-up reinforcing steel ratio, $\rho_{sv} = A_{sv}/s_v b$.

ρ_{pv} = bent-up prestressing tendon ratio, $\rho_{pv} = A_{pv}/s_{pv} b$.

s_v = spacing between two adjacent stirrups.

s_{pv} = spacing between two adjacent vertical prestressing tendons.

A_{sv} = total area of the legs of the stirrup in the same section.

A_{pv} = total area of the legs of prestressing tendon in the same section.

f_{td} = design tensile strength of the concrete.

f_{sd} = design tensile strength of the bent-up reinforcing steel.

f_{pd} = design tensile strength of the bent-up prestressing tendon.

$f_{sd,v}$ = design tensile strength of the stirrup.

$f_{pd,v}$ = design tensile strength of the vertical prestressing tendon.

A_{pb} = area of the bent-up prestressing tendons.

θ_p = angle between the bent-up prestressing tendons and the beam's axis.

A_{sb} = area of bent-up reinforcing steels.

θ_s = angle between the bent-up reinforcing steels and the beam's axis.

Different combinations of the four terms in Eq. (9.41) lead to different formulas and corresponding coefficients to calculate shear strength provided in the codes.

The first combination mode is the same as in Eq. (9.41), which is adopted in the Q/CR 9300-2018 code.

Fig. 7.6 and Eq. (7.50a) show that the contribution of the prestressing to the shear resistance is achieved by reducing the tensile principal stress of concrete, while the transverse stirrups directly provide diagonal tension resistance, so the prestressing tendons, the stirrups, and the concrete should be considered comprehensively to resist shear force, hence, Eq. (9.41) is rewritten as

$$V_u = \alpha_p V_{cs} + V_{sb} + V_{pb} \tag{9.46}$$

where V_{cs} = shear bearing capacity provided by the concrete, transverse stirrups, and vertical prestressing tendons in the oblique section.

α_p = increasing coefficient of the prestressing on the shear bearing capacity.

The second combination mode shown in Eq. (9.46) is adopted in the JTG 3362-2018 code.

In the third combination mode, the concrete and vertical prestressing tendons in the oblique section are taken into account comprehensively to provide the shear resistance, while the nonprestressing stirrups and the prestressing are considered separately, as

$$V_u = V_{cs} + V_{sb} + V_{pb} + V_{p1} \tag{9.47}$$

where V_{cs} = shear bearing capacity provided by the concrete and the stirrups in the oblique section, $V_{cs} = \alpha_{cv} V_c + V_{sv}$

α_{cv} = shear bearing capacity coefficient of concrete.

V_{p1} = shear bearing capacity provided by the prestressing tendons in the oblique section.

Eq. (9.47) is adopted in the GB 50010-2010 code.

9.9.2.2 Shear bearing capacity of the railway beams

The Q/CR 9300-2018 code stipulates that the shear bearing capacity of a prestressed concrete railway beam with rectangular or T-sections can be calculated by Eqs. (9.41)–(9.45), rewritten as

$$V_u = V_c + V_{sv} + 0.8f_{sd}A_{sb}\sin\theta_s + 0.8\sigma_{pe}A_{pb}\sin\theta_p \tag{9.48}$$

$$V_c = \frac{0.6}{m}(1+0.35P)\beta\alpha f_{td}bh_0 \tag{9.49}$$

$$V_{sv} = 0.8(0.35+0.4m)\left(\rho_{sv}f_{sd,v} + \rho_{pv}f_{pd,v}\right)bh_0 \tag{9.50}$$

where m = generalized shear span ratio, $m = 1.0$ if $m < 1.0$ and $m = 3.0$ if $m > 3.0$.

β = increasing coefficient by the prestressing.

α = shear resistance coefficient of the extension part of the compressed flange in the T-section.

ρ = ratio of longitudinal reinforcement in the tension zone in the oblique section

$P = 100\rho$.

β, α, and ρ are given by

$$\beta = 1 + \frac{M_0}{M} = 1 + \lambda \leq 2.0 \tag{9.51}$$

$$\alpha = 1 + \frac{h_f^2}{b_w h_0} \tag{9.52}$$

$$\rho = \frac{A_p + A_{pb} + A_s}{bh_0} \leq 0.03 \tag{9.53}$$

where M_0 = decompression moment, corresponding to the moment counterbalancing the precompressive stress in concrete in the extreme fiber in the tension zone to zero.

M = bending moment due to the frequency combination and the quasi-permanent combination in serviceability limit state.

λ = prestress level, given by Eq. (1.11).

h_f = depth of the compression zone in the flange of the T-section.

b_w = web thickness at the level under consideration, the nominal web thickness, $b_{wn} = b_w - \frac{1}{2}\sum\phi$, should be taken if the tendons with diameter ϕ greater than $b_w/8$ passing through the web.

A_p = area of longitudinal prestressing tendons in the tension zone in the oblique section.

A_s = area of longitudinal reinforcing steels in the tension zone in the oblique section.

For a simply supported beam with rectangular or T-sections, Eq. (9.54) should be satisfied to prevent the oblique section from inclined compression failure:

$$V_d \leq 0.25 f_{cd} bh_0 \qquad (9.54)$$

where V_d = maximum design value of shear force in the section considered.

f_{cd} = design strength of concrete in compression.

In the continuous beam with rectangular or T-sections with a constant depth, the shear bearing capacity in the oblique section in the vicinity of the middle supports may be taken as

$$V_u = \frac{0.34}{0.18 + m}\left(1 + \overline{\lambda_u}\right)(2 + P) f_{td} bh_0 + V_{sv} + 0.8(0.35 + 0.4m)\left(\rho_{sv} f_{sd,v} + \rho_{pv} f_{pd,v}\right) bh_0 \qquad (9.55)$$

$$\overline{\lambda_u} = \frac{\lambda_u^+ + \lambda_u^-}{2} \qquad (9.56)$$

$$\lambda_u^+ = \frac{M_0^+}{2M^+}, \lambda_u^+ = \frac{|M_0^-|}{|2M^-|} \qquad (9.57)$$

where $\overline{\lambda_u}$ = average prestress level.

M_0^+ = decompression moment in the positive bending region, corresponding to the moment counterbalancing the precompressive stress in concrete in the extreme fiber in the tension zone to zero.

M_0^- = decompression moment in the negative bending region.

M^+ = bending moment due to the frequency combination and the quasi-permanent combination in serviceability limit state in the positive bending region.

M^- = bending moment due to the frequency combination and the quasi-permanent combination in serviceability limit state in the negative bending region.

9.9.2.3 Shear bearing capacity of the highway beams

The JTG 3362-2018 code stipulates that the shear bearing capacity of a prestressed concrete highway beam with rectangular or T-sections can be calculated by Eqs. (9.44)−(9.46), where $k_3 = k_4 = 0.75$.

If the design tensile strength of concrete is replaced by the cubic compressive strength, Eq. (9.42) is rewritten as

$$V_c = k_0 \frac{(2 + 0.6P)}{m} \sqrt{f_{cu,k}} bh_0 \qquad (9.58a)$$

where $f_{cu,k}$ = characteristic value of cubic compressive strength of concrete.

ρ = ratio of longitudinal reinforcement in the tension zone in the oblique section,

$\rho = \frac{A_p + A_s}{bh_0}$, $P = 100\rho$.

k_0 = empirical influence coefficient.

The shear resistance provided by the stirrups and vertical prestressing tendons intersecting with the oblique section (oblique cracking plane) is given by

$$V_{sv} = k_s \left(f_{sd,v} \sum A_{sv} + f_{pd,v} \sum A_{pv} \right) \tag{9.58b}$$

where k_s = coefficient of nonuniform stress in the stirrups.

To obtain the total stirrups intersecting with the oblique section, the horizontal projection length C of the oblique section should be derived first. Experimental data show $C \approx 0.6mh_0$, hence

$$V_{sv} = 0.6k_s m \left(f_{sd,v}\rho_{sv} + f_{pd,v}\rho_{pv} \right) bh_0 \tag{9.58c}$$

Add Eq. (9.58a) and Eq. (9.58c) to obtain

$$V_{cs} = V_c + V_{sv} = k_0 \frac{(2 + 0.6P)}{m} \sqrt{f_{cu,k}} bh_0 + 0.6k_s m \left(f_{sd,v}\rho_{sv} + f_{pd,v}\rho_{pv} \right) bh_0 \tag{9.58d}$$

Research shows that the concrete and the transverse reinforcement work together to provide the shear resistance rather than working alone, which is related to the shear span ratio, as shown in Fig. 9.16. Hence, the minimum value of V_{cs} can be found by $\frac{dV_{cs}}{dm} = 0$ from Eq. (9.58d), thereby

$$V_{cs} = 0.45bh_0 \sqrt{(2 + 0.6P)\sqrt{f_{cu,k}}\left(\rho_{sv}f_{sd,v} + 0.6\rho_{pv}f_{pd,v}\right)} \tag{9.58e}$$

As the influences of the prestressing and opposite-sign bending moment are taken into account further, Eq. (9.58e) becomes

$$V_{cs} = 0.45\alpha_1\alpha_2\alpha_3 bh_0 \sqrt{(2 + 0.6P)\sqrt{f_{cu,k}}\left(\rho_{sv}f_{sd,v} + 0.6\rho_{pv}f_{pd,v}\right)} \tag{9.59}$$

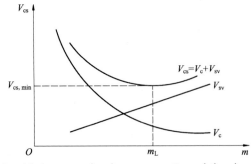

Figure 9.16 Relationship between the shear span ratio and the shear bearing capacity.

where $\alpha_1 =$ coefficient reflecting the influence of the positive and negative bending moment, 1.0 is taken for the oblique section close to the support of the simply supported beams or side support of the continuous beams, while 0.9 is taken for the oblique section close to the middle supports of the continuous beams.

$\alpha_2 =$ increasing coefficient by the prestressing, 1.25 is taken generally for the prestressed concrete flexural members; while, 1.0 is taken when the bending moment due to the resultant of reinforcement and external moment have the same direction, or the flexural members allowing cracking.

$\alpha_3 =$ influence coefficient on compression flange, 1.0 is taken for rectangular section and 1.1 for T-, I-, or box sections.

Eqs. (9.44)–(9.46) and (9.59) where $k_3 = k_4 = 0.75$ are used to calculate the shear bearing capacity of the prestressed concrete highway beams with a constant depth. For continuous beams and cantilever beams with variable section depth, the effect of additional shear stress should be considered.

For a simply supported beam with rectangular, T-, I-, or box sections, Eq. (9.60) should be satisfied to prevent the oblique section from inclined compression failure and to avoid excessive width of the inclined cracks:

$$V_d \leq 0.51bh_0 \sqrt{f_{cu,k}} \tag{9.60}$$

where $V_d =$ maximum design value of shear force in the section considered.

9.9.2.4 Shear bearing capacity of the beams in civil and industrial buildings

The GB 50010-2010 code stipulates that the shear bearing capacity of a prestressed concrete beam with rectangular or T-sections can be calculated by Eqs. (9.44)–(9.45) and (9.47), as

$$V_u = V_{cs} + V_{p1} + 0.8f_{sd}A_{sb} \sin \theta_s + 0.8f_{pd}A_{pb} \sin \theta_p \tag{9.61}$$

$$V_{p1} = 0.05N_{p0} \tag{9.62}$$

where $V_{cs} =$ shear bearing capacity provided by the concrete and the stirrups in the oblique section.

$V_{p1} =$ shear bearing capacity provided by the tendons in the oblique section, when the bending moment due to N_{p0} and external moment have the same direction, or for the flexural members allowing cracking, $V_{p1} = 0$

$N_{p0} =$ resultant of the longitudinal tendons and reinforcing steels when the normal stress of concrete in the section is zero, where the bent-up tendons are disregarded and $N_{p0} \leq 0.3f_{cd}A_0$, f_{cd} is the design strength of concrete in compression and A_0 is the area of the converted section.

For general members, V_{cs} in Eq. (9.61) may be calculated by

$$V_{cs} = \alpha_{cv}f_{td}bh_0 + f_{sd,v}\frac{A_{sv}}{s_v}h_0 \qquad (9.63a)$$

where α_{cv} = shear capacity coefficient of concrete in the oblique section, 0.7 is taken for general flexural members.

For the sections of an independent beam under concentrated loads where the shear force produced by the concentrated loads is more than 75% of the total shear force, the shear span ratio has a significant impact on the shear bearing capacity, in this case, V_{cs} in Eq. (9.61) should be calculated by

$$V_{cs} = \frac{1.75}{m+1}f_{td}bh_0 + f_{sd,v}\frac{A_{sv}}{s_v}h_0 \qquad (9.63b)$$

where m = shear span ratio, given by Eq. (9.39), $m = 1.5$ when $m < 1.5$ and $m = 3$ when $m > 3$.

For flexural members with rectangular, T-, or box sections with an inclined edge in the tension zone (Fig. 9.17), the shear bearing capacity of the oblique section can be obtained by

$$V_u = V_{cs} + V_{sp} + 0.8f_{sd}A_{sb}\sin\theta_s \qquad (9.64)$$

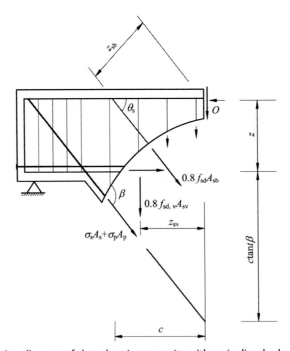

Figure 9.17 Calculation diagram of shear bearing capacity with an inclined edge in the tension zone.

$$V_{sp} = \frac{M_d - 0.8\left(\sum f_{sd,v}A_{sv} + \sum f_{sd}A_{sb}\sin\theta_{sb}\right)}{z + c\tan\beta}\tan\beta \qquad (9.65)$$

$$\sigma_{pe}A_p\sin\beta \leq V_{sp} \leq \left(f_{sd}A_s + f_{pd}A_p\right)\sin\beta \qquad (9.66)$$

where M_d = design value of the bending moment at the section under consideration.

V_{cs} = shear bearing capacity provided by the concrete and the stirrups in the oblique section, given by Eq. (9.63).

V_{sp} = projection of design value of the resultant force of the inclined prestressing tendons and reinforcing steels in the tension zone in the vertical plane.

z_{sv} = distance between the resultant force of the stirrups and the resultant force in the compression zone in the oblique section.

z_{sb} = distance between the resultant force of the bent-up prestressing tendons and the resultant force in the compression zone in the oblique section.

z = distance from the resultant force of the longitudinal prestressing tendons and reinforcing steels in the tension zone to the resultant force in the compression zone in the oblique section, $z \approx 0.9h_0$ is taken approximately.

β = inclination angle between the initially inclined longitudinal prestressing tendons and reinforcing steels in the tension zone in the oblique section.

c = horizontal projection length of the oblique section, $c \approx h_0$ is taken approximately.

To prevent the oblique section from inclined compression failure, the dimensions of the rectangular or T-sections under consideration should satisfy:

When $h_w/b \leq 4$

$$V_d \leq 0.25\beta_c f_{cd}bh_0 \qquad (9.67a)$$

When $h_w/b \geq 6$

$$V_d \leq 0.2\beta_c f_{cd}bh_0 \qquad (9.67b)$$

where V_d = maximum design value of shear force in the section considered.

β_c = influence coefficient of concrete strength, 1.0 is taken for concrete with grade C50 or less, 0.8 is taken for concrete with grade C80, and the values for intermediate strength between C50 and C80 can be obtained by the linear interpolation method.

h_w = web depth, the effective depth is taken for a rectangular section, the effective depth minus the flange depth is taken for a T-section, and the web depth is taken for an I-section.

When $4 < h_w/b < 6$, the intermediate values are determined by the linear interpolation method from Eq. (9.67).

9.10 Flexural bearing capacity of the oblique section

In Fig. 9.14, the flexure failure in the oblique section may occur due to bending from $T_{s,c}$ around point O when $T_{s,c}$ is not large enough. If the longitudinal prestressing tendons and reinforcing steels are bent up in reasonable positions and have sufficient anchor length, this kind of failure can be prevented in most cases.

Consider two typical critical oblique sections in the vicinity of the support of a simply supported beam and the middle support of a continuous beam, respectively, as shown in Fig. 9.18. It is assumed that the tensile stresses in stirrups, longitudinal prestressing tendons

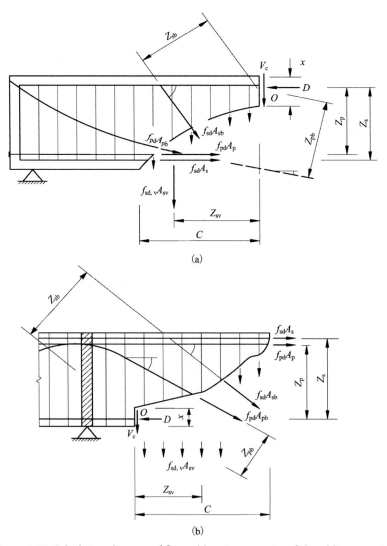

Figure 9.18 Calculation diagram of flexural bearing capacity of the oblique section.

and reinforcing steels, bent-up prestressing tendons, and reinforcing steels intersecting with the oblique section reach their design values at failure.

Establishing the moment equilibrium at a reference point O in Fig. 9.18A, the action point of the resultant of compression in concrete, the flexural bearing capacity of the oblique section can be derived as

$$M_u = f_{sd}A_sZ_s + f_{pd}A_pZ_p + \sum f_{sd}A_{sb}Z_{sb} + \sum f_{pd}A_{pb}Z_{pb} + \sum f_{sd,v}A_{sv}Z_{sv} \quad (9.68)$$

In this condition, the most disadvantageous horizontal projection length of the oblique section can be determined by the trial-and-error method from

$$V_d = \sum f_{sd}A_{sb}\sin\theta_s + \sum f_{pd}A_{pb}\sin\theta_p + \sum f_{sd,v}A_{sv} \quad (9.69)$$

where V_d = maximum design value of shear force in the oblique section.

A_{pb} = area of the inclined prestressing tendons.

A_{sb} = area of the inclined reinforcing steels.

A_p = area of the longitudinal prestressing tendons.

A_s = area of the longitudinal reinforcing steels.

A_{sv} = area of the stirrups.

f_{sd} = design tensile strength of the longitudinal reinforcing steel.

f_{pd} = design tensile strength of the longitudinal prestressing tendon.

$f_{sd,v}$ = design tensile strength of the stirrups.

θ_p = angle between the inclined prestressing tendons and the beam's axis.

θ_s = angle between the inclined reinforcing steels and the beam's axis.

Z_s = distance between the center of the longitudinal reinforcing steels to the resultant of compression in concrete.

Z_p = distance between the center of the longitudinal prestressing tendons to the resultant of compression in concrete.

Z_{sb} = distance between the center of the inclined reinforcing steels to the resultant of compression in concrete.

Z_{pb} = distance between the center of the inclined prestressing tendons to the resultant of compression in concrete.

Z_{sv} = distance between the center of stirrups to the resultant of compression in concrete.

When calculating the flexural bearing capacity of the oblique section within the anchorage length of the prestressing tendons, zero is replaced for f_{pd} at the end of the beam and f_{pd} is taken at the end of anchorage length in Eqs. (9.68) and (9.69), and the values for the intermediate position can be obtained by the linear interpolation method.

9.11 Example 9.4

A simply supported prestressed concrete beam with an effective span of 32.0 m is used in a mixed passenger and freight railway with a design speed of 160 km/h. The materials adopted are the same as in Example 9.1 and the sectional dimensions along the span are the same as shown in Fig. 9.8. In the quarter-span section (I—I section in Fig. 9.19), the stirrups (HPB300, $\phi10$) with two legs in a transverse section are arranged with a spacing of 200 mm, N1 and N2 prestressing tendons bend at 3 degrees to the beam's axis. The design value of shear force and the corresponding bending moment are 6.73×10^2 kN and 3.81×10^3 kN·m, respectively, and the decompression bending moment is 4.58×10^3 kN·m. If 1.2 is taken as the shear span ratio and the effective prestress in the prestressing tendons is 830 MPa, calculate and verify the shear bearing capacity of the quarter-span section according to the Q/CR 9300-2018 code.

9.12 Solution

9.12.1 Parameters

$f_{td} = 1.8$ MPa, $f_{cd} = 22.5$ MPa, $f_{sd,v} = 240$ MPa, $f_{pe} = 830$ MPa

$m = 1.2$, $s_v = 200$ mm, $b = 180$ mm

$A_{sv} = 2 \times \frac{\pi}{4} \times 10^2 = 157.1$ mm^2,

$\rho_{sv} = \dfrac{A_{sv}}{S_{sv}b} = \dfrac{157.1}{200 \times 180} = 0.004$, $A_p = 3 \times 1.26 \times 10^3 = 3.78 \times 10^3$ mm^2,

$A_{pb} = 2 \times 1.26 \times 10^3 = 2.52 \times 10^3$ mm^2

$a_p = \dfrac{3 \times 120 + 2 \times 350}{5} = 212$ mm

$h_0 = h - a_p = 2500 - 212 = 2288$ mm.

9.12.2 Verify the cross-sectional dimension

$0.25 f_{cd} b h_0 = 0.25 \times 22.5 \times 180 \times 2288 = 2316600\text{N} = 23.166 \times 10^2$ kN

Since. $V_d = 6.73 \times 10^2$ kN $< 0.25 f_{cd} b h_0 = 23.166 \times 10^2$ kN

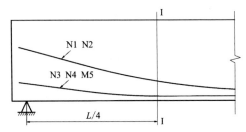

Figure 9.19 Arrangement of longitudinal prestressing tendons.

The cross-section meets the requirement for the minimum dimension prescribed in the code.

9.12.3 Calculate and verify the shear bearing capacity

$$\beta = 1 + \frac{M_0}{M} = 1 + \frac{4580}{3810} = 2.20$$

$\beta = 2.20 > 2.0$, so $\beta = 2.0$ is taken.

$$\rho = \frac{A_p + A_{pb} + A_s}{bh_0} = \frac{(3.78 + 2.52) \times 10^3}{170 \times 2288} = 0.02 < 0.03$$

Therefore, $p = 100\rho = 100 \times 0.02 = 2.0$ is taken.

$$\phi = 100 \text{ mm} < \frac{b_w}{8} = \frac{880}{8} = 110 \text{ mm}$$

Therefore, $b_w = 880$ mm is taken.
Hence,

$$\alpha = 1 + \frac{h_f^2}{b_w h_0} = 1 + \frac{150 \times 150}{880 \times 2288} = 1.01$$

$$V_c = \frac{0.6}{m}(1 + 0.35p)f_{td}bh_0\beta\alpha$$

$$= \frac{0.6}{1.2} \times (1 + 0.35 \times 2) \times 1.8 \times 180 \times 2288 \times 2.0 \times 1.01 = 519.32 \text{ kN}$$

$$V_{sv} = 0.8(0.35 + 0.4m)\rho_{sv}f_{sd,v}bh_0$$

$$= 0.8 \times (0.35 + 0.4 \times 1.2) \times 240 \times 0.004 \times 180 \times 2288 = 262.52 \text{ kN}$$

$$\sigma_{pe}A_{pb}\sin\theta_p = 830 \times 2.52 \times 10^3 \times 3° = 109.47 \text{ kN}$$

$$V_u = V_c + V_{sv} + 0.8f_{sd}A_{sb}\sin\theta_s + 0.8\sigma_{pe}A_{pb}\sin\theta_p = 519.32 + 262.52 + 87.58$$
$$= 869.42 \text{ kN}$$

Since $V_u = 869.42$ kN $> V_d = 673.0$ kN.
The shear bearing capacity of the oblique section in the quarter-span meets the requirement prescribed in the Q/CR 9300-2018 code.

9.13 Example 9.5

A simply supported highway beam with an effective span of 24.2 m with materials and sections as in Example 9.2 is used. In the section away from the support with a distance of half the depth of the section, the dimension and reinforcement at the section are shown in Fig. 9.20, and the stirrups (HPB300 $\phi 10$) with two legs in a transverse section are arranged with a spacing of 100 mm, all prestressing tendons bend at 8 degrees to the beam's axis. The design value of shear force is $V_d = 895.20$ kN. Calculate and verify the shear bearing capacity of this section according to the JTG 3362-2018 code.

9.14 Solution

9.14.1 Parameters

$f_{cu,k} = 40$ MPa, $f_{sd,v} = 195$ MPa

$s_v = 100$ mm, $b = 400$ mm

$A_{pb} = 2763.6$ mm^2, $A_{sv} = 2 \times \frac{\pi}{4} \times 10^2 = 157.08$ mm^2

$a_p = \dfrac{1210 + 400}{2} = 805$ mm

$h_0 = h - a_p = 1750 - 805 = 945$ mm.

9.14.2 Verify the cross-sectional dimension

$$0.51 \times 10^{-3}\sqrt{f_{cu,k}}bh_0 = 0.51 \times 10^{-3} \times \sqrt{40} \times 400 \times 945 = 1219.24 \text{ kN}$$

Since $V_d = 895.2$ kN $< 0.51 \times 10^{-3}\sqrt{f_{cu,k}}bh_0 = 0.51 \times 10^{-3} \times \sqrt{40} \times 400 \times 945 = 1219.24$ kN

The cross-section meets the requirement for the minimum dimension prescribed in the code.

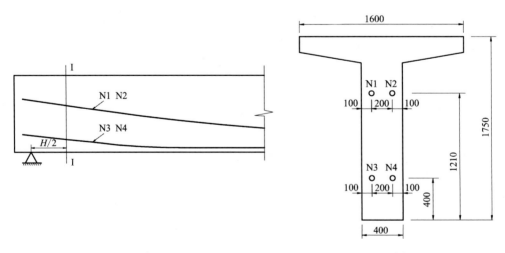

(a) Part of a simply supported highway beam   (b) I–I section

Figure 9.20 Arrangement of longitudinal prestressing tendons.

9.14.3 Calculate and verify the shear bearing capacity

$$p = 100\rho = \frac{100(A_p + A_{pb})}{bh_0} = \frac{100 \times 2763.6}{400 \times 945} = 0.731$$

$$\rho_{sv} = \frac{A_{sv}}{s_v b} = \frac{157.08}{100 \times 400} = 0.003927$$

$$V_{cs} = 0.45\alpha_1\alpha_2\alpha_3 bh_0 \sqrt{(2 + 0.6p)\sqrt{f_{cu,k}}\rho_{sv}f_{sd,v}}$$

$$= 0.45 \times 1.0 \times 1.25 \times 1.1 \times 400 \times 945$$

$$\times \sqrt{(2 + 0.6 \times 0.731)\sqrt{40} \times 0.003927 \times 195 \times 10^{-3}}$$

$$= 803.78 \text{ kN}$$

$$V_{pb} = 0.75 \times f_{pd} \times \sum A_{pb} \sin\theta_p$$

$$= 0.75 \times 1260 \times 4 \times 690.9 \times \sin 8° \times 10^{-3} = 363.46 \text{ kN}$$

$$V_u = V_{cs} + V_{pb} = 803.78 + 363.46 = 1167.24 \text{ kN}$$

Since $V_u = 1167.24$ kN $> V_d = 895.2$ kN.

The shear bearing capacity of the oblique section meets the requirement prescribed in the JTG 3362-2018 code.

9.15 Example 9.6

Consider a post-tensioned beam with a total length of 12 m as in Example 9.3. The dimension and reinforcement in the quarter-span section are shown in Fig. 9.21, the stirrups (HPB300, $\phi 8$) with two legs in a transverse section are arranged with a spacing of 200 mm, and the design value of shear force is 238.5 kN. The prestressing tendon is arranged with a straight line, and the resultant of the longitudinal prestressing tendons and reinforcing steels in the tension zone is $N_{p0} = 612.4$ kN when the normal stress of concrete is zero. Calculate and verify the shear bearing capacity of the quarter-span section according to the GB 50010-2010 code.

9.16 Solution

9.16.1 Parameters

$f_{cu,k} = 40$ MPa, $f_{cd} = 23.1$ MPa, $f_{sd,v} = 210$ MPa, $f_{pd} = 1320$ MPa

$s_v = 200$ mm, $b = 400$ mm, $A_p = 548.0$ mm^2, $A_0 = 128149$ mm^2, $a_p = 60$ mm

$h_0 = h - a_p = 1050 - 60 = 990$ mm.

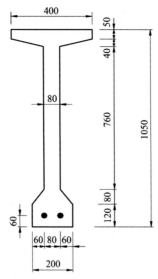

Figure 9.21 Arrangement of longitudinal prestressing tendons.

9.16.2 Verify the cross-sectional dimension

$$\frac{h_w}{b} = \frac{760}{80} = 9.5 > 6$$

$0.2\beta_c f_c b h_0 = 0.2 \times 1.0 \times 23.1 \times 80 \times 990 = 365904$ N.
$0.3 f_{cd} A_0 = 0.3 \times 21.3 \times 128149 = 818872$ N.
Since $V_d = 238.5$ kN $< 0.2\beta_c f_c b h_0 = 365.904$ kN

$$N_{p0} = 612.4 \text{ kN} < 0.3 f_{cd} A_0 = 818.87 \text{ kN}$$

The cross-section meets the requirement for the minimum dimension prescribed in the code.

9.16.3 Calculate and verify the shear bearing capacity

$$V_{cs} = \alpha_{cv} f_{td} b h_0 + f_{sd,v} \frac{A_{sv}}{s_v} h_0$$

$$= 0.7 \times 1.89 \times 80 \times 990 + \frac{210 \times 2 \times 50.3}{200} \times 990 = 209355 \text{ N} = 209.4 \text{ kN}$$

$$V_p = 0.05 N_{p0} = 0.05 \times 612.4 = 30.6 \text{ kN}$$

$$V_u = V_{cs} + V_p = 209.4 + 30.6 = 240.0 \text{ kN}$$

Since $V_u = 240.0$ kN $> V_d = 238.5$ kN.

The shear bearing capacity of the oblique section meets the requirement prescribed in the GB 50010-2010 code.

9.17 Torsion bearing capacity in the beams

9.17.1 General concepts of torsional failure and torsional strength

When a curved continuous prestressed concrete beam comes under traffic loads, or a prestressed concrete flexural member is subjected to eccentric loads about the longitudinal central axis (such as the traffic loads locate at the eccentric lanes), the sections are twisted and additional stresses are generated by the torsion in the section apart from the normal stress and shear stress due to bending moment and shear force.

The torsion can be divided into two different types. One is balanced torsion, in which the torque is independent of the torsional stiffness of the components of the section and it can be calculated directly from the balanced conditions. The other is coordinated torsion, in which the torque relates to the torsional stiffness of components, and the torque will reduce due to internal force redistribution in the section once cracking occurs. The curved continuous prestressed beams in the highway and railway and frame beams in the building are the typical coordinated torsion structures.

Box, rectangular, T-, and I-sections are usually adopted in the prestressed concrete torsional members. Under twisting, shear stress and principal stress are generated in the section. When the maximum tensile principal stress exceeds the tensile strength of concrete, cracking occurs perpendicular to the tensile principal stress. For rectangular or box sections without reinforcement, cracks first develop at about 45° to the beam's axis near the midpoint of the long side and propagate quickly to other short, eventually leading to the spatial warping destroy in which three sides fail due to tension and one side fails due to compression. If the reinforcement is arranged along the direction of tensile principal stress, it can not only delay the onset of cracking but also improve the torsion bearing capacity. Considering the difficulty in arranging the rotating reinforcement, the closed transverse stirrups, and inclined prestressing tendons or reinforcing steels are commonly employed. In addition, the tensile principal stress due to torsion can be reduced by increasing the longitudinal prestressing force effectively, consequently, the torsion bearing capacity can be improved by the longitudinal prestressing, and research shows that it can improve by up to 2.5 times.

Similar to the failure patterns due to the bending moment, a prestressed concrete member subjected to torsion may fail by under-reinforced, balanced, and over-reinforced patterns. When excessive transverse stirrups and longitudinal reinforcements are arranged, the member fails abruptly with very few torsional cracks, which is a typical brittle failure pattern, so the over-reinforced section must be avoided in the design.

Under the combined action of bending moment, shear force, and torsional moment, the mechanism in the section of the prestressed concrete member is complex. Its failure pattern and bearing capacity are not only related to the internal factors such as section shape, reinforcement, prestress level, and material strength, but also to the external loading conditions such as the ratio of the torsional moment to bending moment and the ratio of torsion to shear force, etc. Generally, the normal cracks caused by the bending moment develop first when the torsion-bending ratio is small, finally resulting in bending failure. Meanwhile, the torsional cracking occurs first when the torsion-bending ratio or torsion-shear ratio is large, finally resulting in torsional failure. If both torsional moment and shear force are large, the shear-torsion failure will occur. Experimental results show that bending, shearing, and twisting interact, and it is difficult to use a unified function to describe their contributions to the bearing capacity of the section. Hence, the simplified approach is that the flexural, shear, and torsional bearing capacities are calculated alone, then they are combined by the correlation or superposition method to attain the total bearing capacity of the section being commonly adopted.

For prestressed flexural members subjected to torsion, the torsion bearing capacity and shear bearing capacity should satisfy the requirements specified in the codes:

$$T_d \leq T_u \tag{9.70}$$

$$V_d \leq V_u \tag{9.71}$$

where T_d = design value of the torsional moment.

T_u = torsion bearing capacity of the section considering the effect of shear.
V_d = design value of shear force.
V_u = shear bearing capacity of the oblique section considering the effect of torsion.

In addition, the correlated or superposed equation of T_u and V_u should also satisfy the provisions stipulated in the codes.

9.17.2 Ultimate strength of prestressed concrete members under pure torsion

The bearing capacity of prestressed concrete members with rectangular or box sections under pure torsion is provided by the transverse stirrups, longitudinal reinforcement, and concrete, including the contribution of the prestressing. Comparative experimental studies show that the increase of torsional bearing capacity by the prestressing is about $0.08 \frac{N_{p0}}{A_0} W_t$. Consider the adverse effects due to uneven distribution of actual stress in the section, in which $0.05 \frac{N_{p0}}{A_0} W_t$ is usually taken. The JTG 3362-2018 code specified that the bearing capacity of prestressed concrete members can be obtained by

$$T_u = 0.35 \beta_a f_{td} W_t + 1.2 \sqrt{\zeta} \frac{f_{sd,v} A_{sv1} A_{cor}}{s_v} + 0.05 \frac{N_{p0}}{A_0} W_t \tag{9.72}$$

$$\zeta = \frac{f_{sd}A_{st}s_v}{f_{sd,v}A_{sv1}U_{cor}} \tag{9.73}$$

where ζ = strength ratio of longitudinal reinforcement to transverse stirrups, 1.7 is taken if $e_{p0} \leq h/6$ and $\zeta \geqslant 1.7$.

β_a = influence coefficient of the wall thickness of the section; $\beta_a = 1.0$ for a rectangular section, β_a should be calculated by Eq. (9.74) for a box section.

b = width of a rectangular or box section.

h = depth of a rectangular or box section.

t_1 = wall thickness of the long side of a box section.

t_2 = wall thickness of the short side of a box section.

W_t = polar section modulus of plasticity.

A_{sv1} = area of a single leg of the stirrup in the calculation of pure torsion.

f_{td} = design tensile strength of concrete.

$f_{sd,v}$ = design tensile strength of stirrup.

f_{sd} = design tensile strength of longitudinal reinforcing steel.

A_{st} = total area of longitudinal reinforcing steels arranged symmetrically around the periphery of the section.

A_{cor} = core area of the section surrounded by the inner surface of the stirrups $A_{cor} = b_{cor}h_{cor}$, where b_{cor} and h_{cor} are the short side length and the long side length of the core area, respectively.

U_{cor} = perimeter of the core area of the section, $U_{cor} = 2(b_{cor} + h_{cor})$.

s_v = spacing between two adjacent stirrups.

N_{p0} = generalized prestressing force, $N_{p0} = 0.3f_{cd}A_0$ if $N_{p0} > 0.3f_{cd}A_0$.

e_{p0} = eccentricity of N_{p0} with respect to the centroid of the converted section.

A_0 = area of the converted section.

In Eq. (9.72), if $e_{p0} \leq h/6$ or the value of ζ obtained from Eq. (9.73) is greater than 1.7, the effect of the prestressing should be disregarded, and the member is treated as the reinforced concrete member where $\zeta = 1.7$ is taken for the calculation of the torsion bearing capacity.

The JTG 3362-2018 code specified that $t_2 \geq 0.1b$ and $t_1 \geq 0.1b$ should be satisfied when Eq. (9.72) is used.

The GB 50010-2010 code specified that the influence coefficient of the wall thickness of the section in Eq. (9.72) should be calculated by

$$\beta_a = \frac{2.5t_w}{b_h} \tag{9.74}$$

where t_w = wall thickness of the box section, which should be great than $b_h/7$.

b_h = depth of the rectangular or box section.

The JTG 3362-2018 specifies that the influence coefficient of the wall thickness of the section in Eq. (9.72) can be chosen by:

If $0.1b \leq t_2 \leq 0.25b$ or $0.1h \leq t_1 \leq 0.25h$

$$\beta_a = \min\left(4\frac{t_2}{b}, 4\frac{t_1}{h}\right) \tag{9.75a}$$

If $t_2 > 0.25b$ and $t_1 > 0.25h$

$$\beta_a = 1.0 \tag{9.75b}$$

The polar section modulus of plasticity of a rectangular section (Fig. 9.22A) is given by

$$W_t = \frac{b^2}{6}(3h - b) \tag{9.76}$$

The polar section modulus of the plasticity of the box section (Fig. 9.22B) is given by

$$W_t = \frac{b^2}{6}(3h - b) - \frac{(b - 2t_1)^2}{6}[3(h - 2t_2) - (b - 2t_1)] \tag{9.77}$$

For T- and I-sections under pure torsion, dividing the section into several rectangular components first, then the torsional bearing capacity of each part can be calculated by Eq. (9.72), and the total torsional bearing capacity is obtained by adding them together.

Experiments show that the stress in longitudinal reinforcements and stirrups can almost reach the yield strengths when the section fails due to torsion if the strength ratio

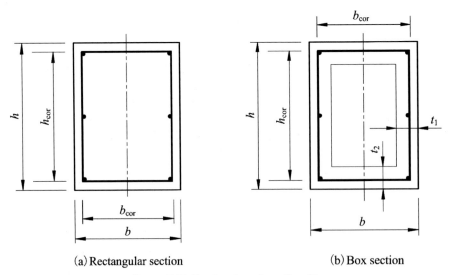

(a) Rectangular section   (b) Box section

Figure 9.22 Torsional sections ($h > b$).

of longitudinal reinforcement to stirrup is within 0.5–2.0. In the codes, ζ is limited to 0.6–1.7 generally, and $\zeta = 1.2$ is commonly adopted in design.

To avoid over-reinforced failure under pure torsion, the ratio of torsional reinforcement should be limited, from another point of view, the section dimension should be sufficiently large, that is

$$\frac{T_d}{W_t} \leq 0.51 \times 10^{-3} \sqrt{f_{cu,k}} \tag{9.78}$$

where $f_{cu,k}$ = characteristic value of cubic compressive strength of concrete.

9.17.3 Ultimate strength of prestressed concrete members under combined bending, shear, and torsion

Under the combined action of bending, shear, and torsion, the bearing capacity of prestressed concrete members with rectangular or box sections can be achieved by the simplified approach, in which the correlation effect is taken into account for concrete resistance under shear-torsion and the contribution of reinforcement is added by superposition.

9.17.3.1 Bearing capacity of prestressed concrete members under shear-torsion

As discussed previously, the approach in calculating the shear bearing capacity in different codes is different, hence, the corresponding formulas to calculate the bearing capacity of the section under shear-torsion are also different.

The correlation of shear-torsion bearing capacity of a pure concrete member

Tests on members without web reinforcement under combined shear and torsion show that the torsion bearing capacity decreases with the increase of shear force. Similarly, the shear bearing capacity decreases with the increase of torsion. Their correlation can be approximately described by a quarter circle (Fig. 9.23), expressed as

$$\left(\frac{V_d}{V_{d0}}\right)^2 + \left(\frac{T_d}{T_{d0}}\right)^2 = 1 \tag{9.79}$$

where V_d = design value of shear force under shear-torsion.

T_d = design value of torsional moment under shear-torsion.

V_{d0} = design value of shear force under pure shear force.

T_{d0} = design value of torsional moment under pure torsion.

To simplify the analysis, the correlation curve in Fig. 9.23 is replaced approximately by three polygons, AB, BC, and CD, namely:

When $\frac{T_d}{T_{d0}} \leq 0.5$ (AB): $\frac{V_d}{V_{d0}} = 1$ (no reduction for shear bearing capacity)

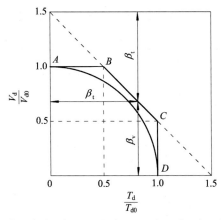

Figure 9.23 Correlation between shear and torsion bearing capacity.

When $0.5 < \frac{T_d}{T_{d0}} < 1$ (BC):

$$T_d = \beta_t T_{d0} \tag{9.80a}$$

$$V_d = \beta_v V_{d0} \tag{9.80b}$$

When $\frac{V_d}{V_{d0}} \leq 0.5$ (CD): $\frac{T_d}{T_{d0}} = 1$ (no reduction for torsion bearing capacity)

Substituting $V_{d0} = 0.7f_{td}bh_0$ obtained from Eq. (9.63a) where $A_{sv} = 0, \alpha_{cv} = 0.7$, and $T_{d0} = 0.35f_{td}W_t$ obtained from Eq. (9.72) into the geometric relation in Fig. 9.23, gives

$$\beta_t = \frac{1.5}{1 + \dfrac{V_d T_{d0}}{V_{d0} T_d}} = \frac{1.5}{1 + 0.5\dfrac{V_d}{T_d}\dfrac{W_t}{bh_0}} \tag{9.81}$$

$$\beta_v = 1 - \beta_t \tag{9.82}$$

where β_t = reduction coefficient for torsion bearing capacity in the pure concrete member under shear-torsion, $\beta_t = 0.5$ if $\beta_t < 0.5$ and $\beta_t = 1.0$ if $\beta_t > 1$.

W_t = polar section modulus of plasticity, it is replaced by $\beta_a W_t$ for a box section under shear-torsion.

b = width of a rectangular or box section.

h_0 = effective width of rectangular or box section parallel to the acting plane of bending moment.

Bearing capacity of prestressed concrete members under shear-torsion in GB 50010

For general prestressed members with rectangular and box sections under shear-torsion, the shear capacity and torsional capacity of the building members can be obtained from Eqs. (9.62), (9.63a), and (9.72), as

$$V_u = (1.5 - \beta_t)\left(0.7 f_{td} b h_0 + 0.05 N_{p0}\right) + f_{sd,v} \frac{A_{sv1}}{s_v} h_0 \tag{9.83}$$

$$T_u = \beta_t \left(0.35 \beta_a f_{td} + 0.05 \frac{N_{p0}}{A_0}\right) W_t + 1.2 \sqrt{\zeta} \, \frac{f_{sd,v} A_{sv1} A_{cor}}{s_v} \tag{9.84}$$

For the sections of an independent beam under concentrated loads where the shear force produced by the concentrated loads is more than 75% of the total shear force, the shear capacity and torsional capacity can be obtained from Eqs. (9.62), (9.63b), and (9.72), as

$$V_u = \left(1.5 - \beta_{t,m}\right)\left(\frac{1.75}{m+1} f_{td} b h_0 + 0.05 N_{p0}\right) + f_{sd,v} \frac{A_{sv1}}{s_v} h_0 \tag{9.85}$$

$$T_u = \beta_{t,m}\left(0.35 \beta_a f_{td} + 0.05 \frac{N_{p0}}{A_0}\right) W_t + 1.2 \sqrt{\zeta} \, \frac{f_{sd,v} A_{sv1} A_{cor}}{s_v} \tag{9.86}$$

$$\beta_{t,m} = \frac{1.5}{1 + 0.2(m+1)\dfrac{V_d}{T_d}\dfrac{W_t}{b h_0}} \tag{9.87}$$

where m = shear span ratio, given by Eq. (9.39), $m = 1.5$ when $m < 1.5$ and $m = 3$ when $m > 3$.

Bearing capacity of prestressed concrete members under shear-torsion in JTG 3362-2018

In the JTG 3362–2018 code, the shear capacity in the oblique section is provided by the concrete and the stirrups respectively, so it is necessary to introduce an equivalent coefficient α_{cs} in Eq. (9.59) to consider the reduction influence for shear capacity by torsion as expressed in Eq. (9.80b), that is

$$\alpha_{cs} V_{cs} = \beta_v V_c + V_{sv} = (1 - \beta_t) V_c + V_{sv}$$

Based on the comparative calculations, the equivalent reduction coefficient is obtained from which

$$\alpha_{cs} = \frac{10 - 2\beta_t}{9} \tag{9.88}$$

Hence, the shear capacity and torsional capacity of the prestressed concrete members with rectangular or box sections under shear-torsion are obtained from Eqs. (9.59), (9.72), and (9.88), as

$$V_u = 0.5 \times 10^{-4}\alpha_1\alpha_2\alpha_3(10 - 2\beta_t)bh_0\sqrt{(2 + 0.69)\sqrt{f_{cu,k}\rho_{sv}f_{sd,v}}} \tag{9.89}$$

$$T_u = \beta_t\left(0.35\beta_a f_{td} + 0.05\frac{N_{p0}}{A_0}\right)W_t + 1.2\sqrt{\zeta}\,\frac{f_{sd,v}A_{sv1}A_{cor}}{S_v} \tag{9.90}$$

9.17.3.2 Bearing capacity of prestressed concrete members under combined bending, shear, and torsion

The comprehensive resistance of prestressed concrete members under combined bending, shear, and torsion can be achieved by adequate section dimensions and reinforcements that provide sufficient flexural bearing capacity, shear bearing capacity, and correlated shear-torsion bearing capacity as stipulated in the codes.

For prestressed members under combined bending, shear, and torsion, the GB 50010 code stipulates that the dimension of the box section with $h_w/b_w \leq 6$, the rectangular or T-sections with $h_w/b \leq 6$ should satisfy the following provisions:

When $\dfrac{h_w}{b} \leq 4$ or $\dfrac{h_w}{b_w} \leq 4$

$$\frac{V_d}{bh_0} + \frac{T_d}{0.8W_t} \leq 0.25\beta_c f_{cd} \tag{9.91}$$

When $\dfrac{h_w}{b} = 6$ or $\dfrac{h_w}{b_w} = 6$

$$\frac{V_d}{bh_0} + \frac{T_d}{0.8W_t} \leq 0.20\beta_c f_{cd} \tag{9.92}$$

where b = width of the rectangular, box, T-, or I-section.

h_0 = effective depth of the section.

β_c = influence coefficient of concrete strength, 1.0 is taken for C50 and lower grade concrete and 0.8 for C80 concrete, respectively, and the values for intermediate strength between C50 and C80 can be obtained by the linear interpolation method.

f_{cd} = design strength of concrete in compression.

h_w = web depth, the effective section depth h_0 is taken for a rectangular section, the effective section depth minus flange thickness is taken for a T-section and net web depth is taken for an I-section.

t_w = wall thickness of the box section, which should be great than $b_h/7$, where b_h is the section width.

When $4 < \dfrac{h_w}{b} < 6$ or $4 < \dfrac{h_w}{b_w} < 6$, the correlation of shear-torsion bearing capacity can be obtained by interpolation method from Eqs. (9.91) and (9.92).

When $\dfrac{h_w}{b} > 6$ or $\dfrac{h_w}{b_w} > 6$, the requirement for section dimension and calculation of bearing capacity should conform to the provisions in the code.

For the members subjected to combined bending, shear, and torsion, the verification for bearing capacity is not required and the reinforcement is arranged only according to the structural requirement if the following equations hold:

$$\frac{V_d}{bh_0} + \frac{T_d}{W_t} \leq 0.7 f_{td} + 0.05 \frac{N_{p0}}{bh_0} \tag{9.93a}$$

Or

$$\frac{V_d}{bh_0} + \frac{T_d}{W_t} \leq 0.7 f_{td} + 0.07 \frac{N}{bh_0} \tag{9.93b}$$

where N = design values of axial force corresponding to the design values V_d and T_d, $N = 0.3 f_{td} A$ is taken if $N > 0.3 f_{td} A$, where A is the section area of the member.

For prestressed concrete members under combined bending, shear, and torsion, the JTG 3362-2018 code stipulates that the dimension of the box, rectangular, or T-section should satisfy:

$$\frac{V_d}{bh_0} + \frac{T_d}{W_t} \leq f_{cv} \tag{9.94}$$

If Eq. (9.95) holds, the verification for torsion bearing capacity is not required and the reinforcement is arranged only according to the structural requirement:

$$\frac{V_d}{bh_0} + \frac{T_d}{W_t} \leq 0.50 \alpha_2 f_{td} \tag{9.95}$$

where f_{cv} = design value of nominal shear stress, $f_{cv} = 0.51 \sqrt{f_{cu,k}}$.

$f_{cu,k}$ = characteristic value of cubic compressive strength of concrete.

f_{td} = design tensile strength of concrete.

α_2 = increasing coefficient by the prestressing, 1.25 is taken generally for the prestressed concrete flexural members; while 1.0 is taken when the bending moment due to the resultants of reinforcement and external bending moment have the same direction, or the flexural members allow cracking.

9.18 Example 9.7

A prestressed concrete member with a rectangular section is subjected to transverse uniform load and torsion. The design value of shear force is 721.1 kN and the design value of torque is 22.3 kN·m. A designed rectangular section with C40 concrete is shown in Fig. 9.24, where two groups of bonded strands $4\phi^S 12.7$ (4-7ϕ4) with an eccentricity of 142 mm and $9\phi 16$ (HRB335) longitudinal reinforcing steel bars are arranged. The

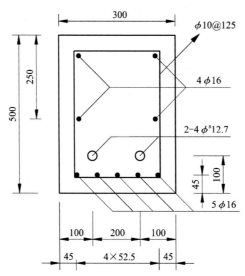

Figure 9.24 Dimension and reinforcement of rectangular section.

closed transverse stirrups (HRB335 ϕ10) with a spacing of 125 mm are arranged around the longitudinal steel bars. The characteristic tensile strength of the strand is 1860 MPa. Calculate and verify the torsion bearing capacity of this section according to the GB 50010-2010 code.

9.19 Solution

9.19.1 Parameters

$f_{cd} = 19.1$ MPa, $f_{td} - 1.71$ MPa, $f_{ptk} = 1860$ MPa, $f_{pd} = 1320$ MPa, $f_{sd} = 300$ MPa
$A_s = 1809$ mm^2, $A_p = 790$ mm^2, $a_p = 100$ mm, $a_s = 45$ mm, $s_v = 125$

$$a = \frac{f_{pd}A_p a_p + f_{sd}A_s a_s}{f_{pd}A_p + f_{sd}A_s} = \frac{1320 \times 790 \times 100 + 300 \times 1809 \times 45}{1320 \times 790 + 300 \times 1809} = 81.17 \text{ mm}$$

$h_0 = h - a = 500 - 81 = 419$ mm.

9.19.2 Calculation and verification of bearing capacity

$$0.35 f_{td} b h_0 = 0.35 \times 1.71 \times 300 \times 419 = 75231 \text{ N}$$

Since $0.35 f_{td} b h_0 = 75231 \text{ N} > V_d = 72100 \text{ N}$.

This member only needs verification of the flexural bearing capacity and the torsional bearing capacity. In this example, only the torsional bearing capacity is chosen to be verified.

$$W_t = \frac{b^2}{6}(3h - b) = \frac{300^2}{6} \times (3 \times 500 - 300) = 1.8 \times 10^7 \text{ mm}^3$$

$$A_{cor} = b_{cor}\Delta h_{cor} = (500 - 37 \times 2) \times (300 - 37 \times 2) = 96276 \text{ mm}^2$$

Longitudinal reinforcing steel bars:

$$A_{st} = 6 \times \frac{\pi}{4} \times 16^2 = 1206 \text{ mm}^2$$

Stirrups:

$$A_{sv1} = \frac{\pi}{4} \times 10^2 = 79 \text{ mm}^2, \, s_v = 125 \text{ mm}$$

$$u_{cor} = 2(b_{cor} + h_{cor}) = 2 \times [(500 - 37 \times 2) + (300 - 37 \times 2)] = 1304 \text{ mm}$$

$$\zeta = \frac{f_{sd} A_{st} S_v}{f_{sd,v} A_{sv1} U_{cor}} = \frac{300 \times 1206 \times 125}{210 \times 79 \times 1304} = 2.09 > 1.7$$

Therefore, $\zeta = 1.7$ is taken.

Since $e_{p0} = 142 \text{ mm} > \frac{h}{6} = \frac{500}{6} = 83.3 \text{ mm}$.

The effect of the prestressing should be disregarded, and the member is treated as the reinforced concrete member where $\zeta = 1.7$ is taken in the calculation of the torsion bearing capacity.

$$T_u = 0.35 f_{td} W_t + 1.2 \sqrt{\zeta} \, \frac{f_{sd,v} A_{sv1} A_{cor}}{S_v}$$

$$= 0.35 \times 1.71 \times 1.8 \times 10^7 + 1.2 \times \sqrt{1.7} \times 210 \times \frac{79 \times 96276}{125}$$

$$= 30765178 \text{ N} \cdot \text{mm} = 30.76 \text{ kN} \cdot \text{m}$$

Since $T_u = 30.76 \text{ kN} \cdot \text{m}, > T_d = 22.3 \text{ kN} \cdot \text{m}$.

The torsion bearing capacity of the member meets the requirement prescribed in the code.

9.20 Bearing capacity of anchorage zone under local compression

9.20.1 Failure mechanism of the local zone under local compression

As shown in Fig. 7.15, the shaded area of concrete under the anchorage device, the local zone, bears huge local compression due to the concentrated pressing force, and the lateral deformation of concrete is confined by the surrounding concrete, namely, the concrete in the local zone is in the state of three-dimensional compression. Research shows that the compressive strength of concrete increases due to the lateral confines. The hooping strengthening theory treats the lateral restraint effect by the surrounding concrete as hooping for the local zone, once the surrounding concrete cracks, the hooping effect will disappear and the local zone concrete finally bursts. The hooping strengthening theory has been used to explain the behavior and failure mechanism of confined concrete under local compression for a long time. However, further investigation shows that the local compression can continue to increase even after cracking occurs in the surrounding concrete, demonstrating that the partial lateral restraint effect still exists by the cracked concrete, that is, the hooping strengthening theory cannot completely and exactly explain the failure mechanism of confined concrete under local compression.

Figs. 7.14D and 7.15 show that the maximum transverse tensile stress of concrete under local compression appears in the region at a certain distance away from the anchorage device, where it will develop bursting cracking first, as shown in Fig. 9.25B. The local compression can still increase even after cracking occurs. As the compression increases, the crack propagates up and downward, but it will bifurcate into two oblique cracks at the top end as the crack extends to a certain height, as shown in Fig. 9.25C. This illustrates that the shear force in the oblique direction provides the resistance for the development of inclined cracks, thus the shear theory is presented to explain the failure mechanism of confined concrete under local compression.

In the analytical model of shear theory, the local zone works as an arch structure with multiple transverse tie rods inside. Under compression, the generated transverse tension in concrete is counterbalanced by the virtual tie rods before cracking which reflects the restraint by the surrounding concrete. As the local compression increases, the tie rods in the bottom region of the vault where the transverse tensile stress is maximum break first, then the vertical crack develops correspondingly, and the position of the resultant of the tie rods moves downward. As the local compression increases further, the effect of tie rods diminishes and the crack propagates up and downward. As shown in Figs. 7.14B and 7.15, the local zone bears compression longitudinally and transversely, thus the bursting crack propagates upward with an inclined direction, as shown in Fig. 9.25C. When the ratio of tension in the tie rods to local compression reaches a certain value, the concrete in the local zone gradually forms a wedge of shear failure and loses the bearing capacity finally.

The hooping strengthening theory is adopted in the Q/CR 9300-2018 code and the shear theory is adopted in the JTG 3362-2018 and GB 50010-2010 codes to calculate the bearing capacity of the local zone under local compression.

9.20.2 Compressive bearing capacity of the local zone

For prestressed concrete members, the bearing capacity of the local zone under local compression should satisfy

$$F_{ld} \leq F_{lu} \tag{9.96}$$

where F_{ld} = design value of the local compression due to the concentrated prestressing force.

F_{lu} = compressive bearing capacity of the local zone.

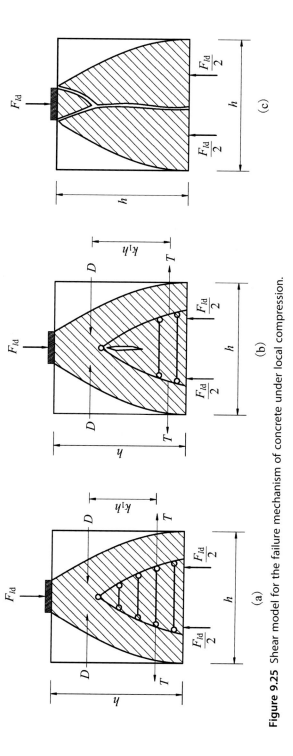

Figure 9.25 Shear model for the failure mechanism of concrete under local compression.

9.20.2.1 Compressive strength of confined concrete under local compression

The design value of the compressive strength of confined concrete under local compression can be obtained by

$$f_{l,cd} = \beta f_{cd} \tag{9.97}$$

where f_{cd} = design strength of concrete in compression.

$f_{l,cd}$ = design strength of confined concrete in compression under local compression.

β = strength-increasing factor of concrete due to lateral restraint.

The strength-increasing factor of concrete can be derived by combining the shear theory or the hooping strengthening theory with experimental data, as

$$\beta = \sqrt{\frac{A_b}{A_l}} \tag{9.98}$$

where A_b = calculated bottom area of concrete under local compression.

A_l = area of concrete directly under local compression.

The determination of A_b should conform to the principle that the centroids of A_b and A_l should coincide and the features of both are consistent.

The Q/CR 9300-2018 code stipulates that A_b can be calculated from Fig. 9.26 (the area enclosed by the dashed lines is A_b), the JTG 3362-2018 and GB 50010-2010 codes stipulate that A_b can be calculated from Fig. 9.27.

9.20.2.2 Compressive bearing capacity of the local zone

Additional helical reinforcements or transverse reinforcements are usually arranged to improve the crack resistance and the compression bearing capacity of the local zone, as shown in Figs. 3.7 and 3.15. These additional reinforcements are called indirect reinforcements. Hence, the local compressive bearing capacity is provided by the concrete and the indirect reinforcements, that is

$$F_l = F_{l,c} + F_{l,s} \tag{9.99}$$

where $F_{l,c}$ = bearing capacity provided by the confined plain concrete.

$F_{l,s}$ = bearing capacity provided by the indirect reinforcements.

The bearing capacity provided by the confined plain concrete can be attained as

$$F_{l,c} = f_{l,cd} A_{ln} = \beta f_{cd} A_{ln} \tag{9.100}$$

where A_{ln} = net area of concrete under the anchorage device, and the area of holes inside the device should be deducted.

The bearing capacity provided by the indirect reinforcements can be attained by the shear theory or the hooping strengthening theory.

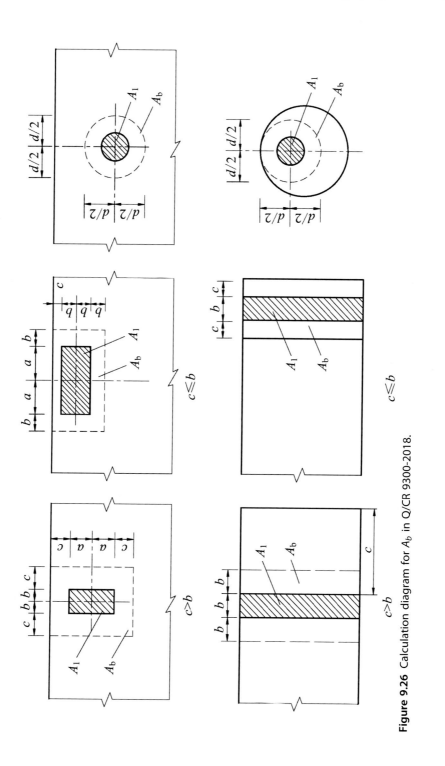

Figure 9.26 Calculation diagram for A_b in Q/CR 9300-2018.

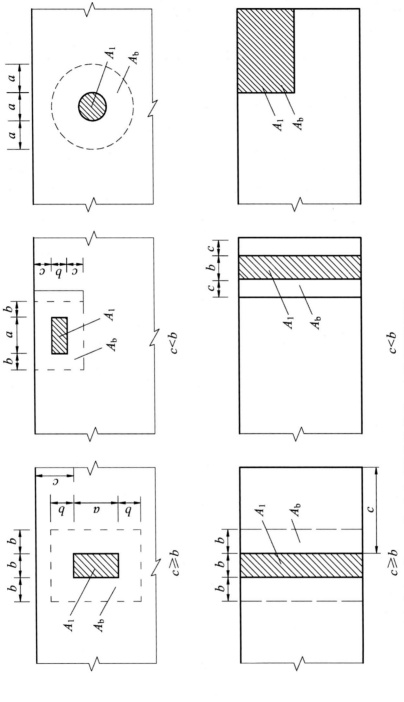

Figure 9.27 Calculation diagram for A_b in JTG 3362-2018 and GB 50010-2010.

Calculation of $F_{l,s}$ by the hooping strengthening theory

Consider a continuous helical reinforcing steel bar arranged in the local zone, as shown in Fig. 9.28A. Cut the helical reinforcement and the surrounding concrete in half along the symmetrical plane, as shown in Fig. 9.28B. It is assumed that the tensile stress in the helical reinforcement reaches the design value in the limit state. On the longitudinal section within one pitch, the tension in helical reinforcement is counterbalanced by the compression in the core concrete confined by the helical reinforcement, that is

$$2A_{ss1}f_{sd} = \sigma_r d_l s \tag{9.101}$$

where d_l = diameter of core concrete confined by the helical reinforcement.

    s = depth of one pitch.

    d_{cor} = diameter surrounded by the helical bar.

    A_{ss1} = area of the helical bar.

    f_{sd} = design tensile strength of the helical bar.

    σ_r = lateral radial pressure on concrete.

    The volume ratio of the helical bar can be obtained in one pitch, as

$$\rho_v = \frac{A_{ss1}\pi d_{cor}}{\frac{1}{4}\pi d_{cor}^2 s} = \frac{4A_{ss1}}{d_{cor}s} \tag{9.102}$$

where ρ_v = volume ratio of the helical bar.

    d_{cor} = diameter surrounded by the helical bar.

    Substituting Eq. (9.102) into Eq. (9.101) yields

$$\sigma_r = \frac{1}{2}\rho_v\beta_{cor}f_{sd} \tag{9.103}$$

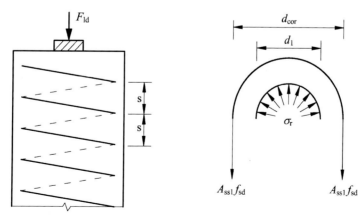

(a) A block subjected to local compression   (b) Analytical model for a helical steel bar

Figure 9.28 Calculation diagram for $F_{l,s}$ by the hooping strengthening theory.

where β_{cor} = increasing coefficient of concrete under local compression due to indirect reinforcement.

f_{sd} = design tensile strength of indirect reinforcement.

β_{cor} can be attained by

$$\beta_{cor} = \sqrt{\frac{A_{cor}}{A_l}} \tag{9.104}$$

where A_{cor} = area of the core concrete confined by the helical bar or the area of additional transverse reinforcement, $A_{cor} = A_b$ if $A_{cor} > A_b$.

Three-dimensional compression tests of a cylindrical specimen show that the compressive strength of confined concrete can increases up to 4 σ_r when the concrete is subjected to radial pressure σ_r, hence, the compression bearing capacity provided by the indirect reinforcement can be attained as

$$F_{l,s} = 4\sigma_r A_{ln} = 2\rho_v \beta_{cor} f_{sd} A_{ln} \tag{9.105}$$

Calculation of $F_{l,s}$ by the shear theory

The compression bearing capacity provided by the indirect reinforcement, $F_{l,s}$, can be attained from the shear theory along with Fig. 9.25. It is assumed that the tensile stress in most of the indirect reinforcements reaches the design value at shear failure, hence

$$T_s = w_s f_{sd} A_s \tag{9.106}$$

where w_s = incomplete coefficient considering that portion of the indirect reinforcement does not reach the design strength, 0.9 is taken generally.

A_s = total area of the indirect reinforcement.

Apply T_s on the sides of the wedge as transverse compression, as shown in Fig. 9.25C. According to the Mohr's theory of strength, the vertical (compression) bearing capacity provided by the indirect reinforcement can be deduced from the wedge subjected to lateral compression, as

$$F_{l,s} = k A_{s0} f_{sd} \tag{9.107}$$

where k = influence coefficient by the indirect reinforcement.

A_{s0} = converted section area of the indirect reinforcement.

When the helical reinforcement is employed, Eqs. (9.103), (9.104), and (9.107) give

$$F_{l,s} = k\rho_v \beta_{cor}^2 f_{sd} A_{ln} \tag{9.108}$$

When the reinforcing steel mats are employed, as shown in Fig. 9.29A, the area of core concrete is $A_{cor} = l_1 l_2$, and the volume ratio of indirect reinforcement is attained as

$$\rho_v = \frac{n_1 A_{s1} l_1 + n_2 A_{s2} l_2}{l_1 l_2 s} \tag{9.109}$$

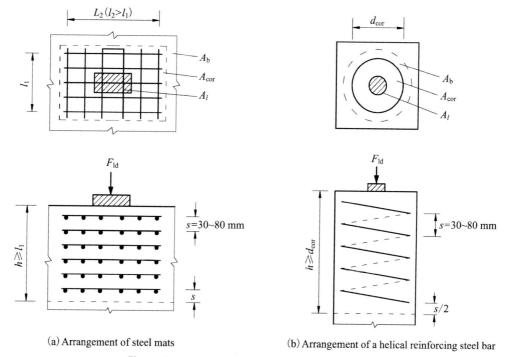

(a) Arrangement of steel mats   (b) Arrangement of a helical reinforcing steel bar

Figure 9.29 Arrangement of indirect reinforcement.

where ρ_v = volume ratio of indirect reinforcement.

n_1 = number of indirect reinforcement along the l_2 direction.

n_2 = number of indirect reinforcement along the l_1 direction.

A_{s1} = area of single indirect reinforcement along the l_1 direction.

A_{s2} = area of single indirect reinforcement along the l_2 direction.

s = layer spacing of indirect reinforcement.

Compressive bearing capacity of the local zone stipulated in the Q/CR 9300-2018 code

Substituting Eqs. (9.100) and (9.105) into Eq. (9.99), the compressive bearing capacity of the local zone of the prestressed concrete railway beams can be obtained:

$$F_{lu} = \frac{\beta f_{cd} A_{ln}}{\gamma_{c2}} + \frac{2\rho_v \beta_{cor} f_{sd} A_{ln}}{\gamma_s} \tag{9.110}$$

where F_{lu} = bearing capacity of the local zone under local compression.

γ_{c2} = comprehensive partial factor of bearing capacity of concrete under local compression, 1.02 is taken generally.

γ_s = comprehensive partial safety factor of bearing capacity of indirect reinforcement, 1.2 is taken generally.

f_{cd} = design strength of concrete in compression.

f_{sd} = design tensile strength of indirect reinforcement.

β = strength increasing factor of concrete due to lateral restraint.

β_{cor} = increasing coefficient due to indirect reinforcement.

ρ_v = volume ratio of indirect reinforcement.

$A_{\ln}$ = net area of concrete under the anchorage device, the area of holes inside the device shall be deducted.

The dimension of the local zone should be large enough to avoid cracking, which requires

$$F_{ld} \leq \frac{\beta f_{cd} A_{\ln}}{\gamma_{c1}} \tag{9.111}$$

where γ_{c1} = comprehensive partial factor of bearing capacity of concrete under local compression, 0.77 is taken generally.

F_{ld} = design value of the local compression due to the concentrated prestressing force.

If Eq. (9.111) cannot be satisfied, the dimension should be enlarged or the grade of concrete should be improved, or the position of the anchorage needs to be adjusted.

Compressive bearing capacity of the local zone stipulated in the JTG 3362-2018 code

Substituting Eqs. (9.100) and (9.108) into Eq. (9.99), the compressive bearing capacity of the local zone of the prestressed concrete highway beams can be obtained:

$$F_{lu} = \beta f_{cd} A_l + k\rho_v \beta_{cor}^2 f_{sd} A_{\ln} \tag{9.112}$$

To simplify the calculation, β_{cor}^2 is replaced by β_{cor}. As the effect of the decrease of concrete plasticity with the increase of strength is taken into account, Eq. (9.112) can be rewritten as

$$F_{lu} = 0.9(\eta_s \beta f_{cd} + k\rho_v \beta_{cor} f_{sd}) A_{\ln} \tag{9.113}$$

where η_s = correction coefficient of concrete strength under local compression, 1.0 is taken for C50 and lower grade concrete and 0.76 for C80 concrete, respectively, and the values for intermediate strength between C50 and C80 are obtained by the linear interpolation method.

k = influence coefficient of indirect reinforcement, 2.0 is taken for C50 and lower grade concrete and 1.7 for C80 concrete, respectively, and the values for intermediate strength between C50 and C80 are obtained by the linear interpolation method.

The dimension of the anchorage zone should satisfy:

$$F_{ld} \leq 1.3\eta_s\beta f_{cd}A_{ln} \qquad (9.114)$$

where F_{ld} = design value of the local compression, it should be taken as 1.2 times the maximum tension at jacking for post-tensioned members.

If Eq. (9.114) cannot be satisfied, the dimension should be enlarged, the grade of concrete should be improved, or the position of the anchorage needs to be adjusted.

Compressive bearing capacity of the local zone stipulated in the GB 50010-2010 code

The compressive bearing capacity of the local zone provided in the GB 50010-2010 code is almost the same as in Eq. (9.113):

$$F_{ld} \leq 0.9(\beta_c\beta f_{cd} + k\rho_v\beta_{cor}f_{sd})A_{ln} \qquad (9.115)$$

where β_c = influence coefficient of concrete strength, 1.0 is taken for C50 and lower grade concrete and 0.8 for C80 concrete, respectively, and the values for intermediate strength between C50 and C80 is obtained by the linear interpolation method.

The dimension of the anchorage zone should satisfy:

$$F_{ld} \leq 1.35\beta_c\beta f_{cd}A_{ln} \qquad (9.116)$$

If Eq. (9.116) cannot be satisfied, the dimension should be enlarged or the grade of concrete should be improved, or the position of the anchorage needs to be adjusted.

9.20.3 Tension bearing capacity in the general zone

The local zone in the anchorage zone is defined as the region with a transverse area of A_b and local compressive length of l_b. A_b can be calculated from Figs. 9.26 and 9.27 and l_b is taken as 1.2 times of the longer-side length of the anchor plate. The general zone is achieved by subtracting the local zone from the anchorage zone, where the tension is generated by the concentrated force in several regions, as shown in Fig. 7.15. The following introduces the calculation of the tension and the required bearing capacity in the general zone specified in the JTG 3362-2018 code.

The tension bearing capacity of each tension part of the general zone should satisfy:

$$T_{w,d} \leq f_{sd}A_s \qquad (9.117)$$

where $T_{w,d}$ = design value of the tension force caused by the concentrated force at each part of the general zone.

f_{sd} = design tensile strength of reinforcing steels.

A_s = area of the reinforcing steels as the tie rod in each part.

When the value of $T_{w,d}$ is obtained, Eq. (9.117) can be used to verify the corresponding bearing capacity in the general zone or obtain the quantity of reinforcement that is required to be arranged in the general zone.

9.20.3.1 Calculation of tension force in the general zone of end anchor
Bursting force under the anchorage

Figs. 7.14D and 7.15 show that the maximum transverse tensile stress of concrete under local compression occurs in the section at a certain distance away from the anchorage device, and the corresponding force is called the bursting force. Fig. 9.30 shows the typical action positions of bursting force under anchorage devices, from which the bursting force and the position can be derived approximately.

For a single anchor plate, when $M_{DE} = T_{b,d} \times 0.25h$ is adopted approximately in Eq. (7.118), the bursting force can be obtained as

$$T_{b,d} = 0.25P_d(1 + \gamma)^2\left[(1 - \lambda) - \frac{a}{h}\right] + 0.5P_d\left|\sin\alpha\right| \qquad (9.118)$$

The horizontal distance between the bursting force and the anchorage is given by

$$d_b = 0.5(h - 2e) + e\sin\alpha \qquad (9.119)$$

where $T_{b,d}$ = bursting force under the anchorage device.

P_d = design value of the concentrated prestressing force, 1.2 times the tension in prestressing tendons at jacking is taken.

a = width of the anchor plate.

h = section depth of the member.

e = eccentricity of the concentrated prestressing force with respect to the centroid of the section.

γ = eccentricity ratio, $\gamma = 2e/h$.

α = inclined angle of the prestressing tendons, generally -5 degree to $+20$ degree.

When numerous anchor plates are arranged together, it is described as a dense arrangement if the distance between adjacent anchor plates is less than two times the plate width (Fig. 9.30B), where all the concentrated prestressing forces should be considered for bursting force simultaneously; otherwise, it is called a nondense arrangement (Fig. 9.30C), and in this case, the maximum bursting force should be obtained first, then the bearing capacity should be checked.

Spalling force around the anchorage

The concrete under the anchor plate bears huge compression while the concrete around the device bears tension which is called the spalling force, as shown in Fig. 9.31.

The spalling force is related to the distance between adjacent anchor plates and the concentrated prestressing force on each plate. When the distance is small (less than half of the end section depth), the spalling force may be taken as

$$T_{s,d} = 0.02\max\{P_{di}\} \qquad (9.120)$$

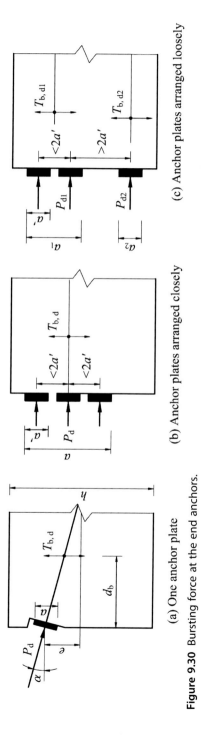

(a) One anchor plate

(b) Anchor plates arranged closely

(c) Anchor plates arranged loosely

Figure 9.30 Bursting force at the end anchors.

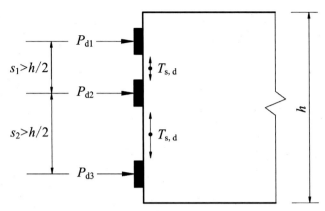

Figure 9.31 Spalling force at the end anchors.

where $T_{s,d}$ = spalling force around anchorage device.

P_{di} = design value of the concentrated prestressing force on each plate.

When the distance between adjacent anchor plates is greater than half the end section depth, the spalling force can be given by

$$T_{s,d} = 0.45\overline{P}_d \cdot \left(\frac{2s}{h} - 1\right) \qquad (9.121)$$

where $\overline{P}_d$ = average design value of two adjacent concentrated prestressing forces, $\overline{P}_d = \frac{P_{d2}+P_{d3}}{2}$, as shown in Fig. 9.31.

s = distance between two adjacent concentrated prestressing forces.

h = depth of the end section of the member.

The value obtained from Eq. (9.121) should be not less than 0.02 times the design value of the maximum concentrated prestressing force.

Tension force at the edge region close to the end section

The transverse tension force at the edge region close to the end section is generated by the concentrated prestressing force on the anchorage device, as shown in Figs. 7.15 and 9.32. Simultaneously, the tension, paralleling the concentrated prestressing force, at the edge region close to the end section is also produced, such as the section $K1$ shown in Fig. 7.14B. The tension force at the edge region due to the concentrated prestressing force can be calculated by

$$T_{et,d} = \frac{(3\gamma - 1)^2}{12\gamma} P_d \qquad (9.122)$$

where γ = eccentricity ratio, $\gamma = 2\frac{e}{h} \geq \frac{1}{3}$.

If $\gamma \leq \frac{1}{3}$, it is not necessary to check the tension resistance against $T_{et,d}$.

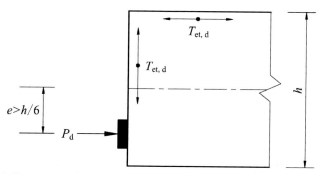

Figure 9.32 Tensions at the edge region close to the end section with an anchorage.

9.20.3.2 Calculation of tension force in the general zone of tooth anchor

As shown in Fig. 7.16, tensions are generated in several regions in the general zone in a triangular tooth block. Fig. 9.33 shows the typical regions where the tension bearing capacity should be calculated and verified.

Bursting force under the anchorage device is given by

$$T_{b,d} = 0.25 P_d \left(1 - \frac{a}{2d} \right) \tag{9.123}$$

Spalling force around the anchorage device may be taken as

$$T_{s,d} = 0.04 P_d \tag{9.124}$$

Traction force in the concrete plate at the back of the tooth block may be taken as

$$T_{tb,d} = 0.20 P_d \tag{9.125}$$

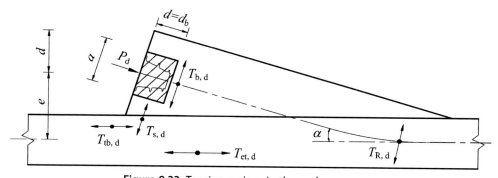

Figure 9.33 Tension regions in the anchorage zone.

Flexural tension in the concrete plate edge opposite the tooth block caused by the bending moment due to the eccentric concentrated prestressing force can be obtained approximately as

$$T_{et,d} = \frac{(2e - d)^2}{12e(e + d)}$$

(9.126)

Radial tension in the concrete plate caused by the bent-up tendons may be taken as

$$T_{R,d} = \alpha P_d$$

(9.127)

where $T_{b,d}$ = design value of bursting force under the anchorage device.

$T_{s,d}$ = design value of spalling force around the anchorage device.

$T_{tb,d}$ = design value of traction in the concrete plate at the back of the tooth block.

$T_{et,d}$ = design value of flexural tension in the concrete plate edge opposite to the tooth block caused by the eccentric concentrated prestressing force.

$T_{R,d}$ = design value of radial tension in the concrete plate caused by the bent-up tendons.

d = vertical distance between the concentrated prestressing force and the edge of the tooth block.

e = distance between the concentrated prestressing force and the centroid of the concrete plate.

α = tangent angle between the concentrated prestressing force and the centroid of the concrete plate before the prestressing tendons bend up, rad.

Suggested readings

[1]   GB 50010: 2015. Code for design of concrete structures. Beijing: China Construction Industry Press; 2015.
[2]   Q/CR 9300: 2018. Code for design on railway bridge and culvert. Beijing: China Railway Publishing House; 2018.
[3]   JTG 3362: 2018. Specifications for design of highway reinforced concrete and prestressed concrete bridges and culverts. Beijing: People's Communications Press; 2018.
[4]   Nawy Edward G. Prestressed concrete: a fundamental approach. 5th ed. Upper Saddle River: Prentice-Hall,Inc; 2006.
[5]   Hu Di. Basic principles of prestressed concrete structure design. 2nd ed. Beijing: China Railway Publishing House; 2019.

CHAPTER 10

Design of prestressed concrete flexural structures

Contents

10.1 Outline for designing a prestressed concrete flexural structure

The section stress analysis and the calculations of deflection and crack width discussed in the earlier chapters are based on the designed prestressed concrete flexural structures. The opposite process for analysis is to design a new member or structure. This chapter focuses on outlining the design contents and the train of thought of designing a flexural structure, mainly including the introduction to durability design, section design, area estimation

Analysis and Design of Prestressed Concrete
ISBN 978-0-12-824425-8, https://doi.org/10.1016/B978-0-12-824425-8.00010-5

and layout of prestressing tendons, and layout reinforcing steel layout in the anchorage zones.

A designed prestressed concrete structure shall meet the following basic functions during construction and in the service period:

(1) It can withstand various actions that may occur simultaneously during normal construction and the service period.

(2) It bears good working performance in the service period.

(3) It has sufficient durability under normal maintenance.

(4) It can maintain the required safety and overall stability under accidental actions.

The above function requirements essentially propose the basic requirements in the design, that is, the structure must have sufficient strength to resist the most disadvantageous actions, meeting the requirements in the ultimate limit state; the member must have sufficient stiffness and crack resistance, controlling the deflections or cambers, crack width, normal stresses, principal stresses, and frequency within the allowable values at service loads, meeting the requirements in serviceability limit state. Also, the structure must have sufficient anticorrosion ability for steels, antideterioration and antiwear ability for concrete, meeting the requirements of durability and service life stipulated in the design codes. Simultaneously, the design scheme needs to consider the economics and aesthetics.

Generally, the typical design steps for a prestressed concrete flexural structure are as follows:

(1) Carry out the durability design according to the required service life, the ambient environment categories, and action characteristics.

(2) Preliminarily determine the type and dimensions of the section according to the purposes of use and overall structure system while also referring to the existing structures and related stuffs.

(3) Obtain the design values of internal forces in the control sections (bending moment, shear, torsion, compression, and tension) by structural analysis based on the preliminary design scheme.

(4) Choose the type of prestressing tendons, estimate the tendon area, and lay out the tendons in the preliminary sections according to the design internal forces and the limitations of deflections and stresses.

(5) Calculate the geometric properties of the sections.

(6) Determine the stretching stress of the tendons, estimate the prestress losses and the corresponding effective stress in tendons at each stage.

(7) Calculate the quantity of the reinforcing steels required in the anchorage zone, initially lay out the reinforcing steels.

(8) Verify the flexural and shear bearing capacity of the critical sections at service loads.

(9) Verify the short-term deflection under the action of a live load.

(10) Verify the stresses in sections at the stages of stretching tendons, transfer, transportation, and installation, and in the service period.

(11) Verify the crack resistance of the critical sections.

(12) Verify the crack width and the long-term deflection.

(13) Verify the fatigue performance for structures subjected to repeated loading.

(14) Verify the natural frequency of the railway structure.

(15) Verify the rotation angle of the railway member ends.

(16) Verify the bearing capacity of the anchorage zone.

In step (4), if the prestressing tendons cannot be reasonably arranged, return to step (2) to adjust the section dimensions. In steps (7)−(16), calculate and verify the design parameters according to the provisions in the design codes. Once it does not meet the requirements, return to step (2) to modify the section dimensions or return to step (4) to modify the prestressing tendon design. An economical, rational, and feasible design often needs several iterations and calculations to be met.

For the members or structures with seismic design requirements, the prestressing tendons and reinforcing steels shall be arranged per the specifications, and verification of earthquake resistance should be finished. For example, the *Codes for design of concrete structures* (GB 50010-2010) stipulates that the prestressed concrete structure used for a seismic fortification intensity of 6−8 degrees, in addition to meeting the requirements of the general ultimate strength limit state and the serviceability limit state, a seismic design must be fulfilled according to the *Building Seismic Design Code* (GB 50011-2018) and *Prestressed Concrete Structure Seismic Design Regulations* (JGJ140-2019) codes.

10.2 Durability design of prestressed concrete members

The durability design of prestressed concrete structures shall be carried out according to the design service life and environmental category specified in the codes, such as the *Guide to Durability Design and Construction of Concrete Structure* (CCES 01-2018), *Code for Durability Design on Concrete Structure of Railway* (TB100005-2010), and *Codes for design of concrete structures* (GB 50010-2010). Generally, it includes the following contents:

(1) Determining the environmental category of the structure.

(2) Putting forward the basic requirements for the durability of concrete materials.

(3) Determining the minimum concrete cover for reinforcements.

(4) Determining the technical measures to guarantee durability under different environmental conditions.

(5) Presenting the inspection and maintenance requirements in the service period.

The GB100005-2010 code also stipulates that the durability design shall be based on the premise that the structure has sufficient bearing capacity and crack resistance, and the time-dependent effect due to concrete shrinkage and creep on the prestress level of the prestressed concrete structures also shall be taken into account.

According to the corrosion mechanism of concrete and steel, the environment can be grouped into carbonation, chloride salt, chemical erosion, sulfate physical attack, salt crystallization destroy, freezing-thaw, and abrasion environments. Under each environment, the impact on the material durability can be divided into several classes by the degree of corrosion influence, therefore, in determinations of the materials for prestressed structures, the class of attack action on the concrete structures by the environment should be considered first.

10.2.1 Concrete durability

The choice of raw materials and concrete mix proportion mixture ratio should be determined according to the durability design class and the corresponding durability indexes. Generally, the total chloride ion content in concrete for prestressed structures shall not exceed 0.06% of the total amount of cementitious materials. Under a corrosive environment, corrosion-resistant measures shall be taken for structures. Other durability indexes of concrete such as resistance class to crack, resistance classes to sulfate physical attack, salt crystallization destroy, freezing–thaw, alkali–aggregate reactivity, and abrasion should satisfy the corresponding requirements specified in the codes according to the durability design grade. When the environmental attack is very severe or extremely severe, in addition to the strict durability requirements of the concrete itself, reliable additional anticorrosion measures should be proposed, such as using epoxy-coated steel bars, adding rust inhibitors or water-soluble polymer emulsion to the concrete, painting or covering with protective material on the concrete surface and setting cathodic protection.

10.2.2 Structural design

The shape of the prestressed concrete structures should be simple and convenient for maintenance and repair. The structural details should be beneficial to reduce any corrosion attack on the structure by the environment, facilitate drainage and ventilation, and help avoid the accumulation of water, vapor, and harmful substances on the concrete surface.

The minimum concrete cover for the prestressing tendons is determined by the design service life and the type of environment. For prestressing steels with continuous anticorrosion sealing sheath inside the concrete, the concrete cover is the distance from the outer edge of the sheath to the concrete surface; the minimum coverage may be the same as that of reinforcing steels while it should exceed half of the sheath diameter. For pretensioned steels without an anticorrosion system, the minimum concrete cover should be 10 mm larger than that of reinforcing steels specified in the codes. The distance from the outer edge of the post-tensioning tendon duct to the concrete surface shall not be less than the duct diameter on the top and side of the section, and shall not be less than 60 mm on the bottom. For prestressing steels arranged outside the concrete, a continuous anticorrosion sealing sheath and other protection measures are required, especially when environmental attack on the steels is very or extremely severe.

10.3 Section design for prestressed concrete flexural structures

10.3.1 Sectional types

Some common section types and section dimensions have been formed for various in engineering practices, which can be used for reference in the design. Fig. 10.1 shows several commonly used cross-sections of prestressed concrete members.

10.3.1.1 Prestressed concrete hollow slab

The hollow core molds with circular, round–ended, or ellipse shapes are used to make the prestressed concrete hollow slabs (Fig. 10.1A) by the pretensioning method in a precast factory or the construction location. The broken lines may be taken for prestressing

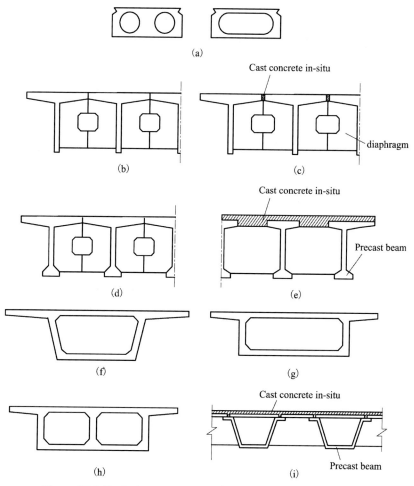

Figure 10.1 Typical cross-sections of prestressed concrete beams.

tendons in a longer span slab. Generally, the suitable span of prestressed concrete hollow slabs is 10—20 m, and as a maximum up to 30 m. Hollow slabs are commonly used in highway prestressed concrete beams with the ratio of section depth to the effective span (depth-to-span ratio) from 1/23 to 1/15. In buildings, prestressed concrete hollow slabs can be used to build a flat plate system.

10.3.1.2 Prestressed concrete T-section

A prestressed concrete T-section (Fig. 10.1B—D) is one of the most broadly used section types in the simply supported prestressed beams, both in highway bridges and railway bridges. The suitable span of a T-section is 16—40 m, which is commonly made by the post-tensioning method. If the span is less than 20 m, a web with constant thickness is commonly adopted (Fig. 10.1B and C). For a longer span with more tendons required at the midspan section, two protruding parts (called a horseshoe block) near the bottom of the web are made to provide space for arranging the tendons, as shown in Fig. 10.1D.

To connect each T-beam into an integrated structure with strong transverse rigidity, the diaphragms arranged at a certain distance are connected by reinforced concrete along with welding of two embedded steel plates at the ends. For better coordination of the bridge deck, portion concrete at the end of the flange plate is cast in the construction location, as shown in Fig. 10.1C.

When the prestressed T-section is used in simply supported highway beams, a reasonable depth-to-span ratio is from 1/25 to 1/15, while from 1/15 to 1/10 is sed in railway beams.

10.3.1.3 Prestressed concrete I-section

To reduce the self-weight of the precast beam to facilitate transportation and installation, the width of the upper flange of the T-section can be shortened to form an I-section, as shown in Fig. 10.1E. After the I-beams are installed in place, the portion between two adjacent flange ends and a whole reinforced concrete plate above the I-beams are cast in the construction place to form a combined T-beam system, also, with reinforcing steels inside, and the diaphragms arranged at a certain distance are connected to provide sufficient transverse rigidity. I-sections can be used in the prestressed beams with an effective span of from 16 to 30 m, in which the reasonable depth-to-span ratio is from 1/18 to 1/15.

10.3.1.4 Prestressed concrete box section

The prestressed concrete box sections (Fig. 10.1F—H) are most widely used in highway bridges and railway bridges with an effective span from 20 to 200 m. The economical span for simply supported prestressed box girders varies from 20 to 50 m, while the span can reach over 70 m when the external prestressing tendons are arranged simultaneously. For bridges with a span from 50 to 200 m, the continuous prestressed concrete box girder is probably the most competitive scheme to some extent, and for the bridges with a span from 200 to 300 m, it also has high competitiveness.

The box section is a closed section with good overall working performance and high flexural rigidity, its torsional stiffness is much larger than that of the open sections (such as a T-section), and the lateral distribution of load is relatively uniform. The webs of the box can be made thinner and the self-weight is lighter. The upper and bottom plates can provide sufficient space to arrange the prestressing tendons to resist the positive and negative bending moment, respectively, and the webs can also provide a rich space to arrange the bending tendons to resist shear force, therefore, prestressed concrete box girders are very suitable for long-span continuous structures.

The web can be made straight or inclined, with the latter being easy to demold and more attractive. According to the numbers of boxes and chambers, the box section can be divided into a single box with a single chamber (Fig. 10.1F and G), a single box with multichamber (Fig. 10.1H), multibox with a single chamber, and multibox with multi-chamber. In the case where the width of the bridge deck is less than 15 m, a single box with a single chamber is preferred.

Prestressed box sections with constant depth are commonly used in simply supported beams and for continuous beams with a span of less than 60 m constructed by the incremental launching method. For simply supported highway box girders, the depth-to-span ratio varies from 1/25 to 1/15.

Generally, the prestressed box sections with variable depth are mainly used in continuous beams with long spans. Table 10.1 shows the typical depth-to-span ratios for continuous highway and railway beams.

10.3.1.5 Prestressed concrete combined box section

The prestressed concrete combined box section is built using a precast channel beam with a precast hollow slab, as shown in Fig. 10.1I. Generally, the channel beams with standard

Table 10.1 Typical ratios for prestressed concrete continuous highway and railway beams.

Items	Continuous beams with constant section depth	Continuous beams with variable section depth	
	Section depth	Section depth at support	Section depth at midspan
High-speed railway beams	$(1/14.0-1/11.0)L$	$(1/16.0-1/12.0)L$	$(1/2.0-1/1.5)H_s$
Common-speed railway beams	$(1/18.0-1/16.0)L$	$(1/16.0-1/12.0)L$	$(1/2.0-1/1.5)H_s$
Highway beams	$(1/20.0-1/15.0)L$	$(1/19.0-1/15.0)L$	$(1/2.5-1/2.0)H_s$

Notes: L is the effective span; H_s is the section depth above the middle support.

design are prefabricated in a factory by the pretensioning method, which is connected with the precast hollow slabs by cast-in-place concrete pavement. The combined section has high flexural rigidity and torsional rigidity, which is suitable for prestressed concrete beams with a span from 16 to 35 m. The reasonable depth-to-span ratio of the prestressed concrete combined box section varies from 1/20 to 1/16.

10.3.2 Section design

In the service period, it is assumed that the whole section of the prestressed concrete flexural members works elastically. The bending moment due to the external loads is balanced by the internal force couple composed of the tension (T_{ps}) provided by the prestressing tendons and reinforcing steels in the tension zone and the compression (C_c) provided by the concrete in the compression zone. As the external bending moment increases, the internal lever arm between T_{ps} and C_c increases, because the tension T_{ps} changes in a relatively low range. This shows that the greater the variation range of the internal lever arm, the higher the flexural strength of the section. For members in which the tensile stress is not allowed, the internal lever arm falls within the core distance in the section. Therefore, the rationality and economy of the section can be reflected by the bending efficiency index of the section, defined as the ratio of the distance between the upper core distance and the lower core distance on the section to the overall depth of the section:

$$\eta_b = \frac{K_u + K_b}{h} \tag{10.1}$$

where η_b = geometric efficiency of the cross-section with respect to bending
$\quad K_u$ = distance from the centroid of the section to the upper limit of the central kern
$\quad K_b$ = distance from the centroid of the section to the lower limit of the central kern
$\quad h$ = overall depth of the section

The larger the value of η_b, the more rational and economical the section is. When the sections are used in their correspondingly reasonable spans, generally, $\eta_b = 1/3$ for prestressed rectangular sections, $\eta_b = 0.4 - 0.55$ for prestressed hollow slabs, and $\eta_b = 0.5$ for prestressed T-sections. Therefore, a section with a large η_b value is preferred in section design.

According to the established formulas of concrete stress and the limited values prescribed in the codes during construction and in the service period, the equations for estimating the geometric parameters of a reasonable section dimension can be derived. As an example, the following introduces how to obtain the section modulus of a rational section of a pretensioned beam.

The normal stress in concrete at transfer can be obtained from Eq. (7.10), rewritten as

$$\sigma_{c,u} = \frac{N_p}{A_c} + \frac{-N_p e_p + M_g}{W_{c,u}} \geq [\sigma_{ct}]_1 \tag{10.2}$$

$$\sigma_{c,b} = \frac{N_p}{A_c} - \frac{-N_p e_p + M_g}{W_{c,b}} \leqslant [\sigma_{cc}]_1 \qquad (10.3)$$

where $\sigma_{c,u}$ = stress of concrete in the top fiber of the section at transfer

$\sigma_{c,b}$ = stress of concrete in the bottom fiber of the section at transfer

N_p = effective prestressing force at transfer

e_p = eccentricity of N_p with respect to the centroid of the converted section

A_c = cross-sectional area of the converted section

$W_{c,u}$ = section modulus of the converted section with respect to the top fiber

$W_{c,b}$ = section modulus of the converted section with respect to the bottom fiber

M_g = bending moment due to the self-weight of the beam

$[\sigma_{ct}]_1$ = limited value of tensile stress in concrete at transfer

$[\sigma_{cc}]_1$ = limited value of compressive stress in concrete at transfer.

The normal stress in concrete at service loads can be obtained from Eq. (7.30), rewritten as

$$\sigma_{c,u} = \frac{\zeta N_p}{A_c} + \frac{-\zeta N_p e_p + M_{\text{design}}}{W_{c,u}} \leqslant [\sigma_{cc}]_2 \qquad (10.4)$$

$$\sigma_{c,b} = \frac{\zeta N_p}{A_c} - \frac{-\zeta N_p e_p + M_{\text{design}}}{W_{c,b}} \geqslant [\sigma_{ct}]_2 \qquad (10.5)$$

where M_{design} = design value of bending moment due to a specified combination of actions.

ζ = ratio of the effective prestressing force after all losses have taken place to the effective prestressing force at transfer.

$[\sigma_{ct}]_2$ = limited value of tensile stress in concrete at service loads.

$[\sigma_{cc}]_2$ = limited value of compressive stress in concrete at service loads.

Subtracting Eq. (10.2) from Eq. (10.4) and subtracting Eq. (10.3) from Eq. (10.5), yields

$$W_{c,u} \geqslant \frac{(1 - \zeta)M_g + (M_{\text{design}} - M_g)}{[\sigma_{cc}]_2 - \zeta[\sigma_{ct}]_1} \qquad (10.6)$$

$$W_{c,b} \geqslant \frac{(1 - \zeta)M_g + (M_{\text{design}} - M_g)}{\zeta[\sigma_{cc}]_1 - [\sigma_{ct}]_2} \qquad (10.7)$$

Eqs. (10.6) and (10.7) can be used to check the initially determined section dimensions, then a reasonable and economical section design could be obtained when the bending efficiency index is further taken into account.

10.4 Area estimation for prestressing tendons

Based on the preliminary cross-sectional dimensions of concrete and the design load along with the limitations for working stress in the serviceability limit state and bearing capacity requirements in the ultimate limit state, the quantity of prestressing tendons can be estimated.

10.4.1 Estimation based on the concrete stress limits during stretching and in the service period

From Eqs. (10.2)−(10.5), the Mangnel's inequalities in a critical section are obtained:

$$\frac{1}{A_p} \geq \frac{\sigma_{pe}}{\dfrac{M_g}{W_{c,u}} - [\sigma_{ct}]_1} \left(\frac{e_p}{W_{c,u}} - \frac{1}{A_c} \right) \tag{10.8}$$

$$\frac{1}{A_p} \geq \frac{\sigma_{pe}}{[\sigma_{cc}]_1 + \dfrac{M_g}{W_{c,u}}} \left(\frac{1}{A_c} + \frac{e_p}{W_{c,b}} \right) \tag{10.9}$$

$$\frac{1}{A_p} \geq \frac{\zeta\sigma_{pe}}{[\sigma_{cc}]_2 - \dfrac{M_{design}}{W_{c,u}}} \left(\frac{1}{A_c} - \frac{e_p}{W_{c,u}} \right) \tag{10.10}$$

$$\frac{1}{A_p} \leq \frac{\zeta\sigma_{pe}}{[\sigma_{ct}]_2 + \dfrac{M_{design}}{W_{c,h}}} \left(\frac{1}{A_c} + \frac{e_p}{W_{c,b}} \right) \tag{10.11}$$

Eqs. (10.8)−(10.11) show that $1/A_p$ changes with e_p linearly, as shown in Fig. 10.2, in which lines E to H correspond to Eqs. (10.8)−(10.11). The shaded area envelops the

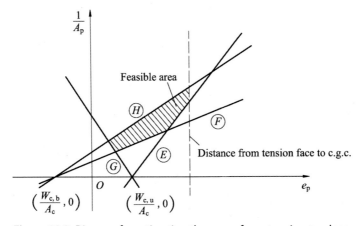

Figure 10.2 Diagram for estimating the area of prestressing tendons.

selectable range of $1/A_p$ and e_p. When estimating, the effective stress in prestressing tendons can be taken as $\sigma_{pe} = 0.4 - 0.6f_{ptk}$ (f_{ptk} is the characteristic tensile strength). Considering the reasonable values of e_p according to the concrete cross-section and the selected types of prestressing tendons, the total cross-sectional area of the prestressing tendons can be determined from Fig. 10.2, then the corresponding number of prestressing tendons is derived.

If the prestressing tendons are required on both the upper and lower edges of the section, the contribution of all prestressing tendons to the section stresses should be comprehensively taken into account in Eqs. (10.2)–(10.5). Accordingly, Eqs. (10.8)–(10.11) should be modified.

10.4.2 Estimation based on the ultimate strength limit state

Take the flexural rectangular section as an example (Fig. 9.4) to discuss how to estimate the area of the prestressing tendons according to the bearing capacity. To simplify the calculation, the contribution of the reinforcing steels to the bearing capacity is neglected.

In the ultimate limit state, the bearing capacity of the normal section of a flexural rectangular section is given by Eqs. (9.21) and (9.22):

$$M_d = f_{cd}bx\left(h_0 - \frac{x}{2}\right) \tag{10.12}$$

$$f_{pd}A_p = f_{cd}bx \tag{10.13}$$

From which

$$x = h_0 - \sqrt{h_0^2 - \frac{2M_d}{f_{cd}b}} \tag{10.14}$$

$$A_p = b\left(h_0 - \sqrt{h_0^2 - \frac{2M_d}{f_{cd}b}}\right)\frac{f_{cd}}{f_{pd}} \tag{10.15}$$

where M_d = design value of bending moment due to a specified combination of actions

f_{cd} = design strength of concrete in compression

f_{pd} = design tensile strength of prestressing tendon

A_p = cross-sectional area of prestressing tendons in the tension zone

b = width of the rectangular section

b = width of the rectangular section

x = depth of the neutral axis

h_0 = effective depth of the section, $h_0 = h - a_p$

a_p = depth of prestressing tendons in the tension zone from tension face.

When the prestressing tendons are arranged in the compression zone, the results from Eqs. (10.14) and (10.15) are too large, while it is still a reasonable estimation in most cases because the quantity of prestressing tendons in the compression zone is relatively small. If the section is subject to a positive−negative bending moment, it can be regarded as a single-strength section, the areas of prestressing tendons in the tension and compression zones can be estimated by Eqs. (10.14) and (10.15) according to the positive and negative moments, respectively.

For prestressed concrete members that allow cracking, the nominal concrete tensile stress corresponding to the allowable crack width can be obtained first from Eqs. (8.53), (8.55), and (8.56), then the required quantity of prestressing tendons can be estimated from Eqs. (10.8)−(10.11). Finally, the total prestressing tendons are determined when Eqs. (10.14) and (10.15) are considered further. The replenished quantity and arrangement of reinforcing steels by the crack resistance and other provisions stipulated by the codes can also be obtained.

10.5 Selection and layout of prestressing tendons

10.5.1 Selection of prestressing tendons

The selection of prestressing tendons should take into account the prestress loss values, spaces for layout and anchor of tendons, the required eccentric prestressing force, cross-sectional dimensions of concrete, and the characteristics of the prestressing system. Generally, the prestressed concrete flexural structures with a large or medium span are constructed by post-tensioning and the members with a short span are constructed by pretensioning, while some medium-span members may also be constructed by pretensioning. To counterbalance the bending moment due to service loads, the prestressing force and its eccentricity are required to change along the length of the member, in this case, the prestressing steel strands which can be bent at any section are most widely used as the longitudinal tendons, as shown in Fig. 6.7. For pretensioned flexural members, the prestressing steel strands, indented steel wires, and threaded steels can be used as the longitudinal tendons. When the smooth steel wires are used, measures should be taken to ensure reliable anchoring of wires in concrete.

In some cases, prestressing FRP can be used for the structures in a severe ambient environment.

When the required prestressing tendons are short, whether by post-tensioning or by pretensioning, the prestressing threaded steel bars are the best choice in most cases, since the nut anchorage bears the lowest value of tendon retraction due to anchorage set. Therefore, the threaded steel bars are broadly used in the short members (such as the pretensioned sleepers for a ballastless track system in a high-speed railway, as shown in Fig. 10.3A), and used as the vertical tendons in the webs and horizontal tendons in the flange of box girders, as shown in Fig. 10.4. Considering that the vertical tendons in the webs are fixed at the bottom ends and stretched when the concrete of webs is

(a) The threaded steel bars in the
pretensioned sleepers before stretching

(b) The threaded steel bars are stretching by
the automatic intelligent tensioning system

(c) The sleeper made by pretensioning

Figure 10.3 The threaded steel bars in the pretensioned sleepers for a ballastless track system in a high-speed railway.

cast and hardens enough, the threaded steel bars could be the best choice as the vertical tendons because no additional steel meshes are needed to keep the tendons in positions vertically.

When the top plate of a box girder has a large span cantilever or the distance between two webs is relatively large, with which a large transverse bending moment will occur under the action of the service loads, the horizontal tendons should be bent transversely to counterbalance the positive moment at the middle of the top plate between two webs and the negative moment in the top plate above the webs. In this condition, the prestressing steel strands should be adopted.

10.5.2 Reasonable position of the prestressing tendons

From Eqs. (10.2)−(10.5), a rational range of the position of the prestressing tendons can be derived:

$$e_{p1} \leqslant k_b + \frac{M_g - W_{c,u}[\sigma_{ct}]_1}{N_p} \tag{10.16}$$

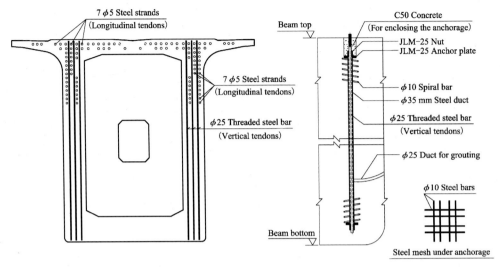

(a) Vertical tendons in the webs of a box section   (b) Details for vertical tendons

Figure 10.4 Schematic diagram of the vertical tendons in a box section.

$$e_{p2} \geqslant -k_u + \frac{M_g + M_d + M_L + W_{c,b}[\sigma_{ct}]_2}{\zeta N_p} \tag{10.17}$$

where e_{p1}, e_{p2} = distance from the point of prestressing force to the c.g.c. of the converted section.

k_u, k_b = upper and bottom core distances, $k_u = \dfrac{W_{c,b}}{A_c}$, $k_b = \dfrac{W_{c,u}}{A_c}$.

Eqs. (10.16) and (10.17) give the boundary positions of the prestressing tendons, called the tendon boundary (or cable boundary). Therefore, the rational position of the resultant of the prestressing force should be within these boundaries, as

$$-k_u + \frac{M_g + M_d + M_L + W_{c,b}[\sigma_{ct}]_2}{\zeta N_p} \leqslant e_p \leqslant k_b + \frac{M_g - W_{c,u}[\sigma_{ct}]_1}{N_p} \tag{10.18}$$

The schematic diagram of the tendon boundary in a typical prestressed concrete T-beam is shown in Fig. 10.5.

As long as the point of the prestressing force falls within the tendon boundaries, namely, the eccentricity satisfies Eq. (10.18), the tensile stresses in the upper and lower edges in a section do not exceed the limit values stipulated in the codes during construction and in the service period, which provides overall guidance for the layout of prestressing tendons. It should be noted that it is permissible to arrange some of the prestressing tendons outside the boundaries.

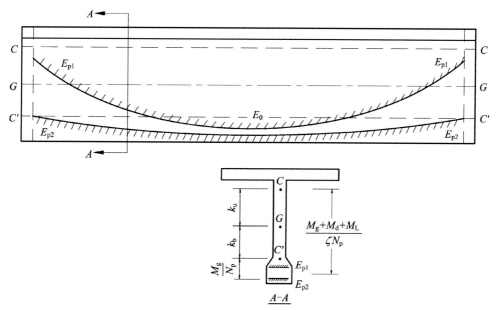

Figure 10.5 Schematic diagram of the tendon boundary in a T-beam.

10.5.3 Principles of the prestressing tendon layout

Generally, the principles of the prestressing tendon layout include the following aspects:
(1) The point of the prestressing force in a section falls within the tendon boundaries.
(2) The change of eccentricity of the prestressing tendons along the length of the structure should be followed with the bending moment caused by the service loads.
(3) The prestressing tendons should bend to match with the change of the shear forces in the sections caused by the service loads.
(4) Using the concept of concordant tendon and the linear transformation principle along with the requirements in (1)−(3) to layout the tendons preliminary.
(5) The positions of the prestressing tendon ends should be dispersed to facilitate the arrangement of the anchorages and reduce the local compression in the vicinity of the anchorage zone.
(6) For thinner flange and bottom plate of a box section, the longitudinal prestressing tendons should be placed as close as possible to the web regions to prevent transverse cracking.
(7) Structural details for layout, such as the concrete cover for tendons and net spacing for anchorages, should satisfy the requirements stipulated in the design codes.

10.5.4 Layout of prestressing tendons in the pretensioned members

The prestressing tendons in the pretensioned members must have sufficient anchorage length, which can be calculated from Eq. (7.117). The design values of the anchorage length for railway bridges can be adopted from Table 7.12, while for highway bridges they are adopted from Table 7.13.

The net spacing between prestressed tendons shall be determined according to the requirements of pouring concrete, prestressing, and anchoring. The minimum net distance shall not be less than 1.5 times their nominal diameters or equivalent diameters. The Q/CR 9300-2018 code stipulates that the minimum net distance should not be less than 30 mm. The JTG 3362-2018 code stipulates that the minimum net distance should not be less than 25 mm for $7\phi5$ strands and 15 mm for wires.

The minimum concrete cover shall not be less than the nominal diameter of the prestressing tendon. In addition, the Q/CR 9300-2018 code also stipulates that the minimum concrete covers should not be less than 50 mm for tendons in the upper and side, and 60 mm for tendons in the bottom. The JTG 3362-2018 code stipulates that the minimum concrete covers for straight tendons should also meet the requirements compiled in Table 10.2.

10.5.5 Layout of prestressing tendons in the post-tensioned structures

The longitudinal prestressing tendons generally begin to bend at the regions from the quarter to one-third span to match the shear force variation. The bending angle should not be greater than 20 degree to reduce the frictional loss when the prestressing tendons are stretched. When the bending angle is large (for example, the prestressing tendons are bent out of the beam top and the corresponding angle is possibly greater than 25 degree), the measures to reduce frictional loss must be taken. The radius in the bending segment

Table 10.2 Minimum concrete covers for straight tendons for highway bridges (JTG 3362-2018) (mm).

Types of member	Beam, plate, tower, and arch		Abutment and pier		Foundation and bearing platform	
Service life (years)	100	50, 30	100	50, 30	100	50, 30
Class I—General environment	20	20	25	20	40	40
Class II—freezing–thaw environment	30	25	35	30	45	40
Class III—chloride salt environment	35	30	45	40	65	60
Class IV—other chloride environment (deicing salts)	30	25	35	30	45	40
Class V—salt crystallization environment	30	25	40	35	45	40
Class VI—chemical erosion	35	30	40	35	60	55
Class VII—abrasion environment	35	30	45	40	65	60

Note: For precast members, the values in the table can be reduced by 5 mm, while the values should be not less than 20 mm.

should be not too large and should consider the magnitude of the frictional loss at stretching and the ease of bending and arranging. A circular arc, a parabola, or a catenary is commonly adopted in the bent portion of the prestressing tendons.

The radius of curvature of the prestressing tendons should meet the following requirements:

(1) The radius of curvature shall not be less than 4 m when the steel wire diameter in a strand or bundle of wires is equal to or less than 5 mm, and the minimum radius should be 6 m if the steel wire diameter is greater than 5 mm.

(2) The radius of curvature shall not be less than 12 m when the threaded steel bar diameter is equal to or less than 25 mm, otherwise, the radius should be not less than 15 m.

The dimensions and positions of the reserved ducts for prestressing tendons and the distance from the duct outer wall to the same side edge shall comply with the requirements in Table 10.3. In general, the inner diameter of the ducts or the diameter of the holes formed by the pipe former should be 10 mm larger than the nominal diameter of the strands or wire bundles.

Table 10.3 Parameters for ducts of prestressing tendons.

Item	Duct dimension (mm)	Minimum net spacing (mm)	Minimum concrete cover (mm)
Q/CR 9300–2018	$d_d - d_p \geqslant 10$	$\geqslant 40 \ (d_p < 55)$ $\geqslant d_p \ (d_p \geqslant 55)$	$\max(50, d_d)$ (beam top, beam side) $\geqslant 60$ (beam side)
JTG 3362–2018	$A_d \geqslant 2A_p$	Straight tendons: $\max(40, 0.6d_d)$ Curved tendons: $\max(40, 0.6d_d)$	Straight tendons: C_s can be adopted from Table 10.1 Curved tendons: $\max(C_{in}, C_s)$ (concave side) $\max(C_{out}, C_s)$ (convex side)
GB 50010–2010	$d_d - d_p \geqslant 10 - 15$	Precast members: $\geqslant 50$ Frame beams: $\geqslant d_d$ (vertical) $\geqslant 1.5d_d$ (horizontal)	Precast members: $\max(30, 0.5d_d)$ Frame beams: $\geqslant 50$ (beam's bottom) $\geqslant 40$ (beam side)

Notes: d_d is the duct diameter; d_p is the tendon diameter; A_d is the cross-sectional area of duct; A_p is the cross-sectional area of prestressing tendons; C_s is the minimum concrete cover for straight tendons; C_{in} is the minimum concrete cover for ducts in the concave side of a curved plane; C_{out} is the minimum concrete cover for ducts in the convex side of a curved plane.

When the duct diameter is not more than 55 mm, the net pipe distance between ducts should not be less than 40 mm; when the duct diameter is greater than 55 mm, the net pipe distance should not be smaller than the duct diameter.

For vertical tendons in the webs of a box section, the longitudinal distance should be 500–1000 mm.

The JTG 3362-2018 code also stipulates that the minimum concrete covers for straight tendons should be greater than half of the duct diameters, and meet the requirements in Table 10.1. The minimum concrete cover for ducts on the concave and convex sides of a curved plane can be calculated by the following.

Minimum concrete cover for ducts in the concave side:

$$C_{in} \geqslant \frac{P_d}{0.266r\sqrt{f'_{cu}}} - \frac{d_s}{2} \qquad (10.19)$$

Minimum concrete cover for ducts on the convex side:

$$C_{out} \geqslant \frac{P_d}{0.266\pi r\sqrt{f'_{cu}}} - \frac{d_s}{2} \qquad (10.20)$$

where C_{in} = minimum concrete cover for ducts in the concave side of a curved plane

C_{out} = minimum concrete cover for ducts in the convex side of a curved plane

P_d = design value of tensioning force, which can be derived by multiplying the tension force obtained by subtracting the prestress losses due to anchor ring, friction, and tendon retraction from the jacking force by 1.2

r = duct radius of curvature, mm

d_s = duct diameter in the convex side, mm

f_{cu} = cubic strength of concrete in compression.

If the minimum concrete cover obtained from Eq. (10.19) or (10.20) is less than the value in Table 10.1, the value in Table 10.1 should be adopted.

10.6 Requirements for reinforcing steels in design

10.6.1 General requirements for reinforcing steels in design

For the flexural members that allow cracking or tensile stress in concrete in the tension zone at service loads, the Q/CR 9300-2018 code stipulates that the reinforcing steels (nonprestressing steels) must be arranged in the tension zone and their cross-sectional area should not be less than 0.3% of the concrete area in tension, the JTG 3362-2018 code stipulates that the cross-sectional area of the reinforcing steels should not be less than 0.003 bh_0, where b and h_0 are the section width and effective depth of the section, respectively. The deformed steel bars with small diameters should be preferred as the longitudinal steels outside the prestressing tendons in the tension zone and should be arranged closely.

For the flexural members in which tensile stress in concrete is not allowed, the Q/CR 9300-2018 code stipulates that the reinforcing steels also must be arranged in the tension zone, and the diameter of steel bars should not be less than 8 mm, and the distance between the steel bars should not be less than 100 mm.

When the prestressing tendons are concentratedly arranged in the upper or bottom plate of the section, sufficient stirrups or vertical prestressing tendons should be arranged in these regions.

For pretensioned members, a spiral steel bar with a length of not less than 150 mm shall be arranged at the end of a single prestressed tendon; when a plurality of prestressing tendons is arranged, three to five layers of steel mesh within the length of 10 times the tendon diameter should be employed.

In the end anchorage zone of the post-tensioned members, a bearing plate with a bell mouth below the anchorage should be adopted. Indirect steel bars are also required to arrange under the anchor plate, and their volume ratio shall not be less than 0.5%.

The stirrups should be determined by the requirements for shear bearing capacity, shear stress, and crack resistance (such as cracking in the webs due to concrete shrinkage). Generally, the stirrup diameter should not be less than 8 mm, and hot-rolled ribbed steel bars should be used.

The Q/CR 9300-2018 code stipulates that the distance between the stirrups should not exceed 200 mm in the webs. When the longitudinal tendons are arranged in the upper flange, the closed stirrups should be employed and the minimum distance is 100 mm. Within 500 mm of the beam end, the distance between the stirrups should be 80−100 mm. In a torsion member, closed stirrups are required.

The JTG 3362-2018 code stipulates that the stirrups with a diameter 10−20 mm should be arranged in the webs of the T-section, I-section, and box section with a distance not greater than 120 mm. If the length is greater than the section depth from the bearing center, closed stirrups with a distance not greater than 120 mm should be employed. In the horseshoe at the bottom of the T- or I-section, additional closed stirrups with a distance not greater than 200 mm are needed.

10.6.2 Design of reinforcing steels in the anchorage zone
10.6.2.1 Calculation of reinforcing steels in the anchorage zone
In the post-tensioned structures, the reinforcing steels in the local anchorage zone against the high local compression force due to the prestressing force and the overall zone against the tensions shall be calculated according to the design jacking force and characteristics of the anchorage zone.

Estimation of indirect reinforcement volume ratio in the local compression zone
Within a certain distance from the anchor plate, the concrete is subjected to the local compression force due to the concentrated prestressing force where the indirect

reinforcement (a spiral steel bar or square steel meshes) is required to be arranged to improve the local compressive bearing capacity.

For railway bridges, the indirect reinforcement volume ratio can be estimated by

$$\rho_v = \frac{\gamma_s F_{ld} - \dfrac{\beta f_{cd} A_{ln}}{\gamma_{c2}}}{2\beta_{cor} f_{sd} A_{ln}} \qquad (10.21)$$

where ρ_v = volume ratio of indirect reinforcement

F_{ld} = design value of the local compression due to the concentrated prestressing force

γ_s = comprehensive partial safety factor of bearing capacity of indirect reinforcement, 1.2 is taken generally

γ_{c2} = comprehensive partial factor of bearing capacity of concrete under local compression, 1.02 is taken generally

f_{cd} = design strength of concrete in compression

f_{sd} = design tensile strength of indirect reinforcement

β = strength increasing factor of concrete due to lateral restraint

β_{cor} = increasing coefficient due to indirect reinforcement.

A_{ln} = net area of concrete under the anchorage device, the area of holes inside the device shall be deducted.

For highway bridges, the indirect reinforcement volume ratio can be estimated by

$$\rho_v = \frac{\dfrac{F_{ld}}{0.9 A_{ln}} - \eta_s \beta f_{cd}}{k \beta_{cor} f_{sd}} \qquad (10.22)$$

where η_s = correction coefficient of concrete under local compression, 1.0 is taken for C50 and lower grade concrete and 0.76 for C80 concrete, respectively, and the values for intermediate strength between C50 and C80 are obtained by the interpolation method.

k = influence coefficient of indirect reinforcement, 2.0 is taken for C50 and lower grade concrete and 1.7 for C80 concrete, respectively, and the values for intermediate strength between C50 and C80 are obtained by the linear interpolation method.

Estimation of reinforcing steels in the overall zone

Section 7.7 discussed the tension regions in the overall zone due to the concentrated prestressing force, as shown in Figs. 7.14–7.16. Generally, the reinforcing steels required to resist the tensions in the overall zone can be obtained by

$$A_s = \frac{T_{w,d}}{f_{sd}} \qquad (10.23)$$

where $T_{w,d}$ = design value of the tension caused by the concentrated prestressing force at each part of the overall zone, including $T_{b,d}$, $T_{s,d}$, $T_{et,d}$, $T_{tb,d}$ and $T_{R,d}$ given by Eqs. (9.118)–(9.127)

f_{sd} = design tensile strength of reinforcing steels

A_s = area of reinforcing steels as the tie rod in each part.

10.6.2.2 Layout of reinforcing steels in the anchorage zone
Layout of reinforcing steels in the anchorage zone of pretensioned members

For a single prestressing tendon, a spiral steel bar with a length not less than 150 mm over four turns must be arranged at the end of the tendon. For scattered prestressing tendons, three to five layers of steel mats perpendicular to the tendons should be employed in a length of 10 times the nominal tendon diameter from the tendon end.

Layout of reinforcing steels in the anchorage zone of post-tensioned members

In a post-tensioned simply supported beam, most of the anchorages are arranged at the beam ends (Fig. 7.13A), some may be arranged at the bottom or upper plate. In a post-tensioned continuous box girder, most of the anchorages for longitudinal tendons in the upper flange are arranged at the segment ends when the balanced cantilever method is adopted, most of the anchorages for longitudinal tendons in the bottom plate are arranged in the triangular tooth blocks (Fig. 7.13B) protruding from the bottom plate, and some are arranged at the beam ends. If necessary, some are arranged in the triangular tooth blocks (Fig. 7.13B) protruding from the webs and upper plate.

The indirect reinforcement volume ratio can be calculated from Eq. (10.21) or (10.22) with a minimum of 0.5%. The indirect steels shall be arranged within the depth specified in Fig. 9.29, where at least four turns of spiral steel bar or four layers of steel mats shall be employed. Under the anchor plate, a steel bearing plate with a thickness not less than 16 mm or with a bell tube shall be provided (Fig. 3.7). In the overall zone under and around the anchorage, the transverse steels to resist transverse splitting force, and the closed stirrups with a spacing less than 120 mm shall be arranged.

The prestressing tendons anchored on the tooth blocks should adopt a large bending radius, and the reinforcing steels should be set according to the concentrated prestressing force and the specifications. The indirect reinforcement volume ratio in the local compression zone of the tooth block can be calculated from Eq. (10.21) or (10.22), and the reinforcing steels in the overall zone can be calculated from Eq. (10.23). The JTG 3362-2018 code stipulates that the layout of reinforcing steels in the overall zone should adhere to the following specifications:

(1) Under the anchorage, the closed or U-shaped stirrups with a distance not greater than 150 mm must be arranged within a length of not less than 1.2 times the tooth depth ($A1$ in Fig. 7.16) to resist transverse splitting force.

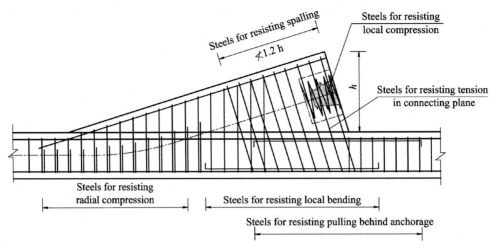

Figure 10.6 Schematic diagram of reinforcing steels in the anchorage zone.

(2) In the plan between the tooth block and the concrete plate (bottom plate, upper flange, or web of the box girder) (A3 in Fig. 7.16), the stirrups perpendicular to the plane extending into the tooth block and the concrete plate must be arranged.

(3) In the pulling region of the concrete plate behind the anchorage (A5 in Fig. 7.16), the longitudinal reinforcing steels must be arranged in the concrete plate behind the anchorage while being close to the inner side to resist the pull force. The steels should be set in a length not less than 1.5 m and within three times the width of the bearing plate laterally.

(4) On the outside of the tooth block, the longitudinal reinforcing steels should be arranged in the concrete plate to resist the local bending moment due to compression in the inner side caused by the prestressing force (A6 in Fig. 7.16). The steels should be set in a length of not less than 1.5 m and within three times the width of the bearing plate laterally.

(5) In the radial compression region of the curved tendons (A4 in Fig. 7.16), the vertical stirrups and U-shaped steels should be arranged in the concrete plate within the length of the curved segment to resist the radial compression and disintegration.

The layout of reinforcing steels in the tooth block anchorage area is shown in Fig. 10.6.

Suggested readings

[1] CCES 01: 2018. Durability design and construction of concrete structure. Beijing: China Construction Industry Press; 2018.
[2] TB100005: 2010. Code for durability design on concrete structure of railway. Beijing: China Railway Publishing House; 2010.

[3]   JTG 3362: 2018. Specifications for design of highway reinforced concrete and prestressed concrete bridges and culverts. Beijing: People's Communications Press; 2018.

[4]   Q/CR 9300: 2018. Code for design on railway bridge and culvert. Beijing: China Railway Publishing House; 2018.

[5]   TB10621: 2014. Code for design of high-speed railway bridge. Beijing: China Railway Publishing House; 2014.

[6]   GB 50010: 2015. Code for design of concrete structures. Beijing: China Construction Industry Press; 2015.

[7]   JGJ/T140: 2019. Standard for seismic design of prestressed concrete structures. Beijing: China Construction Industry Press; 2019.

[8]   Magnel G. Prestressed concrete. 3rd ed. London: Concrete Publication Ltd; 1954.

[9]   Lin TY, Burns NH. Design of prestressed concrete structures. 3rd ed. New York: John Wiley and Sons; 1981.

[10]   Naaman Antoine E. Prestressed concrete structure analysis and design. 2nd ed. New York: McGraw Hill; 2004.

[11]   Hu Di. Basic principles of prestressed concrete structure design. 2nd ed. Beijing: China Railway Publishing House; 2019.

CHAPTER 11

Analysis and design of tension and compression members

Contents

11.1 Analysis and design of tension members

Prestressed concrete tension members are divided into axial tension members and eccentric tension members, the latter being subdivided again into small eccentric tension members and large eccentric tension members. When the action point of the tension force coincides with the centroid of the cross-section of the member, it is called the axial tension member. There are few absolute axial tension members in practical engineering,

Analysis and Design of Prestressed Concrete
ISBN 978-0-12-824425-8, https://doi.org/10.1016/B978-0-12-824425-8.00011-7

since deviation of the setting action point of axial force is inevitable during construction. In an eccentric tension member, the load could be either an eccentric tension force or an axial tension force combined with a transverse bending moment.

When the external load exerted on a prestressed concrete tension member increases from zero to the magnitude that causes the member to fail, the behavior can be distinguished into three states: approximate elastic working state before cracking, working state after cracking, and failure state. The ultimate strength of tension members is obtained based on the failure state.

In the following sections, the prestressed concrete tension members with which the prestressing tendons and the concrete are bonded are discussed.

11.1.1 Stress and deformation of axial tension members

The normal stress of concrete caused by the effective prestressing force at transfer is given by

$$\sigma_{pc,0} = \frac{N_p}{A_i} \tag{11.1}$$

where N_p = effective prestressing force, given by Eq. (11.2).

A_i = cross-sectional area, converted section (A_0) is taken for pretensioned members and net section (A_n) is taken for post-tensioned members.

The effective prestressing force at transfer is given by

$$N_p = \sigma_{p0} A_p \tag{11.2}$$

where σ_{p0} = stress in prestressing tendons when the normal stress of concrete is zero, $\sigma_{p0} = \sigma_{con} - \sigma_{l,I} + \sigma_{l4}$ for pretensioned members, $\sigma_{p0} = \sigma_{con} - \sigma_{l,I}$ for post-tensioned members.

A_p = cross-sectional area of prestressing tendons.

σ_{con} = control stress in prestressing tendons at stretching.

$\sigma_{l,I}$ = first stage prestress loss, calculated by Eq. (5.81) or (5.83).

σ_{l4} = prestress loss due to elastic shortening.

Therefore, the deformation of the member at transfer can be derived from Eq. (11.1):

$$\Delta l_{pc} = \frac{N_p l_0}{E_c A_i} \tag{11.3}$$

where Δl_{pc} = deformation of the member at transfer due to the effective prestressing force.

l_0 = effective length of the axial tension member.

E_c = modulus of the elasticity of concrete.

In the service period, the prestress losses due to concrete shrinkage and creep and tendon relaxation can be derived from Eqs. (5.62) and (5.67), or obtained from the formulas provided in the design codes. If the assumption that the stress change in reinforcing steels due to concrete shrinkage and creep is taken as σ_{l6}, the change of the prestressing force after transfer can be obtained as

$$\Delta N_{ps} = \Delta N_p + \Delta N_s = A_p[\sigma_{l5}(t) + \sigma_{l6}(t)] + A_s\sigma_{l6} \tag{11.4}$$

where A_p = cross-sectional area of the prestressing tendons.

A_s = cross-sectional area of the reinforcing steels.

σ_{l5} = prestress loss due to tendon relaxation.

σ_{l6} = prestress loss due to concrete shrinkage and creep.

Therefore, the deformation change due to long-term prestress loss can be obtained as approximately:

$$\Delta l_{loss} = \frac{\Delta N_{ps}l_0}{E_c A_0} \tag{11.5}$$

And the stress change in concrete due to long-term prestress loss is obtained as approximately:

$$\Delta\sigma_{pc} = \frac{\Delta N_{ps}}{A_0} \tag{11.6}$$

The stress in concrete in the service period due to the effective prestressing force can be obtained from Eqs. (11.1) and (11.6):

$$\sigma_{pc} = \frac{N_p}{A_i} - \frac{\Delta N_{ps}}{A_0} \tag{11.7}$$

From Eqs. (11.3) and (11.5), the final deformation of the tension member due to the effective prestressing force is derived as

$$\Delta l_c = \frac{N_p l_0}{E_c A_i} - \frac{\Delta N_{ps}l_0}{E_c A_0} \tag{11.8}$$

The effective stress in prestressing tendons in the service period is

$$\sigma_{pe} = \sigma_{con} - \sigma_{l,I} - \sigma_{l,II} \tag{11.9}$$

Under the action of external loads, the deformation of the tension member and the stresses in the section are given by

$$\Delta l_c = \frac{F l_0}{E_c A_0} - \left(\frac{N_p l_0}{E_c A_i} - \frac{\Delta N_{ps}l_0}{E_c A_0}\right) \tag{11.10}$$

$$\sigma_{ct} = \frac{F}{A_0} - \left(\frac{N_p}{A_i} - \frac{\Delta N_{ps}}{A_0}\right) \tag{11.11}$$

$$\sigma_p = \sigma_{con} - \sigma_{l,I} - \sigma_{l,II} + \alpha_{EP}\frac{F}{A_0} \tag{11.12}$$

$$\sigma_s = \alpha_{ES}\left[\frac{F}{A_0} - \left(\frac{N_p}{A_i} - \frac{\Delta N_{ps}}{A_0}\right)\right] \tag{11.13}$$

where F = external axial load.

$\sigma_{l,II}$ = second-stage prestress loss, $\sigma_{l,II} = \sigma_{l5}(t) + \sigma_{l6}(t)$.

α_{ES} = modulus ratio of reinforcing steel to concrete.

α_{EP} = modulus ratio of tendon to concrete.

For prestressed axial tension members in the railway, the *Code for Design on Railway Bridge and Culvert* (Q/CR 9300-2018) prescribes that the normal stress of concrete in the service period should be controlled to a limited value. For the members in which tensile stress is not allowed, the maximum normal tensile stress of concrete should satisfy:

$$\sigma_{ct} \leqslant 0 \tag{11.14}$$

For the members in which tensile stress is allowed while cracking is not allowed, the maximum normal tensile stress of concrete should satisfy:

$$\sigma_{ct} \leqslant 0.7 f_{ctk} \tag{11.15}$$

where σ_{ct} = normal tensile stress of concrete.

f_{ctk} = characteristic tensile strength of concrete.

For prestressed axial tension members in the highway in which cracking is not allowed in the service period, the *Specifications for Design of Highway Reinforced Concrete and Prestressed Concrete Bridges, and Culverts* (JTG 3362-2018) prescribes that the maximum tensile stress in prestressing tendons should satisfy the following.

Wires and strands:

$$\sigma_p \leqslant 0.65 f_{ptk} \tag{11.16}$$

Threaded steel bars:

$$\sigma_p \leqslant 0.75 f_{ptk} \tag{11.17}$$

If the external tension force is large enough, the tension member will experience cracking, and the stresses in prestressing tendons and reinforcing steels should be calculated based on the cracked section. In this condition, the maximum crack width should be controlled. The crack width can be derived from Eqs. (8.53), (8.55), or (8.56).

11.1.2 Bearing capacity of tension members

11.1.2.1 Bearing capacity of axial tension members

The bearing capacity of axial tension members should satisfy:

$$N_d \leqslant N_u \tag{11.18}$$

where N_d = design value of axial tension load.

N_u = tension bearing capacity of the section.

Under the action of axial tension load, the reinforcing steels and prestressing tendons cannot yield simultaneously in most cases, because they have different tensile strains corresponding to their yield points. Considering that the section tension bearing capacity is calculated by design strength stipulated in the design codes rather than the yield strength, it is generally assumed that both the reinforcing steels and prestressing tendons reach their design tensile strengths at failure. Therefore, the tension bearing capacity of the normal section of prestressed concrete members under axial tension load can be given by

$$N_u = f_{sd}A_s + f_{pd}A_p \tag{11.19}$$

where f_{pd} = design tensile strength of prestressing tendon.

f_{sd} = design tensile strength of reinforcing steel.

A_p = cross-sectional area of prestressing tendons.

A_s = cross-sectional area of reinforcing steels.

11.1.2.2 Bearing capacity of eccentric tension members

The loads exerted on the eccentric tension members can be equivalent to an axial tension load and its corresponding eccentricity. According to the magnitude of the eccentricity, eccentric tension members can be grouped into small eccentric tension members and large eccentric tension members. Their working mechanism and failure patterns are different, and the corresponding calculation formulas of bearing capacity are also different.

The bearing capacity of the eccentric tension members should satisfy Eq. (11.18) and the following equation:

$$N_d e \leqslant M_u \tag{11.20}$$

where N_d = design value of axial tension load.

e = eccentricity of the axial tension.

M_u = flexural bearing capacity of the section.

The position of the axial tension load is used as the criterion to judge the small or large eccentric tension. When the tension force is in the vicinity of one side, that is, close to the center of tendons A_p and reinforcing steels A_s in one side (or A'_s and A'_p in the other side), the tension force is almost borne by the tendons and reinforcing steels in the same side while the stresses in prestressing tendons and reinforcing steels in the other side are almost zero at failure. This is the critical position distinguishing small eccentric tension and large eccentric tension.

For ease of description, set N_{ps} as the resultant provided by A_p and A_s, and N'_{ps} as the resultant provided by A'_p and A'_s, respectively.

Fig. 11.1 is a sketch illustrating the section bearing capacity of small eccentric tension and large eccentric tension members with prestressing tendons and reinforcing steels. When the eccentric tension force acts between N_{ps} and N'_{ps}, the section is subjected to small eccentric tension, as shown in Fig. 11.1A. Under the action of eccentric tension, the concrete near the tension side cracks first. As the tension force increases, the concrete opposite to the tension side also develops cracking, consequently, the tension force is entirely borne by the prestressing tendons and reinforcing steels. At tension failure, the tendons and reinforcing steels near the tension side reach the design tensile strengths,

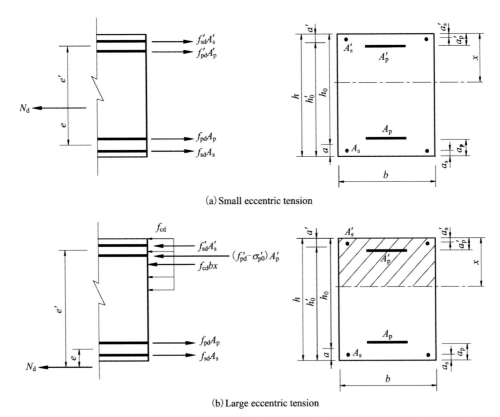

(a) Small eccentric tension

(b) Large eccentric tension

Figure 11.1 Typical prestressed concrete rectangular section subjected to eccentric tension force.

therefore, the bearing capacity of small eccentric tension members with a rectangular section can be obtained by

$$N_d e \leqslant M_u = f_{sd} A_s' \left(h_0 - d_s' \right) + f_{pd} A_p' \left(h_0 - d_p' \right) \tag{11.21}$$

$$N_d e' \leqslant M_u = f_{sd} A_s \left(h_0' - a_s \right) + f_{pd} A_p \left(h_0' - a_p \right) \tag{11.22}$$

where f_{pd} = design tensile strength of the prestressing tendon.

f_{sd} = design tensile strength of the reinforcing steel.

A_p = cross-sectional area of the prestressing tendons in one side.

A_p' = cross-sectional area of the prestressing tendons in the other side.

A_s = cross-sectional area of the reinforcing steels in one side.

A_s' = cross-sectional area of the reinforcing steels in the other side.

h_0 = effective depth of the section, $h_0 = h - a$.

a_p = depth of the prestressing tendons from the same side face.

d_p' = depth of the prestressing tendons from the same side face.

a_s = depth of the reinforcing steels from the same side face.

d_s' = depth of the reinforcing steels from the same side face

$$a = \frac{a_s A_s + a_p A_p}{A_s + A_p}, d' = \frac{A_s' d_s' + A_p' d_p'}{A_s' + A_p'}$$

When the eccentric tension force does not act between N_{ps} and N_{ps}', the member is subjected to large eccentric tension, as shown in Fig. 11.1B. Under the action of large eccentric tension, the concrete near the tension side bears tension while there is compression on the other side. As the tension force increases, the concrete near the side with tension cracks first. At failure, the tendons and reinforcing steels near the tension side reach their tensile strengths, simultaneously, the concrete on the other side ruptures when the quantity of the compression reinforcement is in a reasonable range. If the compression side is overreinforced, the compression concrete does not rupture and the compression reinforcements cannot reach the yield strength. If the compression side is underreinforced, the compression concrete ruptures suddenly before the tendons and reinforcing steels reach their tensile strengths, which is a typical brittle failure.

The calculations of bearing capacity of large eccentric tension members with a rectangular section are based on the balanced section. In Fig. 11.1B, taking the force equilibrium in the axial direction of the member and the moment equilibrium at the reference point of the resultant provided by the tendons and reinforcing steels, the formulas of bearing capacity and verification are derived as follows:

$$N_d \leq f_{sd} A_s + f_{pd} A_p - f_{sd}' A_s' - \left(f_{pd}' - \sigma_{p0}' \right) A_p' - f_{cd} bx \tag{11.23}$$

$$N_d e \leq f_{cd} b x \left(h_0 - \frac{x}{2} \right) + f'_{sd} A'_s \left(h_0 - d'_s \right) + \left(f'_{pd} - \sigma'_{p0} \right) A'_p \left(h_0 - d'_p \right) \tag{11.24}$$

The depth of the compression zone should satisfy:

$$x \leq \xi_b h_0 \tag{11.25}$$

When the tendons and reinforcing steels are arranged in the compression zone, and the tendons are in compression, the depth of neutral axis should also satisfy

$$x \geq 2d' \tag{11.26}$$

If only reinforcing steels are arranged in the compression zone, or σ'_p is tensile (negative) when both the tendons and reinforcing steels are arranged in the compression zone, the depth of the neutral axis should also satisfy

$$x \geq 2d'_s \tag{11.27}$$

11.2 Analysis and design of compression members

During the construction or in the service period, excessive tensile stress may occur in the critical sections of the reinforced concrete compression members, in this case, introducing the compressive prestressing into the tension zone rather than enlarging the sectional dimensions is a reasonable and economical solution. For example, each segment of the concrete pile with a diameter of 300–1000 mm and a length of 8–12 m is easy to be pressed into the soil and they are easily connected in the construction place to form a long pile foundation. Meanwhile, the segmental concrete piles can easily experience cracking during pressing, which seriously affects the durability in the service period. Therefore, the piles have to be made of prestressed concrete. Nowadays, in China, prestressed concrete pipe piles are widely used in the delta area where long foundations are required. Considering that the prestressing is mainly introduced to resist the tension in concrete in most cases, the prestressed concrete eccentric compression members are discussed in the following sections.

Actually, the section of the prestressed flexural member can be treated as a prestressed eccentric compression section to some extent, where an equivalent eccentric compression load acts on the prestressed section. Therefore, the stress analysis of prestressed eccentric compression sections is very similar to that of flexural sections in Chapter 7.

11.2.1 General issues on the ultimate bearing capacity of eccentric compression members

Before calculating the section bearing capacity of the compression members, the general issues, such as failure patterns, are discussed as these are the bases of establishing the bearing capacity formulas.

11.2.1.1 Failure patterns and characteristics of eccentric compression members

The experimental results show that the failure patterns of the prestressed concrete members subjected to eccentric compression load are similar to that of reinforced concrete eccentric compression members. The eccentric compression members can be divided into two groups, the members with a small eccentricity and the members with a large eccentricity, which hold different failure patterns.

Under the action of eccentric compression load, section failure may be caused by crushing of concrete in the side of compression load or by the yield of tensile prestressing tendons at the far side; the former occurs in the small eccentric compression members while the latter occurs in the large eccentric compression members.

Under compression load with a small eccentricity, the entire depth of the concrete section bears compression, or a small portion of the concrete bears tension at the far side from the compression load, as shown in Fig. 11.2A and B. In Fig. 11.2A, only part of the prestressing tendons and reinforcing steels may yield when the concrete strain reaches the ultimate value in compressing at compression failure, and the reinforcements at the far side cannot reach yield strength. In Fig. 11.2B, the compression force locates close to the centroid of the converted concrete section, the failure occurs once the compressive concrete strain reaches the ultimate value. These two kinds of failure initiate from the crushing of concrete, also called the initial compression failure.

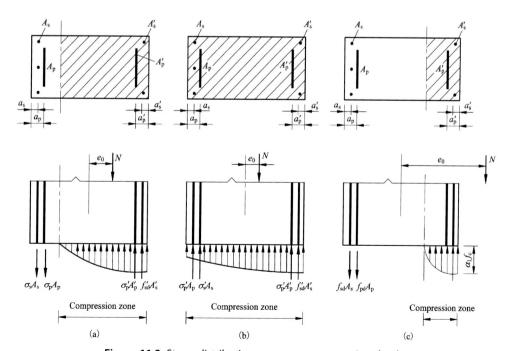

Figure 11.2 Stress distribution across concrete section depth.

Under a compression load with large eccentricity, the concrete at the loading side bears compression while the prestressing tendons and reinforcing steels on the far side bear tension, as shown in Fig. 11.2C. In a balanced section, the concrete crushes when the prestressing tendons and reinforcing steels at the far side yield, which is a typical plastic failure pattern. If a large amount of prestressing tendons and reinforcing steels is arranged in the loading side, the compressive strain in concrete at the loading side cannot reach the ultimate value, this is an overreinforced failure mode which is an obvious brittle failure pattern. In general, the section failures under compression load with a large eccentricity initiate from the yielding of the tensile reinforcement, which is also called the initial tension failure.

11.2.1.2 Basic assumptions for calculating the bearing capacity of eccentric compression members

The failures of prestressed concrete members subjected to eccentric compression are ultimately related to the crushing of concrete, although they are initiated from the crushing of concrete or from yielding of reinforcement, which is similar to that of flexural members. Therefore, the approach of bearing capacity computations for flexural sections can be used for eccentric sections, and the basic assumptions are also similar. The method for distinguishing the small eccentric compression section and the large eccentric compression section can be carried out by means of the relative depth of the compression zone as for flexural sections.

11.2.1.3 Distinguishing failure patterns due to small or large eccentric compression

Let ε_p be the strain in prestressing tendons at the far side from the compression load, the small or large eccentric compression failure of the member can be distinguished according to the following relations.

Failure due to small eccentric compression:

$$\varepsilon_p < \varepsilon_{pd}, \quad x > x_b = \xi_b h_0, \quad \xi > \xi_b$$

Boundary failure (failure in the balanced section):

$$\varepsilon_p = \varepsilon_{pd}, \quad x = x_b = \xi_b h_0, \quad \xi = \xi_b$$

Failure due to large eccentric compression

$$\varepsilon_p > \varepsilon_{pd}, \quad x < x_b = \xi_b h_0, \quad \xi < \xi_b$$

where ε_{pd} = strain in the prestressing tendons corresponding to the design tensile strength, $\varepsilon_{pd} = f_{pd} / E_p$.

f_{pd} = design tensile strength of the prestressing tendons.

ξ_b = relative boundary depth of the compression zone.

ξ = relative depth of the compression zone of the section under eccentric compression.

x = depth of the equivalent rectangular stress block.

x_b = critical depth of the neutral axis.

h_0 = effective depth of the section.

The relative boundary depth of the compression zone ξ_b can be derived from Eq. (9.17) for prestressing threaded steel bars, or from Eq. (9.19) for prestressing wires and strands.

11.2.1.4 Stresses in prestressing tendons and reinforcing steels

When the section failure occurs due to small eccentric compression, the stresses in prestressing tendons and reinforcing steels at the far side are less than their design strengths ($\sigma_p < f_{pd}$ and $\sigma_s < f_{sd}$), which can be obtained from the assumption of the plane section and strain compatibility.

For reinforcing steels:

$$\sigma_{si} = \varepsilon_{cu} E_s \left(\frac{\beta_1 h_{oi}}{x} - 1 \right) \tag{11.28}$$

$$-f'_{sd} \leqslant \sigma_{si} \leqslant f_{sd} \tag{11.29}$$

where σ_{si} = stress in reinforcing steels in the ith layer.

ε_{cu} = ultimate strain of concrete in compression.

E_s = modulus of elasticity of reinforcing steel.

f_{sd} = design tensile strength of reinforcing steel.

f'_{sd} = design strength of reinforcing steel in compression.

x = neutral axis depth.

β_1 = ratio of the depth of stress block to the actual neutral axis depth.

h_{oi} = distance from the level of the ith layer reinforcement to the edge in which the compression is larger.

When σ_{si} is the tensile stress and its value is greater than the design strength, $\sigma_{si} = f_{sd}$ is taken; when σ_{si} is the compressive stress and its absolute value is greater than the design strength, $\sigma_{si} = -f'_{sd}$ is taken.

For prestressing tendons:

$$\sigma_{pi} = \varepsilon_{cu} E_p \left(\frac{\beta_1 h_{oi}}{x} - 1 \right) + \sigma_{poi} \tag{11.30}$$

$$-\left(f'_{pd} - \sigma_{poi} \right) \leqslant \sigma_{pi} \leqslant f_{pd} \tag{11.31}$$

where σ_{pi} = stress in prestressing tendons in the ith layer.

f_{pd} = design tensile strength of the prestressing tendon.

f'_{pd} = design strength of the prestressing tendon in compression.

E_p = modulus of elasticity of the prestressing tendon.

σ_{poi} = stress in prestressing tendons in the ith layer corresponding to the state of zero normal stress in concrete in the same level.

When σ_{pi} is the tensile stress and its value is greater than the design strength, $\sigma_{pi} = f_{pd}$ is taken; when σ_{pi} is the compressive stress and its absolute value is greater than $\left(f'_{pd} - \sigma_{poi}\right)$, $\sigma_{pi} = -\left(f'_{pd} - \sigma_{poi}\right)$ is taken.

In Eqs. (11.28) and (11.30), the strains in reinforcing steels and prestressing tendons depend on their moduli and the ultimate strain of concrete in compression. For simplification, the stresses in reinforcing steels and prestressing tendons can be approximately calculated by the following formulas.

For reinforcing steels:

$$\sigma_{si} = \frac{f_{sd}}{\xi_b - \beta_1}\left(\frac{x}{h_{oi}} - \beta_1\right) \tag{11.32}$$

For prestressing tendons:

$$\sigma_{pi} = \frac{f_{pd} - \sigma_{poi}}{\xi_b - \beta_1}\left(\frac{x}{h_{oi}} - \beta_1\right) \tag{11.33}$$

11.2.1.5 Increasing factor of eccentricity

For eccentric compression members with a high slenderness ratio, the bending will be generated by the eccentric compression load. This additional bending will increase the eccentricity of the compression load, which must be taken into account in the calculation of bearing capacity in the case of $l_0/i \geq 17.5$, where l_0 is the effective length of the member and i is the radius of gyration of section. The influence of increased eccentricity can be described by an increasing factor for eccentricity, η.

The Q/CR 9300-2018 code stipulates that the increasing factor of eccentricity is given by

$$\eta = \left(1 + \frac{1}{1400e_i/h_0}\right)\left(\frac{l_0}{h}\right)^2 \zeta_1 \zeta_2 \tag{11.34}$$

$$\zeta_1 = 0.2 + 2.7\frac{e_i}{h_0} \leqslant 1.0 \tag{11.35}$$

$$\zeta_2 = 1.15 - 0.01\frac{l_0}{h} \leqslant 1.0 \tag{11.36}$$

$$e_i = e_a + e_0 \tag{11.37}$$

where l_0 = effective length of the member.

e_0 = initial eccentricity of the compression load with respect to the centroid of the converted section, $e_0 = M_d/N_d$.

e_a = additional eccentricity due to bending, the larger value of 20 mm and 1/30th of the maximum dimension in section along the direction of the compression load is adopted.

h_0 = effective depth of the section.

h = overall depth of the section.

ζ_1 = influence coefficient for load eccentricity on the sectional curvature.

ζ_2 = influence coefficient for slenderness ratio on the section curvature.

The JTG 3362-2018 code stipulates that the increasing factor of eccentricity is given by

$$\eta = \left(1 + \frac{1}{1300\frac{e_0}{h_0}}\right)\left(\frac{l_0}{h}\right)^2 \zeta_1 \zeta_2 \tag{11.38}$$

11.2.2 Ultimate bearing capacity of eccentric compression members

The commonly used cross-sections of eccentric compression members include rectangular, T, I, box, circular, and annular shapes. For prestressed rectangular, T, I, and box sections under eccentric compression, it is easy to establish the bearing capacity calculation formulas by referring to those in flexural members according to the basic assumptions and equilibrium equations. Meanwhile, the calculation is rather cumbersome for circular and annular sections if the equilibrium equations are established directly according to the basic assumptions, so the trapezoidal stress distribution of the reinforcements around the section is commonly simplified into an equivalent rectangular stress block.

Along the direction of the eccentric compression load, the bearing capacity of normal sections under eccentric compression should satisfy:

$$N_d \leqslant N_u \tag{11.39a}$$

$$N_d e \leqslant M_u \tag{11.39b}$$

where N_d = design value of compression load.

N_u = compression bearing capacity of the normal section.

e = eccentricity of the compression load.

M_u = flexural bearing capacity of the normal section.

For eccentric compression members with rectangular, T, or I section, in addition to calculating the flexural bearing capacity given by Eq. (11.39b), the axial compression bearing capacity perpendicular to the bending moment plane should also be checked in which the stability coefficient should be taken into account.

For slender members, the axial compression bearing capacity perpendicular to the bending moment plane should satisfy:

$$N_d \leqslant \varphi N_u \tag{11.40}$$

where N_u = axial compression bearing capacity of the normal section.

φ = stability coefficient of the member in the plane perpendicular to the bending moment.

11.2.2.1 Bearing capacity of rectangular section under eccentric compression

The calculation diagram of the normal section bearing capacity of a rectangular section under eccentric compression is shown in Fig. 11.3.

Establishing an equilibrium equation along the axial direction of the member:

$$N_d \leqslant N_u = f_{cd}bx + f'_{sd}A'_s + \left(f'_{pd} - \sigma'_{po}\right)A'_p - \sigma_s A_s - \sigma_p A_p \tag{11.41}$$

Establishing a moment equilibrium equation about the point of the resultant provided by the prestressing tendons and reinforcing steels at the far side from the compression load:

$$N_d e \leqslant M_u = f_{cd}bx\left(h_0 - \frac{x}{2}\right) + f'_{sd}A'_s\left(h_0 - d'_s\right) + \left(f'_{pd} - \sigma'_{po}\right)A'_p\left(h_0 - d'_p\right) \tag{11.42}$$

where h_0 = effective depth of the section.

e = final eccentricity of the compression load considering the effect of bending.

f_{cd} = design strength of concrete in compression.

f'_{pd} = design strength of the prestressing tendon in compression.

f'_{sd} = design strength of the reinforcing steel in compression.

σ'_p = stress in prestressing tendons at the loading side, $\sigma'_p = f'_{pd} - \sigma'_{p0}$.

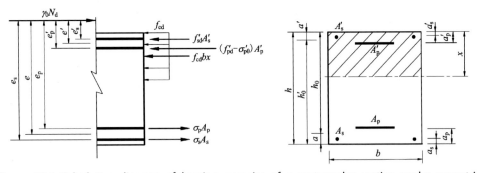

Figure 11.3 Calculation diagram of bearing capacity of a rectangular section under eccentric compression.

σ'_{p0} = stress in the prestressing tendons at the loading side corresponding to zero concrete strain at the level of tendons.

σ_s = stress in the reinforcing steels at the far side, given by Eq. (11.32).

σ_p = stress in the prestressing tendons at the far side, given by Eq. (11.33).

A_p = cross-sectional area of the prestressing tendons at the far side.

A'_p = cross-sectional area of the prestressing tendons at the loading side.

A_s = cross-sectional area of the reinforcing steels at the far side.

A'_s = cross-sectional area of the reinforcing steels at the loading side.

d'_p = depth of the prestressing tendons at the loading side from the same side face.

d'_s = depth of the reinforcing steels at the loading side from the same side face.

For railway columns, the Q/CR 9300-2018 code stipulates that the final eccentricity e in Eq. (11.42) is given by

$$e = \eta e_i + \gamma_{sp} \tag{11.43}$$

$$e_i = e_0 + e_a \tag{11.44}$$

where e_0 = initial eccentricity of the compression load with respect to the centroid of the converted section, $e_0 = M_d/N_d$.

e_a = additional eccentricity due to bending, $e_a = h/30$ can be taken.

γ_{sp} = distance between the point of the resultant provided by the prestressing tendons and reinforcing steels to the centroid of the converted section.

η = increasing factor for eccentricity, given by Eq. (11.34)

For highway columns, the JTG 3362-2018 code stipulates that the final eccentricity e in Eq. (11.42) is given by

$$e = \eta e_0 + \frac{h}{2} - a \tag{11.45}$$

where η = increasing factor for eccentricity, given by Eq. (11.38).

a = distance from the centroid of the prestressing tendons and reinforcing steels at the far side to the same side face, $a = \dfrac{a_s A_s + a_p A_p}{A_s + A_p}$.

σ_s and σ_p in Eq. (11.41) can be determined as follows:

(1) When $\xi \leqslant \xi_b$, the failure due to large eccentric compression, $\sigma_s = f_{sd}$ and $\sigma_p = f_{pd}$ are taken.

(2) When $\xi > \xi_b$, the failure due to small eccentric compression, σ_s and σ_p are calculated from Eqs. (11.28)−(11.31) or from Eqs. (11.32) and (11.33).

The depth of the neutral axis should satisfy:

$$x \geqslant 2d', \quad x \geqslant 2d'_s$$

For small eccentric compression members, when the compression load acts between the resultants provided by A'_s, A'_p and by A_s, A_p, respectively, the compression bearing capacity should also satisfy:

$$N_d e' \leqslant M_u = f_{cd} bh\left(h'_0 - \frac{h}{2}\right) + f_{sd} A_s \left(h'_0 - a_s\right) + \left(f_{pd} - \sigma_{po}\right) A_p \left(h'_0 - a_p\right) \qquad (11.46)$$

$$e' = \frac{h}{2} - d' - (e_0 - e_a) \qquad (11.47)$$

where $h'_0 = h - d'$

$$d' = \frac{A'_s d'_s + A'_p d'_p}{A'_s + A'_p}$$

11.2.2.2 Bearing capacity of the T-section under eccentric compression

When the eccentric compression load acts on the flange side of the T-section and the depth of the compression zone falls within the flange, the bearing capacity of the normal section can be calculated by the formulas for rectangular sections. When the depth of the compression zone falls outside the flange ($x > h'_f$), the compression bearing capacity can be obtained by establishing an equilibrium equation along the axial direction of the member (Fig. 11.4):

$$N_d \leqslant N_u = f_{cd}\left[bx + \left(b'_f - b\right)h'_f\right] + f'_{sd} A'_s + \left(f'_{pd} - \sigma'_{po}\right) A'_p - \sigma_s A_s - \sigma_p A_p \qquad (11.48)$$

where h'_f = flange thickness in a T-section.

b'_f = flange width in a T-section.

b = web width in a T-section.

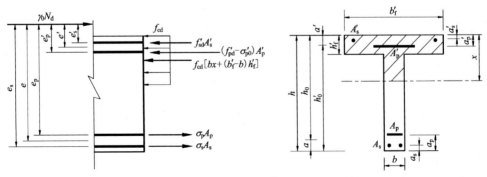

Figure 11.4 Calculation diagram of bearing capacity of a T-section under eccentric compression.

In Fig. 11.4, establishing a moment equilibrium equation about the point of resultant provided by the prestressing tendons and reinforcing steels at the far side from the compression load gives

$$N_d e \leqslant M_u = f_{cd}\left[bx\left(h_0 - \frac{x}{2}\right) + \left(b_f' - b\right)h_f'\left(h_0 - \frac{h_f'}{2}\right)\right]$$
$$+ f_{sd}'A_s'\left(h_0 - d_s'\right) + \left(f_{pd}' - \sigma_{po}'\right)A_p'\left(h_0 - d_p'\right) \tag{11.49}$$

For small eccentric compression members, when the compression load acts between the resultants provided by A_s', A_p' and by A_s A_p, respectively, the bearing capacity should also satisfy:

$$N_d e' \leqslant M_u = f_{cd}\left[bh\left(h_0' - \frac{h}{2}\right) + \left(b_f' - b\right)h_f'\left(\frac{h_f'}{2} - d'\right)\right]$$
$$+ f_{sd}'A_s\left(h_0' - a_s\right) + \left(f_{pd}' - \sigma_{po}\right)A_p\left(h_0' - a_p\right) \tag{11.50}$$

For members under small eccentric compression, when the compression load acts on the side far from the flange, the bearing capacity should also satisfy:

$$N_d e' \leqslant M_u = f_{cd}\left[bh\left(h_0' - \frac{h}{2}\right) + \left(b_f - b\right)h_f\left(h_0' - \frac{h_f}{2}\right)\right]$$
$$+ f_{sd}'A_s\left(h_0' - a_s\right) + \left(f_{pd}' - \sigma_{po}\right)A_p\left(h_0' - a_p\right) \tag{11.51}$$

where h_f = flange thickness in a T-section at the far side.

b_f = flange width in a T-section at the far side.

e' = distance between the compression load and the resultant provided by A_s' and A_p', given by Eq. (11.47).

11.2.2.3 Bearing capacity of circular and annular sections under eccentric compression

For eccentric compression members with an annular section arranged with longitudinal reinforcements uniformly along the periphery (Fig. 11.5), if $r_1/r_2 \geq 0.5$, the normal section bearing capacity can be calculated by

$$N_d \leqslant N_u = \alpha f_{cd}A - \sigma_{po}A_p + \alpha f_{pcd}A_p - \alpha_t\left(f_{ptd} - \sigma_{po}\right)A_p \tag{11.52}$$

$$N_d \eta e_i \leqslant M_u = f_{cd}A\frac{r_1 + r_2}{2}\cdot\frac{\sin\pi\alpha}{\pi} + f_{pcd}A_p r_p\frac{\sin\pi\alpha}{\pi} + \left(f_{ptd} - \sigma_{po}\right)A_p r_p\frac{\sin\pi\alpha_t}{\pi} \tag{11.53}$$

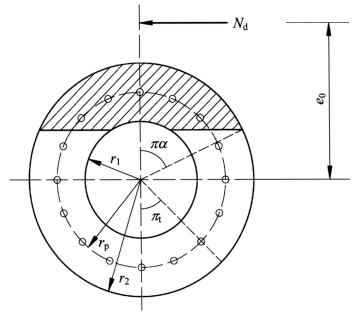

Figure 11.5 Calculation diagram of the bearing capacity of the annular section under eccentric compression.

$$\alpha_t = 1 - 1.5\alpha \tag{11.54}$$

$$e_i = e_0 + e_a \tag{11.55}$$

where A = cross-sectional area of the annular section.

A_p = cross-sectional area of prestressing tendons.

α = ratio of the compressed concrete area to the entire cross-sectional area.

α_t = ratio of prestressing tendon area in the compression zone to entire prestressing tendon area, $\alpha_t = 0$ if $\alpha > 2/3$.

r_1 = inner radius of the annular section.

r_2 = outer radius of the annular section.

r_p = radius of the prestressing tendons.

e_0 = eccentricity of the compression load with respect to the centroid of the converted section.

e_a = additional eccentricity, $e_a = 0.03(r_2 + r_s)$ or $e_a = 0.03(r_2 + r_p)$ is taken.

If $\alpha < \arccos\dfrac{\dfrac{2r_1}{r_1 + r_2}}{\pi}$, the bearing capacity of the annular section can be calculated by the circular section under eccentric compression, as

$$N_d \leq N_u = \alpha f_{cd} A\left(1 - \frac{\sin 2\alpha}{2\pi\alpha}\right) - (\alpha - \alpha_t)f_{pcd}A_p \tag{11.56}$$

$$N_d \eta e_i \leq M_u = \frac{2}{3} f_{cd} A \frac{\sin^3 \pi\alpha}{\pi} + f_{pcd} A_p r_p \frac{\sin \pi\alpha + \sin \pi\alpha_t}{\pi} \qquad (11.57)$$

$$\alpha_t = 1.25 - 2\alpha \qquad (11.58)$$

$$e_i = e_0 + e_a \qquad (11.59)$$

Only when the number of longitudinal prestressing tendons in the annular or circular section exceeds 6 can Eqs. (11.52)–(11.59) be used, otherwise, the equations should be adjusted.

11.2.2.4 Section bearing capacity under biaxial eccentric compression

When the section with two mutually perpendicular symmetrical axes is subjected to biaxial (two-direction) eccentric compression loads, the bearing capacity of the normal sections can be achieved approximately by the correlation relation:

$$N_u = \frac{1}{\dfrac{1}{N_{ux}} + \dfrac{1}{N_{uy}} - \dfrac{1}{N_{uo}}} \qquad (11.60)$$

where N_{uo} = design value of the axial compression bearing capacity of the section.

N_{ux} = design value of the eccentric compression bearing capacity of the section in the x direction corresponding to the compression load with eccentricity of $\eta_x e_{ix}$, in which all prestressing tendons and reinforcing steels in the section are taken into account.

N_{uy} = design value of the eccentric compression bearing capacity of section in the y direction corresponding to the compression load with eccentricity of $\eta_y e_{iy}$, in which all prestressing tendons and reinforcing steels in the section are taken into account.

η_x = eccentricity increasing factor in the x direction, given by Eq. (11.34) or (11.38).

η_y = eccentricity increasing factor in the y direction, given by Eq. (11.34) or (11.38).

e_{ix} = initial eccentricity in the x direction, $e_{ix} = M_{dx}/N_{dx}$.

e_{iy} = initial eccentricity in the y direction, $e_{iy} = M_{dy}/N_{dy}$.

M_{dx} = design value of the bending moment.

N_{dx} = design value of the compression load in the x direction.

M_{dy} = design value of the bending moment.

N_{dy} = design value of the compression load in the y direction.

11.2.2.5 Compression bearing capacity and stability of prestressed concrete members at transfer

For post-tensioned members, the maximum compression force on the section equals the jacking force if all tendons are stretched simultaneously. If the tendons are stretched in batches, the maximum compression force appears when the final batch of tendons is stretched. For pretensioned members, the maximum eccentric compression force due to the prestressing force occurs at transfer. Under maximum eccentric compression, the slender eccentric compression members including the prestressed flexural members with high span−depth ratio possibly achieve overall stability. The Q/CR 9300-2018 code stipulates that the strength and overall stability of the members shall be verified during stretching or at transfer.

Axial compression force on the section during stretching or at transfer

For pretensioned members subjected to axial compression or eccentric compression with small eccentricity $(x > 0.55h'_0)$, the axial compression force exerted on the concrete section can be calculated by

$$N_p = \left(\sigma_{con} - \sigma_{l,t} - f'_{pd} \right) A_p + \left(\sigma'_{con} - \sigma'_{l,t} \right) A'_p \tag{11.61}$$

where N_p = axial compression force exerted on the concrete section at transfer.

A_p = cross-sectional area of the prestressing tendons at the far side.

A'_p = cross-sectional area of the prestressing tendons at the loading side.

f'_{pd} = design strength of the prestressing tendons in compression.

σ_{con} = control stress in the tendons at stretching at the far side, given by Eq. (5.2).

σ'_{con} = control stress in the tendons at stretching at the loading side, given by Eq. (5.2).

$\sigma_{l,t}$ = prestress loss in the tendons at the far side during stretching, where σ_{l4} is deducted.

$\sigma'_{l,t}$ = prestress loss in the tendons at the loading side during stretching, where σ'_{l4} is deducted.

For pretensioned members subjected to eccentric compression with a large eccentricity $(x < 0.55h'_0)$, the axial compression force exerted on the concrete section can be calculated by

$$N_p = \left(\sigma_{con} - \sigma_{l,t} - f'_{pd} \right) A_p \tag{11.62}$$

For post-tensioned members in which all tendons are stretched simultaneously, the axial compression force exerted on the concrete section can be calculated by

$$N_p = \sigma_{con} A_p + \sigma'_{con} A'_p \tag{11.63}$$

where N_p = axial compression force exerted on the concrete section during stretching.

For post-tensioned members subjected to axial compression or eccentric compression with small eccentricity in the case of stretching the tendons in batches, the axial compression force exerted on the concrete section can be calculated by

$$N_p = \left(\sigma_{con} - \sigma_{l,t} - \sigma_{ps}\right)\left(A_p - A_{pm}\right) + \left(\sigma'_{con} - \sigma'_{l,t}\right)\left(A'_p - A'_{pm}\right) + \sigma_{con}A_{pm} + \sigma'_{con}A'_{pm}$$

(11.64)

$$\sigma_{ps} = f'_{pd}\frac{A_p - A_{pm}}{A_p} \leqslant 300 \text{ (MPa)}$$

(11.65)

where N_p = axial compression force exerted on the concrete section during stretching.

A_p = total cross-sectional area of the prestressing tendons at the far side.

A'_p = total cross-sectional area of the prestressing tendons at the loading side.

A_{pm} = lcross-sectional area of the prestressing tendons at the far side stretched in the last batch.

A'_{pm} = cross-sectional arca of the prestressing tendons at the loading side stretched in the last batch.

σ_{ps} = stress reduction in the prestressing tendons prior to failure, MPa.

For post-tensioned members subjected to eccentric compression with a large eccentricity in the case of stretching the tendons in batches, the axial compression force exerted on the concrete section can be calculated by

$$N_p = \left(\sigma_{con} - \sigma_{l,t} - \sigma_{ps}\right)\left(A_p - A_{pm}\right) + \sigma_{con}A_{pm} + \sigma'_{con}A'_{pm}$$

(11.66)

Verification for stability

For slender members subjected to eccentric compression loads during stretching or at transfer, the stability in the plane perpendicular to the bending moment can be verified by

$$N_p \leqslant \frac{1}{\gamma_d}\left[\varphi_s\left(f_{ck}A_n + f'_{sd}A_s\right)\right]$$

(11.67)

where γ_d = comprehensive partial factor of stability during stretching or at transfer, 1.6 can be taken.

φ_s = stability coefficient of the member in the plane perpendicular to the bending moment, given by Table 11.1.

N_p = axial compression force exerted on the concrete section during stretching or at transfer.

A_n = cross-sectional area of the concrete section.

A_s = cross-sectional area of the reinforcing steels.

f_{ck} = characteristic strength of the concrete in compression.

f'_{sd} = design strength of the reinforcing steels in compression.

For pretensioned members, 1.0 is taken as the stability coefficient in Eq. (11.67).

Table 11.1 Stability coefficient.

l_0/b	<4	4	6	8	10	12	14	16	18	20	22	24	26	28	30
l_0/i	<14	14	21	28	35	42	48	55	62	69	76	83	90	97	104
φ_s	1.00	0.98	0.96	0.91	0.86	0.82	0.77	0.72	0.68	0.63	0.59	0.55	0.51	0.47	0.44

Notes: l_0 is the effective length of the member; b is the section width, and i is the radius of gyration of the section.

Verification for bearing capacity

For members subjected to large eccentric compression during stretching or at transfer, the bearing capacity should satisfy

$$N_p \leqslant \frac{1}{\gamma_k} \left(f_{ck} A_c + f'_{sd} A'_s - f_{sd} A_s - f_{pt} A_p \right) \qquad (11.68)$$

If the self-weight of the member needs to be taken into account, the bearing capacity should also satisfy

$$N_p e' \pm M_g \leqslant \frac{1}{\gamma_k} \left[f_{ck} A_c Z + f'_{sd} A_s \left(h'_0 - a_s \right) - f_{pt} A'_p \left(a'_p - a'_s \right) \right] \qquad (11.69)$$

where M_g = bending moment due to the self-weight of the member.

e' = distance between N_p and the centroid of A'_s.

Z = distance between the concrete compressive stress resultant and the centroid of A'_s.

γ_k = adjusting coefficient of the bearing capacity during stretching or at transfer, 1.8 is taken generally.

f_{pt} = calculating value of the tensile strength of the prestressing tendons, $f_{pt} = 0.9 f_{ptk}$ can be taken.

f_{ptk} = characteristic tensile strength of the prestressing tendon.

f_{sd} = design tensile strength of the reinforcing steel.

f'_{sd} = design strength of the reinforcing steel in compression.

A_c = area of concrete in compression.

A_p = cross-sectional area of the prestressing tendons at the far side.

A_s = cross-sectional area of the reinforcing steels at the far side.

A'_s = cross-sectional area of the reinforcing steels at the loading side.

a'_p = depth of the prestressing tendons at the loading side from the same side face.

a'_s = depth of the reinforcing steels at the loading side from the same side face.

For members subjected to small eccentric compression during stretching or at transfer, the bearing capacity should satisfy

$$N_p \leqslant \frac{N_0}{\left[1 + \left(\dfrac{N_0}{N_j} - 1 \right) \dfrac{e'_0}{e'_j - c'} \right] \gamma_k} \qquad (11.70)$$

$$N_0 = f_{ck} A_c + f'_{sd} \left(A_s + A'_s \right) \qquad (11.71)$$

$$N_j = 0.55 f_{ck} b h'_0 + f_{ck} A_f + f'_{sd} A_s - f_{sd} A'_s \qquad (11.72)$$

$$e'_j = \frac{0.4 f_{ck} b h'^2_0 + f_{ck} A_f Z_f + f'_{sd} A_s \left(h'_0 - a_s \right)}{N_j} \qquad (11.73)$$

$$c' = \frac{0.5 f_{ck} b h_0'^2 + f_{ck} A_f Z_f + f_{sd}' A_s \left(h_0' - a_s \right)}{N_0} \tag{11.74}$$

where $N_0 =$ axial compression force corresponds to uniform compression failure in the whole section.

$N_j =$ compression force when the section experiences eccentric compression failure in the boundary state of large and small eccentricity.

$A_f =$ cross-sectional area of the overhanging part of the flange in compression, A_f in T- or I-section can be neglected if A_f is located at the far side.

$e_0' =$ distance between N_p and the centroid of the concrete cross-section, if necessary the effect of self-weight of the member should be taken into account.

$e_i' =$ distance between N_j and the centroid of A_s'.

$Z_f =$ distance between compression $f_{ck} A_f$ and the centroid of A_s'.

$c' =$ distance between N_0 and the centroid of A_s'.

11.3 Example 11.1

A pretensioned column with an effective length of 5 m has a rectangular section with a width of 220 mm and an overall depth of 80 mm. C40 concrete, three prestressing threaded steel bars with a diameter of 12 mm and $f_{ptk} = 785$ MPa, HRB400 reinforcing steels with $2\Phi14$ for A_s' and $2\Phi10$ for A_s (reinforcing steels are structural reinforcements which do not contribute to the bearing capacity) are used. The dimension and reinforcement layout are shown in Fig. 11.6. In the service period, the effective stress in threaded steels is 450.2 MPa, a compression load with 1.230×10^3 kN and an eccentricity of 175 mm are applied on the top of the column. Calculate and verify the bearing capacity of the column according to the JTG 3362-2018 code.

11.4 Solution

11.4.1 Parameters

$\sigma_{pe} = 450.2$ MPa, $N_d = 1.230 \times 10^3$ kN, $e_0 = 175$ mm

$f_{cd} = 18.4$ MPa, $f_{pd} = 650$ MPa, $f_{pd}' = 400$ MPa, $f_{sd}' = 330$ MPa

$E_p = 2.0 \times 10^5$ MPa, $E_s = 2.0 \times 10^5$ MPa, $E_c = 3.25 \times 10^4$ MPa

$\varepsilon_{cu} = 0.0033$, $\beta = 0.80$

$l_0 = 5$ m, $b = 220$ mm, $h_0 = 680$ mm

$a = a_p = 60$ mm, $d' = d_s' = 35$ mm

$h_0 = h - a = 680 - 60 = 620$ mm

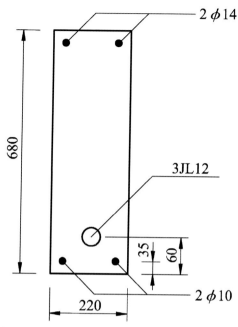

Figure 11.6 Arrangement of prestressing tendons and reinforcing steels.

$$A_p = 3 \times \frac{\pi}{4} \times 12^2 = 339 \text{ mm}^2$$

$$A'_s = 2 \times \frac{\pi}{4} \times 14^2 = 308 \text{ mm}^2$$

$$\alpha_{EP} = \frac{E_p}{E_c} = 5.49, \ \alpha_{ES} = \frac{E_s}{E_c} = 5.63$$

11.4.2 Calculate the eccentricity increasing factor

$$\varsigma_1 = 0.2 + 2.7\frac{e_0}{h_0} = 0.2 + 2.7 \times \frac{175}{620} = 0.962$$

$$\varsigma_2 = 1.15 - 0.01\frac{l_0}{h} = 1.15 - 0.01 \times \frac{5000}{680} = 1.076 > 1.0, \ \varsigma_2 = 1.0 \text{ is adopted}$$

$$\eta = 1 + \frac{1}{1400\dfrac{e_0}{h_0}}\left(\frac{l_0}{h}\right)^2 \times \varsigma_1\varsigma_2$$

$$= 1 + \frac{1}{1400 \times 175/620} \times (5000/680)^2 \times 0.962 \times 1.0 = 1.132$$

11.4.3 Judge the type of eccentric compression

$e_0 = 175 \text{ mm} < 0.3h_0 = 0.3 \times 620 = 186 \text{ mm}$

The value of eccentricity is relatively small, so the trial calculation shall be carried out based on the small eccentric compression.

Taking an equilibrium equation along the axial direction of the member:

$$f_{cd}bx(e - h_0 - x) = \sigma_p A_p e_p + \sigma_s A_s e_s$$

where $\sigma_p = \varepsilon_{cu}E_p\left(\dfrac{\beta h_0}{x} - 1\right) + \sigma_{p0}$ which should satisfy $-\left(f'_{pd} - \sigma_{p0}\right) \leqslant \sigma_p \leqslant f_{pd}$.

Considering $A_s = 0$, from which

$$f_{cd}bx\left(e_s - h_0 + \frac{x}{2}\right) + f'_{sd}A'_s e'_s = \left[\varepsilon_{cu}E_p\left(\frac{\beta h_0}{x} - 1\right) + \sigma_{p0}\right]A_p e_p$$

The above equation is simplified into

$$\frac{1}{2}f_{cd}bx^3 + f_{cd}b(e - h_0)x^2 + \left(f'_{sd}A'_s e'_s + \varepsilon_{cu}E_p A_p e_p - \sigma_{p0}A_p e_p\right)x - \varepsilon_{cu}E_p\beta h_0 A_p e_p = 0$$

When all parameters are input into the above equation, the following equation is obtained:

$$x^3 - 284x^2 + 21356x - 2.621 \times 10^7 = 0$$

Solving
$x = 397 \text{ mm}$

$$\xi = \frac{x}{h_0} - \frac{397}{620} = 0.640 > \xi_b = 0.614$$

So the column is under the action of small eccentric compression, which is consistent with the hypothesis. Thus

$$e = \eta e_0 + \frac{h}{2} - a = 1.132 \times 175 + \frac{680}{2} - 60 = 478 \text{ mm}$$

$e_p = e = 478 \text{ mm}$

$$e'_s = \frac{h}{2} - \eta e_0 - d' = \frac{680}{2} - 1.132 \times 175 - 35 = 107 \text{ mm}.$$

For prestressing threaded steel bars, the relative boundary depth of the compression zone can be derived from Eq. (9.17), where $\sigma_{p0} = \sigma_{pe}$, giving

$$\xi_b = \frac{\beta}{1 + \dfrac{f_{pd} - \sigma_{p0}}{E_p\varepsilon_{cu}}} = \frac{0.8}{1 + \dfrac{650 - 450.2}{2.0 \times 10^5 \times 0.0033}} = 0.614$$

11.4.4 Calculate and verify the bearing capacity

In a section under small eccentric compression, the stress in prestressing steels can obtain by

$$\sigma_p = \varepsilon_{cu}E_p\left(\frac{\beta h_0}{x} - 1\right) + \sigma_{p0} = 0.0033 \times 2.0 \times 10^5 \times \left(\frac{0.8 \times 620}{397} - 1\right)$$

$$+ 450.2 = 614.8 \text{ MPa}.$$

Since $\sigma_p = 614.8 \text{ MPa} < f_{pd} = 650 \text{ MPa}$, the requirement prescribed in the code is met.

The section bearing capacity is given by

$$N_u = f_{cd}bx + f'_{sd}A'_s - \sigma_pA_p = 18.4 \times 220 \times 397 + 280 \times 308 - 614.8 \times 339$$

$$= 1.485 \times 10^3 \text{ kN}.$$

$$M_u = f_{cd}bx\left(h_0 - \frac{x}{2}\right) + f'_{sd}A'_s\left(h_0 - d'_s\right) = 18.4 \times 220 \times 397 \times \left(620 - \frac{397}{2}\right)$$

$$+ 280 \times 308 \times (620 - 35) = 7.278 \times 10^2 \text{ kM.m}.$$

Since

$$N_u = 1.485 \times 10^3 \text{ kN} > \gamma_0 N_d = 1.230 \times 10^3 \text{ kN}$$
$$M_u = 727.8 \text{ kN.m} > \gamma_0 N_d e = 587.9 \text{ kN.m}.$$

The bearing capacity of the column under eccentric compression meets the specifications in the code.

Suggested readings

[1] Q/CR 9300:2018. Code for design on railway bridge and culvert. Beijing: China Railway Publishing House; 2018.
[2] JTG 3362:2018. Specifications for design of highway reinforced concrete and prestressed concrete bridges and culverts. Beijing: People's Communications Press; 2018.
[3] GB 50010: 2015. Code for design of concrete structures. Beijing: China Construction Industry Press; 2015. culverts. Beijing: People's Communications Press, 2015.
[4] Hu Di. Basic principles of prestressed concrete structure design. 2nd ed. Beijing: China Railway Publishing House; 2019.
[5] Naaman Antoine E. Prestressed concrete structure analysis and design. 2nd ed. New York: McGraw Hill; 2004.

CHAPTER 12

Analysis and design of unbonded prestressed concrete flexural structures

Contents

12.1 General concepts on unbonded prestressed concrete flexural structures

The concrete structures in which the prestressing tendons are inside the concrete and not in contact with the concrete directly are called unbonded prestressed concrete structures, and the corresponding tendons are called unbonded prestressing tendons. An unbonded tendon is a single or bundle of tendons wrapped with a special anticorrosion lubricating coating and plastic sheath. The commonly used materials for unbonded tendons are prestressing steel wires, strands, threaded bars, fiber-reinforced polymer (FRP) strands, and FRP rods. During construction, the unbonded tendons are arranged as reinforcing steels before casting in concrete. After the concrete is set and hardens to a sufficient strength, they are stretched and anchored.

Generally, two ends of the unbonded tendons are anchored close to the end of the concrete structures. If there is no or a low quantity of reinforcing steels in the concrete, the bending behavior of this unbonded prestressed concrete structure is similar to that of a plain concrete structure prestressed with unbonded tie rods. Considering a simply supported beam, once one or a few flexural cracks develop in the midspan region, the cracks

Analysis and Design of Prestressed Concrete
ISBN 978-0-12-824425-8, https://doi.org/10.1016/B978-0-12-824425-8.00012-9

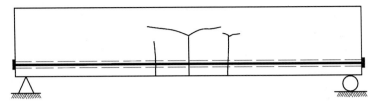

Figure 12.1 Cracking of an unbonded prestressed concrete beam without reinforcing steels.

propagate upward quickly and extend to both sides, then fail suddenly as the external loads increase slightly, as shown in Fig. 12.1. Therefore, a large amount of reinforcing steels must be arranged in the unbonded prestressed concrete flexural structures.

The *Technical specifications for concrete structures prestressed with unbonded tendons* (JGJ 92–2016) code stipulates that the minimum area of longitudinal reinforcing steels of an unbonded prestressed concrete beam shall be the larger of

$$A_s = \frac{1}{3}\left(\frac{\sigma_{pu}h_p}{f_{sd}h_s}\right)A_p \tag{12.1}$$

$$A_s = 0.003bh \tag{12.2}$$

where σ_{pu} = design value of tensile stress of the unbonded prestressing tendon in the calculation of normal bearing capacity.

f_{sd} = design tensile strength of the reinforcing steel.

h_p = distance from the level of the unbonded prestressing tendons in the tension zone to the extreme compressive fiber of concrete.

h_s = distance from the level of the reinforcing steels in the tension zone to the extreme compressive fiber of concrete.

A_p = area of unbonded prestressing tendons in the tension zone.

A_s = area of longitudinal reinforcing steels.

b = width of the section.

h = depth of the section.

For unbonded prestressed concrete slabs under unidirectional force, the minimum area of longitudinal reinforcing steels is

$$A_s = 0.002bh \tag{12.3}$$

For two-way slabs with peripheral supports, the minimum reinforcing steel ratio in the longitudinal and transverse direction is 0.05%, respectively. For two-way slabs in the slab-column structural system, the minimum ratio and layout of reinforcing steels should be determined by the structural requirements and crack resistance requirements combined with the load conditions.

The ACI-318 code specifies that the minimum area of longitudinal reinforcing steels of unbonded prestressed concrete structures is

$$A_{s,min} = 0.004A_{ct} \tag{12.4}$$

where $A_{s,min}$ = minimum area of longitudinal reinforcing steels.

A_{ct} = area of that part of the cross-section between the center of gravity of the cross-section to the extreme tension fiber of the concrete.

Generally, unbonded prestressed concrete structures have the following characteristics:

(1) The structure has a light self-weight. No duct is required in the unbonded prestressed concrete which not only saves the ducts but also provides better flexibility for tendon layout. Hence, compared with the bonded flexural structures with the same span under the same external loads, the sectional dimension can be reduced and the quantity of concrete is decreased also, leading to lighter self-weight.

(2) There is no need to arrange the ducts and grouting for the ducts, consequently, the construction process is simple, then the construction period is shortened and the corresponding construction cost is saved.

(3) The unbonded tendon is coated with anticorrosion grease and wrapped with a plastic sheath, so the unbonded tendon has strong corrosion resistance.

(4) High seismic resistance. When the structure generates large repeated displacements due to seismic load, the stress in unbonded tendons changes within a small range, and the tendons always work elastically, since the large displacement is absorbed evenly in the entire length. In addition, a large quantity of reinforcing steels arranged in the concrete provides sufficient seismic resistance.

(5) Long prestressing tendons can be used efficiently. The prestress loss due to friction at stretching is relatively small since the unbonded tendons do not directly contact the concrete (or ducts), resulting in an almost uniform distribution of stress over the entire length. Furthermore, the stress in tendons is mostly maintained at a constant value approximately due to automatic slip adjustment no matter the loads applied to the structure, making the strength of whole tendons be utilized fully. Therefore, the long unbonded tendons can be used efficiently in continuous structures with multiple long spans, such as continuous beams and continuous flat plates.

(6) The stress in unbonded tendons cannot reach the design strength when the flexural failure occurs. The strain in tendons in the critical section does not change coordinately with the concrete, since the strain change in tendons is the average value between two anchorages.

12.2 Anchorage system for unbonded tendons

The anchorages for unbonded prestressing tendons are classified as stretching-end and fixing-end anchorages, which are commonly provided as the finished anchorage assembly. The anchorage shall be selected according to the type of unbonded tendons, tension value, and ambient environment. When the prestressing steel strands are used as the unbonded tendons, circular sleeve anchorages and integrated-bearing-plate anchorages can be adopted for stretching ends, as shown in Figs. 12.2 and 12.3, and embedded extrusion anchorages and integrated-bearing-plate anchorages can be adopted for fixing ends, as shown in Fig. 12.4.

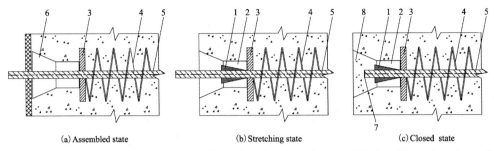

Figure 12.2 Structural diagram of typical circular sleeve anchorage: (1) clip; (2) anchor ring; (3) bearing plate; (4) spiral reinforcing steel bar; (5) unbonded strands; (6) cave mold; (7) plastic cap; (8) microexpansive fine aggregate concrete or nonshrinkage cement mortar.

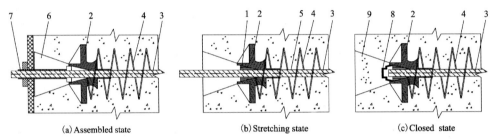

Figure 12.3 Structural diagram of typical integrated-bearing-plate anchorage: (1) clip; (2) integrated-bearing-plate; (3) unbonded strands; (4) spiral reinforcing steel bar; (5) plastic sealing sleeve; (6) cave mold; (7) sealing connector and nut; (8) sealing cover; (9) microexpansive fine aggregate concrete or nonshrinkage cement mortar.

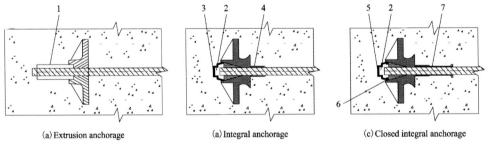

Figure 12.4 Structural diagram of typical fixing-end anchorage (1) extrusion anchorage; (2) special anticorrosive grease; (3) sealing cover; (4) plastic sealing sleeve; (5) pressure sealing cover; (6) sealing ring; (7) thermoplastic pressure sealing sleeve.

The anchorage system at the stretching end should be arranged on the concave concrete surface. A circular sleeve anchorage consists of an anchor ring, clips, bearing plate, and indirect reinforcement (Fig. 12.2). An integrated-bearing-plate anchorage consists of a bearing plate, clips, cave mold, sealing connector, nut, indirect reinforcement, sealing cover, and plastic sealing sleeve (Fig. 12.3).

A typical extrusion anchorage is composed of an extruded anchor, bearing plate, and indirect reinforcement (Fig. 12.4A), and the special sleeve shall be extruded and assembled at the end of the steel strands with equipment. An integral anchorage is composed of an integrated-bearing plate, clips, sealing cover, plastic sealing sleeve, and indirect reinforcement (Fig. 12.4B). A fully enclosed integral anchorage is composed of an integrated-bearing plate, clips, indirect reinforcement, pressure metal sealing cover, sealing ring, and thermoplastic pressure sealing sleeve (Fig. 12.4C). During the installation of integral anchorages at the fixing end, a wedge tightener is used to compress the clip tightly into the anchor plate with a jacking force not less than 0.75 times the characteristic value of the tensile strength of the strand, and the sealing cover is installed finally.

The anchorages are required to bear reliable anchoring performance, sufficient bearing capacity, and good applicability. The unbonded prestressing tendon–anchorage assembly needs to pass the performance test by fatigue loading stipulated in the *Anchorage, grip, and coupler for prestressing tendons* (GB/T 14370-2015). For unbonded prestressing steels, the efficiency coefficient and total elongation of tendon–anchorage assembly should satisfy Eqs. (3.2) and (3.3), while Eqs. (3.7) and (3.8) should be satisfied for unbonded prestressing FRP.

12.3 Flexural behavior of unbonded prestressed concrete structures

The flexural behavior of unbonded structures arranged with sufficient reinforcing steels under loading from zero to the ultimate load is similar to that of the bonded post-tensioned structures to some extent. The flexural behavior can be distinguished into three stages, before cracking, after cracking, and failure stage. The first stage can be divided further into the elastic working stage before the maximum tensile stress of concrete reaches the tensile strength and is then followed by the transition stage to cracking.

To simplify the analysis, the relation between the deflection and the maximum bending moment in the critical section can be approximated to three segments, as shown in Fig. 12.5 (the camber due to jacking force is not drawn), which is similar to Fig. 7.2. The first turning point appears when the cracking occurs in the critical section and the second turning point corresponds to the yield of outermost tensile reinforcing steels. As the external loads increase further, the cracking develops and the section flexural rigidity reduces, so the segment between M_{cr} and M_y should be curved. Considering that the stress in unbonded tendons increases gradually in a small range during this period, the deflection changes almost linearly. When the external loads increase continuously, obvious deflection and many cracks are produced and flexural failure finally occurs.

Consider two simply supported beams arranged with unbonded and bonded tendons, respectively, having the same section and moment of inertia, and the same initial prestressing force and same eccentricity of the tendons. The stress changes in unbonded and bonded tendons with external loads are illustrated in Fig. 12.6, respectively. Due

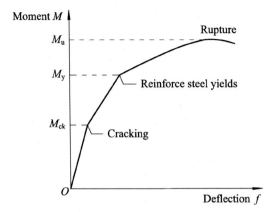

Figure 12.5 Bending moment versus deflection of an unbonded prestressed concrete flexural member.

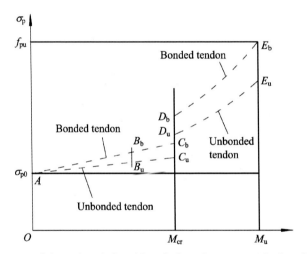

Figure 12.6 Stress of the unbonded and bonded tendons versus the bending moment.

to free slip between the unbonded tendons and the concrete, the rate of stress change in the unbonded tendons is lower than that in the bonded tendons. At cracking, the tension in concrete is transformed to the bonded tendons and reinforcing steels in the tension zone directly, so the stress in bonded tendons jumps from point C_b to point D_b (Fig. 12.6). For unbonded tendons, although the stress also jumps from point C_u to point D_u, the change is relatively smaller since the longitudinal deformation due to cracks is almost evenly distributed over the entire length. At rupture, the stress in bonded tendons reaches the ultimate tensile strength, while the stress in unbonded tendons cannot reach the ultimate tensile strength.

12.4 Calculation of stress in unbonded tendons

12.4.1 Effective stress in unbonded tendons in the service period

Referring to the calculation of effective stress for bonded prestressing tendons in Chapter 7, the calculation of effective stress in unbonded tendons is concisely introduced in the following.

The frictional loss occurs at stretching of the unbonded tendons, although the value is far less than that in the bonded tendons. The prestress loss due to friction is obtained as

$$\sigma_{l1}(x) = \sigma_{con} - \sigma(x) = \sigma_{con}\left[1 - e^{-(\mu\theta + \kappa x)}\right] \tag{12.5}$$

where σ_{con} = control stress in the unbonded tendons at jacking.

μ = frictional coefficient between the unbonded tendons and the sheath.

κ = coefficient of wobble friction.

x = tendon length between the end of stressing anchorage and the point under consideration.

θ = sum of the angular change over a tendon length x, for spatial tendons it can be calculated from Eqs. (5.12) and (5.13).

μ, κ should be determined by in situ tests. If reliable data are not available, the values in Table 12.1 for unbonded steel strands can be adopted. For unbonded FRP rods, the values in Table 12.1 may be used.

If $\mu\theta + \kappa x \leqslant 0.3$, $\sigma_{l1}(x)$ can be calculated approximately by Eq. (5.11).

Generally, the control stress in the unbonded prestressing steels at stretching in Eq. (12.5) should be less than $0.75f_{ptk}$ and in no case shall it exceed $0.8f_{ptk}$. For unbonded prestressing FRP, the control stress should conform to the provisions in the JGJ 92-2016 code as listed in Table 12.2.

The prestress loss due to anchorage set is given by

$$\sigma_{l2}(x) = \frac{\Delta l}{l}E_p \tag{12.6}$$

Table 12.1 The values of μ and κ.

Unbonded prestressing tendon	μ	κ
Steel strands with $d \leqslant 15.2$ mm	0.09	0.004

Note: the data in this table are reproduced from the JGJ 92-2016 code.

Table 12.2 The control stress in the unbonded prestressing FRP.

Type of prestressing FRP	Minimum of σ_{con}	Maximum of σ_{con}
Prestressing carbon fiber	$0.40f_{tpk}$	$0.65f_{tpk}$
Prestressing aramid fiber	$0.35f_{tpk}$	$0.55f_{tpk}$

Note: the data in this table are reproduced from the JGJ 92-2016 code.

where Δl = retraction of the unbonded tendons due to anchorage set.

l = effective length of the unbonded tendons.

E_p = modulus of elasticity of unbonded tendon.

For unbonded steel strands, 5 mm is taken for Δl when there is precompression for the clips or wedges and 6—8 mm in the case of no precompression. For unbonded prestressing FRP, 1—2 mm is taken for adhesive anchorages and 8 mm for wedge-type anchorages.

When the unbonded tendons are stretched in batches, the previously stressed tendons shorten when the subsequent tendons are stretched. $\sigma_{L4}(x)$ can be calculated by Eq. (5.55) or (5.56).

The prestress loss of the unbonded steel strands due to relaxation (σ_{l5}) can be calculated from Table 2.25 (Grade II relaxation), the loss due to concrete shrinkage and creep (σ_{l6} and σ'_{l6}) can be calculated from Eqs. (5.79) and (5.80).

For unbonded prestressing FRP, the loss due to relaxation can be calculated from Eq. (2.38). The loss due to shrinkage and creep can be calculated by

$$\sigma_{l6}(x, t \rightarrow \infty) = \frac{55 + 300\dfrac{\sigma_{pc}(t_0)}{f_{cu,k}}}{1 + 15\rho} \cdot \frac{E_{fp}}{E_p} \tag{12.7}$$

$$\rho = \frac{A_{fp} + A_s}{A_n} \tag{12.8}$$

where $\sigma_{pe}(t_0)$ = normal stress of concrete at the level of the prestressing FRP due to the prestressing force and the self-weight of the structure at time t_0.

$f_{cu,k}$ = characteristic value of cubic compressive strength of concrete at time t_0.

A_{fp} = area of unbonded prestressing FRP.

A_s = area of reinforcing steels.

ρ = ratio of prestressing FRP and reinforcing steels.

E_{fp} = modulus of elasticity of prestressing FRP.

A_n = converted cross-sectional area of concrete and reinforcing steels minus the duct area.

Hence, the effective stress in unbonded tendons in the service period is obtained as

$$\sigma_{pe}(x) = \sigma_{con} - [\sigma_{l1}(x) + \sigma_{l2}(x) + \sigma_{l4}(x) + \sigma_{l5}(x) + \sigma_{l6}(x, t)] \tag{12.9}$$

The JGJ 92-2016 code specifies that the prestress losses of the unbonded tendons should be not less than 80 MPa.

12.4.2 Stress change in unbonded tendons under loading

When the external loads are applied to the unbonded flexural structure, deflection and longitudinal deformation are generated. Considering the relative slip between the unbonded tendons and the concrete, the incremental stress in unbonded tendons due to external loads can be approximately obtained by

$$\Delta\sigma_p = \frac{\Delta l_p}{l_0}E_p \tag{12.10}$$

where $\Delta\sigma_p$ = mean stress change in unbonded tendons over the entire length due to external loads.

Δl_p = length change of the unbonded tendons due to external loads.

l_0 = initial length of the unbonded tendons before loading.

According to the deformation coordination, the length change in the unbonded tendons equals the cumulative deformation of surrounding concrete over the entire length, giving

$$\Delta l_p = \Delta l_c = \int_0^{l_0} \varepsilon(x)dx = \int_0^{l_0} \frac{M(x)e_p(x)}{E_c I(x)}dx \tag{12.11}$$

where $M(x)$ = bending moment due to external loads in the section under consideration.

$e_p(x)$ = eccentricity of the unbonded tendons.

$I(x)$ = moment of inertia of section converted to concrete, $I(x) = I_n(x) + \alpha_{EP}A_p e_p^2(x)$.

$I_n(x)$ = moment of inertia of the net concrete section.

A_p = area of unbonded tendons.

E_c = modulus of elasticity of concrete.

α_{EP} = modulus ratio of unbonded tendon to concrete.

The stress in unbonded tendons in the service period can be obtained from Eqs. (12.9)−(12.11), as

$$\sigma_p = \sigma_{pe} + \Delta\sigma_p = \sigma_{pe} + \frac{\alpha_{EP}}{l_0}\int_0^{l_0}\frac{M(x)e_p(x)}{I(x)}dx \tag{12.12}$$

where $\sigma_{pe}(x)$ = effective stress in unbonded tendons in the service period, given by Eq. (12.9).

$M(x)$ = bending moment due to external loads in the section under consideration.

Approximately, the stress in unbonded tendons in the service period can be estimated by

$$\sigma_p = \sigma_{pe} + \Delta\sigma_p = \sigma_{pe} + \alpha_{EP}\sigma_{p,c} \tag{12.13}$$

where $\sigma_{p,c}$ = concrete stress caused by the external loads at the level of unbonded tendons.

For a simply supported beam with constant section and straight tendons under uniformly distributed load, the stress change in unbonded tendons can be obtained approximately from Eqs. (12.9)−(12.11):

$$\Delta l_p = \int_0^{l_0} \left[\frac{q}{2}(l_0 - x)x \frac{e_p}{E_c I} dx \right] = \frac{q l_0^3 e_p}{12 E_c I} = \frac{2 M_{\max} l_0 e_p}{3 E_c I} \tag{12.14}$$

where $M_{\max}$ = maximum bending moment at midspan, $M_{\max} = \frac{q l_0^2}{8}$.

l_0 = length of the unbonded tendons; can be taken as the length of beam approximately.

Hence, the incremental stress in unbonded tendons is obtained from Eq. (12.14):

$$\Delta \sigma_p = \frac{2}{3} \frac{\alpha_{EP} M_{\max} e_p}{I} = \frac{2}{3} \Delta \sigma_{p,b} \tag{12.15}$$

where $\Delta \sigma_p$ = stress change in unbonded tendons in a simply supported beam with straight tendons due to uniformly distributed load.

$\Delta \sigma_{p,b}$ = stress change in bonded tendons in a simply supported beam with the same effective span and the same constant section under the same uniformly distributed load, $\Delta \sigma_{p,b} = \alpha_{EP} \frac{M_{\max} e_p}{I}$.

Eq. (12.15) shows that for the bonded and unbonded prestressed simply supported beams with the same effective span, section, and reinforcements under the same uniformly distributed load, the stress change in unbonded tendons at midspan is about two-thirds of that in bonded tendons. For prestressed continuous beams, the similar conclusion that the stress change in unbonded tendons is smaller can also be drawn.

12.4.3 Ultimate stress in unbonded tendons at flexural failure

Theoretically, the stress in tendons in the unbonded flexural structures under loading can be derived from Eq. (12.12). Considering that the strain change in the unbonded tendons depends on the longitudinal deformation as shown in Eq. (12.11), which is not only related to the sectional properties, tendon eccentricity, tendon and reinforcing steel ratios but also related to the structural system, the formula of ultimate stress in unbonded tendons at flexural failure is difficult to derive, so the estimation formula based on the theoretical analysis and experimental results is commonly adopted.

The JGJ 92-2016 code specifies that the ultimate stress in unbonded steel strands in the calculation of flexural bearing capacity may be taken as

$$\sigma_{pu} = \sigma_{pe} + \left(240 - 335 \zeta_p \right) \left(0.45 + 5.5 \frac{h}{l_0} \right) \frac{l_2}{l_1} \tag{12.16}$$

$$\zeta_p = \frac{\sigma_{pe}A_p + f_{sd}A_s}{f_{cd}bh_p} \tag{12.17}$$

where σ_{pu} = design value of tensile stress of the unbonded steel strands in the calculation of the flexural bearing capacity.

ζ_p = comprehensive reinforcement index, which should be less than 0.4. For continuous beams and slabs, the average value at support and midspan sections shall be taken.

l_0 = effective span.

l_1 = total length of the unbonded tendon between two anchorages.

l_2 = sum of loading span length determined by the most dangerous layout of live load related to l_1.

σ_{pe} = effective stress in unbonded steel strands, given by Eq. (12.9).

f_{cd} = design strength of concrete in compression.

f_{sd} = design tensile strength of reinforcing steel.

h_p = distance from the level of unbonded tendons to the extreme compressive fiber of concrete.

For flexural members with a T- or I-section in which the flange is in the compression zone and the depth of compression zone is greater than the flange thickness, the comprehensive reinforcement index is given as

$$\zeta_p = \frac{\sigma_{pe}A_p + f_{sd}A_s - f_c\left(b_f' - b\right)h_f'}{f_{cd}bh_p} \tag{12.18}$$

where h_f' = flange thickness of a T- or I-section.

b_f' = flange width of a T- or I-section.

b = web width of a T- or I-section.

The JGJ 92-2016 code specifies that the ultimate stress in unbonded prestressing FRP in the calculation of the flexural bearing capacity may be taken as

$$\sigma_{fpu} = \sigma_{fpe} + \Delta\sigma_{fp} \tag{12.19}$$

$$\Delta\sigma_{fp} = \left(240 - 335\zeta_{0f}\right)\left(0.45 + 5.5\frac{h}{l_0}\right)\frac{l_2}{l_1}\cdot\frac{E_{fp}}{E_p} \tag{12.20}$$

$$\zeta_{0f} = \frac{\sigma_{fpe}A_{fp} + f_{sd}A_s}{f_{cd}bh_{0,fp}} \tag{12.21}$$

where σ_{fpu} = design value of tensile stress in unbonded prestressing FRP in the calculation of the flexural bearing capacity.

σ_{fpe} = effective stress in unbonded prestressing FRP, given by Eq. (12.9).

$\Delta\sigma_{fp}$ = stress change in unbonded prestressing FRP in the calculation of flexural bearing capacity.

ζ_{0f} = comprehensive reinforcement index, which should be less than 0.4. For continuous beams and slabs, the average value at the support and midspan sections shall be taken.

f_{cd} = design strength of concrete in compression.

$h_{0,fp}$ = distance from the level of unbonded prestressing FRP to the extreme compressive fiber of concrete.

Anyway, the design value of tensile stress in unbonded prestressing FRP should be greater than the effective stress σ_{fpe}, while it must be less than the design tensile strength of unbonded prestressing FRP.

12.5 Bearing capacity of the unbonded prestressed concrete sections

For unbonded prestressed concrete flexural structures, the flexural bearing capacity of the normal sections and the shear bearing capacity of the oblique sections should be verified in the design, as expressed in Eqs. (9.1) and (9.2). In calculations of bearing capacity, the basic assumptions adopted are almost the same as those for bonded post-tensioned flexural members.

12.5.1 Flexural bearing capacity of the normal sections

The JGJ 92-2016 code specifies that the flexural bearing capacity of the normal sections of the unbonded flexural structures should be greater than the cracking moment:

$$M_u \geqslant M_{cr} \tag{12.22}$$

$$M_{cr} = \left(\sigma_{pc} + \gamma f_{ctk}\right) W_0 \tag{12.23}$$

where M_u = flexural bearing capacity of the unbonded prestressed concrete section.

M_{cr} = cracking moment of the section under consideration.

W_0 = section modulus of converted section.

f_{ctk} = characteristic tensile strength of concrete.

σ_{pc} = compressive stress of concrete produced by the prestressing force at the extreme fiber in the tension zone.

γ = correction coefficient considering the influence of plasticity of concrete, given by Eq. (7.66).

In an unbonded prestressed balanced section at flexural failure, the reinforcing steel in the tension zone reaches its design tensile strength and the concrete in the compression zone reaches its design value of compressive strength simultaneously, and the ultimate stress in unbonded tendons is given by Eq. (12.16) or (12.19). Therefore, the flexural bearing capacity of the rectangular section (referring to Fig. 9.4) can be obtained as

$$M_u = \sigma_{pu}A_p\left(h_p - \frac{x}{2}\right) + f_{sd}A_s\left(h_s - \frac{x}{2}\right) \tag{12.24}$$

where the depth of the neutral axis can be derived from

$$\sigma_{pu}A_p + f_{sd}A_s = f_{cd}bx + f'_{sd}A'_s + \sigma'_pA'_p \tag{12.25}$$

And the depth of the neutral axis should satisfy:

$$x \leqslant \xi_b h_0 \tag{12.26}$$

When the unbonded tendons and reinforcing steels are arranged in the compression zone, and the unbonded tendons are in compression, the depth of the neutral axis should also satisfy

$$x \geqslant 2d' \tag{12.27}$$

If only reinforcing steels are arranged in the compression zone, or the unbonded tendons in the compression zone are in tension, the depth of the neutral axis should also satisfy

$$x \geqslant 2d'_s \tag{12.28}$$

where σ_{pu} = design value of tensile stress of the unbonded tendons, given by Eq. (12.16) or (12.19).

f_{cd} = design strength of concrete in compression.

f_{sd} = design tensile strength of reinforcing steel.

f'_{sd} = design strength of reinforcing steel in compression.

σ'_p = stress in unbonded tendons in the compression zone, $\sigma'_p = f'_{pd} - \sigma'_{p0}$.

σ'_{p0} = stress in unbonded tendons in the compression zone corresponding to zero concrete strain at the level of tendons.

A_p = cross-sectional area of unbonded tendons in the tension zone.

A'_p = cross-sectional area of unbonded tendons in the compression zone.

A_s = cross-sectional area of reinforcing steels in the tension zone.

A'_s = cross-sectional area of reinforcing steels in the compression zone.

b = width of rectangular section.

h_0 = effective height of the section, $h_0 = h - \dfrac{\sigma_{pu}A_pa_p + f_{sd}A_sa_s}{\sigma_{pu}A_p + f_{sd}A_s}$.

$h_p = h - a_p$.

$h_s = h - a_s$.

a_p = depth of unbonded tendons in the tension zone from the tension face.

d'_p = depth of unbonded tendons in the compression zone from the compression face.

a_s = depth of reinforcing steels in the tension zone from the tension face.

d'_s = depth of reinforcing steels in the compression zone from the compression face.

12.5.2 Shear bearing capacity of the oblique sections

The JGJ 92-2016 code specifies that the shear bearing capacity of the oblique sections of the unbonded prestressed concrete flexural structures can be calculated from

$$V_u = V_{cs} + V_{p1} + 0.8 f_{sd} A_{sb} \sin \theta_s + 0.8 \sigma_{pu} A_{pb} \sin \theta_p \tag{12.29}$$

where V_{cs} = shear bearing capacity provided by the concrete and the stirrups in the oblique section, given by Eq. (9.63).

V_{p1} = shear bearing capacity provided by the unbonded tendons in the oblique section, $V_{p1} = 0.05 N_{p0}$, the value of N_{p0} is determined from Eq. (9.62).

A_{pb} = cross-sectional area of unbonded bending tendons.

θ_p = angle between the bent-up unbonded tendons and the beam's axis.

A_{sb} = cross-sectional area of bending reinforcing steels.

θ_s = angle between the bending reinforcing steels and the beam's axis.

12.6 Calculation and control of deflection and crack

The approaches to calculating and controlling the deflection and crack of the unbonded prestressed concrete flexural structures are similar to those for the bonded flexural structures. Nevertheless, the free slip occurring between the unbonded tendons and the concrete under loading results in deformation disharmony between the tendons and the concrete in a section, therefore, the contribution of the unbonded tendons to the sectional flexural rigidity is different to that from the bonded tendons, and the relation between the bending moment and deflection in the bonded flexural members does not apply completely to the unbonded members. Generally, the effect of free slip of unbonded tendons can be taken into account approximately by reducing the contribution of unbonded tendons to the flexural rigidity.

12.6.1 Deflection of the unbonded prestressed concrete flexural structures

Comparing Fig. 12.2 with Fig. 8.3, the short-term deflection of the unbonded prestressed concrete flexural structures under loading can be calculated by the bilinear method:

When $M(x) \leqslant M_{cr}$

$$f_s = \int_0^L \frac{\overline{M}(x) \cdot M(x)}{B_1} dx \tag{12.30}$$

When $M(x) > M_{cr}$

$$f_s = \int_0^L \overline{M}(x) \left[\frac{M_{cr}(x)}{B_1} + \frac{M(x) - M_{cr}(x)}{B_2} \right] dx \qquad (12.31)$$

where $M(x) =$ bending moment caused by the applied load.

$\overline{M}(x) =$ bending moment due to unit force acting on the considered position.

$M_{cr}(x) =$ cracking moment.

$B_1 =$ short-term flexural rigidity of the uncracked section.

$B_2 =$ short-term flexural rigidity of the cracked section.

$I_c =$ moment of inertia of the uncracked section.

$I_{cr} =$ moment of inertia of the cracked section.

$L =$ effective span of the structure.

The JGJ 92-2016 code specifies that the short-term flexural rigidity should be calculated as follows:

For uncracked sections:

$$B_s = 0.85 E_c I_0 \qquad (12.32)$$

For cracked sections:

$$B_s = \frac{0.85 E_c I_0}{k_{cr} + w(1 - k_{cr})} \qquad (12.33)$$

$$w = \left(1.0 + \frac{0.21}{\alpha_E \rho} \right) \left(1 + 0.45 \gamma_f \right) - 0.7 \qquad (12.34)$$

$$\gamma_f = \frac{(b_f - b) h_f}{b h_0} \qquad (12.35)$$

where $M_k =$ bending moment caused by the characteristic combination of actions.

$M_{cr} =$ cracking moment.

$k_{cr} = \frac{M_{cr}}{M_k}$, $k_{cr} = 1.0$ is taken if $k_{cr} > 1.0$.

$\rho =$ reinforcement ratio, given by Eq. (12.36) or (12.37).

$E_c =$ modulus of elasticity of concrete.

$I_0 =$ moment of inertia of the converted section.

$b_f =$ flange width in the tension zone.

$h_f =$ flange thickness in the tension zone.

$b =$ width of rectangular section or the web width of the T- or I-section.

$h_0 =$ effective depth of the section.

$\alpha_E =$ modulus ratio of reinforcement to concrete.

For flexural members with unbonded prestressing steel strands or wires, the reinforcement ratio in Eq. (12.34) should be calculated by

$$\rho = \frac{0.3A_p + A_s}{bh_0} \tag{12.36}$$

For flexural members with unbonded prestressing FRP rods, the reinforcement ratio should be calculated by

$$\rho = \frac{\dfrac{0.3E_{fp}A_{fp}}{E_p} + A_s}{bh_0} \tag{12.37}$$

where E_{fp} = modulus of elasticity of prestressing FRP.

A_{fp} = cross-sectional area of prestressing FRP.

The JGJ 92-2016 code specifies that the long-term flexural rigidity should be calculated by

$$B = \frac{M_k}{M_q(\theta - 1) + M_k}B_s \tag{12.38}$$

where M_q = bending moment produced by the quasi-permanent combination of actions.

θ = incremental influence coefficient by the long-term loading; generally $\theta = 2.0$ is taken.

12.6.2 Crack width of the unbonded prestressed concrete flexural structures

A certain quantity of longitudinal reinforcing steels are used in unbonded prestressed concrete structures, and some of them are distributed in the tension zone. Control of the crack width is an important measure to protect reinforcing steels from corrosion.

For unbonded prestressed flexural members in which cracking is allowed, the JGJ 92-2016 code specifies that the maximum crack width should be calculated by

$$w_{\max} = \alpha_{cr}\psi\frac{\sigma_{sk}}{E_s}\left(1.9c_c + 0.08\frac{d_{eq}}{\rho_{s,ef}}\right) \tag{12.39}$$

$$\psi = 1.1 - 0.65\frac{f_{ctk}}{\rho_{s,ef}\sigma_{sk}} \tag{12.40}$$

where the nominal diameter of the reinforcement and effective ratio of longitudinal reinforcement are given by

$$d_{eq} = \frac{\sum n_i d_i^2}{\sum n_i \nu_i d_i^2} \tag{12.41}$$

$$\rho_{s,ef} = \frac{A_s}{A_{c,ef}} \tag{12.42}$$

where α_{cr} = coefficient related to the type of member, 1.5 is taken for flexural members.

ψ = coefficient of nonuniformity distribution of the tensile stress in reinforcement, 0.2 is taken if $\psi < 0.2$, 1.0 is taken if $\psi > 1.0$, and 1.0 is taken for members directly subjected to repeated loading.

σ_{sk} = equivalent stress of longitudinal tensile reinforcement in the cracked section caused by the characteristic combination of actions.

n_i = number of bars (or wires) with diameter d_i.

ν_i = coefficient of relative bond property of reinforcement, listed in Table 8.5.

c_c = concrete cover.

$\rho_{s,ef}$ = effective ratio of longitudinal reinforcement, 0.1 is taken if $\rho_{te} > 0.1$ and 0.01 is taken if $\rho_{te} \leqslant 0.01$.

$A_{c,ef}$ = effective area of concrete in tension, $A_{te} = 0.5bh + (b_f - b)h_f$, mm^2.

The equivalent stress of longitudinal tensile reinforcement in Eq. (12.39) can be obtained from

$$\sigma_{sk} = \frac{M_k \pm M_{p2} - N_{p0}(z - e_p)}{(0.3A_p + A_s)z} \tag{12.43}$$

$$e = e_p + \frac{M_k \pm M_{p2}}{N_{p0}} \tag{12.44}$$

$$e_p = \gamma_{ps} - e_{p0} \tag{12.45}$$

$$z = \left[0.87 - 0.12\left(1 - \gamma'_f\right)\left(\frac{h_0}{e}\right)^2\right]h_0 \leqslant 0.87h_0 \tag{12.46}$$

$$N_{p0} = \sigma_{p0}A_p + \sigma'_{p0}A'_p - \sigma_{l6}A_s - \sigma'_{l6}A'_s \tag{12.47}$$

$$\sigma_p = \sigma_{pe} + \Delta\sigma_p = \sigma_{pe} + \frac{\alpha_{EP}}{l_0}\int_0^{l_0}\frac{M(x)e_p(x)}{I(x)}dx \tag{12.48}$$

where M_k = bending moment due to the characteristic combination of actions.

M_{p2} = secondary bending moment due to stretching of the tendons in the statically indeterminate structures.

N_{p0} = generalized prestressing force corresponding to zero normal stress in concrete.

A_p = cross-sectional area of unbonded tendons in the tension zone; for unbonded prestressing steels, it is taken as the cross-sectional area of unbonded tendons; for unbonded prestressing FRP, it is taken as $E_{fp}A_{fp}/E_p$.

A_p' = cross-sectional area of unbonded tendons in the compression zone; for unbonded prestressing steels, it is taken as the cross-sectional area of unbonded tendons; for unbonded prestressing FRP, it is taken as $E_{fp}A_{fp}'/E_p$.

A_s = cross-sectional area of reinforcing steels in the tension zone.

A_s' = cross-sectional area of reinforcing steels in the compression zone.

σ_{p0} = stress in unbonded tendons in the tension zone when the normal stress of concrete at the level of tendons is zero, $\sigma_{p0} = \sigma_{pe}(x) + \sigma_{l4}$, $\sigma_{pe}(x)$ is given by Eq. (12.9).

σ_{p0}' = stress in unbonded tendons in the compression zone when the normal stress of concrete at the level of tendons is zero, $\sigma_{p0}' = \sigma_{pe}'(x) + \sigma_{l4}'$.

z = distance between the resultants in the tension zone and compression zone, respectively.

$e_{p,s}$ = distance between N_{p0} and the resultant of the longitudinal prestressing tendons and reinforcing steels in the tension zone.

y_{ps} = eccentricity of N_{p0}.

e_{p0} = eccentricity of prestressing tendons and reinforcing steels in the tension zone.

γ_f' = ratio of the section area of the compression (flange) to the effective section area of the web, given by Eq. (8.52).

Under the characteristic combination of the actions, considering the long-term effect, the JGJ 92-2016 stipulates that the maximum crack width calculated by Eq. (12.39) should not exceed the allowable values specified in Table 8.7 or Table 8.8.

Suggested readings

[1]   JGJ 92: 2016. Technical specifications for concrete structures prestressed with unbonded tendons. Beijing: China Construction Industry Press; 2016.
[2]   GB/T 26752: 2020. PAN-based carbon fiber. Beijing: China Construction Industry Press; 2020.
[3]   GB/T 26743:2011. Fiber reinforced composite bars for civil engineering. Beijing: China Construction Industry Press; 2011.
[4]   GB/T 14370: 2015. Anchorage, grip, and coupler for prestressing tendons. Beijing: China Standards Press; 2015.
[5]   Hu Di. Basic principles of prestressed concrete structure design. 2nd ed. Beijing: China Railway Publishing House; 2019.
[6]   Lin TY, Burns NH. Design of prestressed concrete structures. 3rd ed. New York: John Wiley and Sons; 1981.

CHAPTER 13

Analysis and design of externally prestressed concrete structures

Contents

13.1 General concepts on externally prestressed concrete structures

The reinforced concrete members or structures with prestressing tendons outside the concrete section which are constructed by the post-tensioning method are called externally prestressed concrete structures, and the tendons adopted here are called external prestressing tendons. The external tendons easily corrode in the natural environment, so special anticorrosion measures must be taken. The external tendons along with the corresponding anticorrosion and protection structure are known as the external prestressing cables, abbreviated to external cables. The external cables are arranged as a straight or broken line, and the ends can be anchored at any position. For a broken line cable, the steering devices are required to be set on the diaphragms or protruding concrete or steel structures at the turning positions in a flexural member. Fig. 13.1 shows the typical layouts

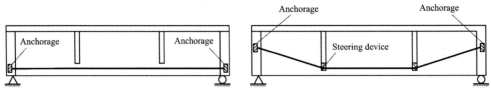

(a) Straight external cable in a simply supported beam   (b) Broken line external cable in a simply supported beam

Figure 13.1 Typical layouts of external cable in a simply supported beam.

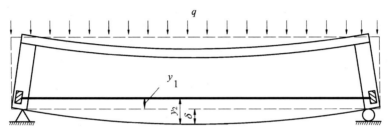

Figure 13.2 Deformation relation between the external cable and beam.

of broken line external cables in an externally prestressed simply supported beam. In most cases, the external cables can slide freely at the steering devices.

One distinctive feature of the externally prestressed beams under loading is that the eccentricity of the external cable changes with the deflection. Consider a simply supported beam with a straight cable under the uniformly distributed load, as shown in Fig. 13.2, where the dashed and solid lines represent the beam positions before and after deflection occurs, respectively. δ is the maximum deflection at midspan. When the external load is applied to the beam, the cable remains straight although the concrete beam bends, so the eccentricity of the cable at the midspan region reduces to ($y_1 < y_2$), resulting in a reduction of the bending moment caused by the prestressing force. This phenomenon is called the secondary effect of prestressing leading to a reduction of flexural resistance.

Another distinctive feature is that the cable strain is disproportional with the concrete strain at the same section. Since the cables are anchored at two ends with the beam and they can slide freely at the steering devices, the strain change in cables is an average value due to the total longitudinal deformation of the beam between two anchorages. In some special cases, such as that shown in Fig. 13.2, the strain change in the cable is very small no matter how large the beam deflects. Therefore, when the critical section fails in flexure, the stress in cables cannot reach the design strength.

If the external cables are treated as tied rods, as discussed for unbonded prestressed structures, the concrete beam should be reinforced or prestressed with a certain reinforcement ratio. Otherwise, an externally prestressed flexural structure may have a brittle failure as in the case of a low amount of reinforcing steels or bonded prestressing steels in concrete.

Generally, the externally prestressed concrete structures have the following characteristics:

(1) The structure has a light self-weight. No duct is required in the concrete, resulting in a smaller sectional dimension and lower amount of concrete, accordingly the self-weight of the structure reduces.

(2) The construction is simple and the construction period is shortened. Although the fabrication of anchoring devices and steering devices is required during construction, the construction is simplified by eliminating the process for assembling reinforcement meshes for ducts and putting the tendons through the ducts.

(3) The segments of reinforced concrete can be precast in the factory, then they are assembled into a structure in situ by the external cables. Standardized segmental prefabrication and construction for concrete beams can reduce construction costs and increase construction efficiency.

(4) The cable layout is flexible, especially when high-strength prestressing tendons and large tonnage anchorage systems are used, also, the number of cables can be reduced, leading to simplification of the design and construction.

(5) The frictional loss is low at stretching tendons since most parts of the cables do not contact the concrete, so the long prestressing tendons can be used efficiently, also the cost of couplers along with the corresponding construction are reduced in cases in which the long prestressing tendons are required.

(6) In an externally prestressed concrete structure with a long span, the secondary effect of prestressing is obvious, and the stress in cables cannot reach the design strength at flexural failure.

(7) The safety detection for the structure is convenient and the structural maintenance is easy carry out. For instance, once some external cables are broken, compared with the bonded prestressed structures, the damaged tendons are easily detected and replaced.

The externally prestressed concrete is widely used to construct new structures, especially for simply supported beams and continuous beams in railway and highway bridges. Also, the external prestressing technology is widely used to adjust the deflection or improve the bearing capacity of the existing flexural structures. This chapter introduces the general knowledge on the externally prestressed concrete and the newly constructed flexural members.

13.2 External prestressing system and external cable assembly

13.2.1 External prestressing system

A typical external prestressing system is composed of the following parts: an external prestressing cable and its anticorrosion system, anchorage system, steering devices, damping devices, and positioning structures. The anchorages, an external cable, steering devices (steering gears), damping devices, conduits, sealing devices, and joints are collectively referred to as an external cable assembly, as shown in Fig. 13.3. The anchor devices are divided into stretching-end anchorages and fixing-end anchorages, which may be

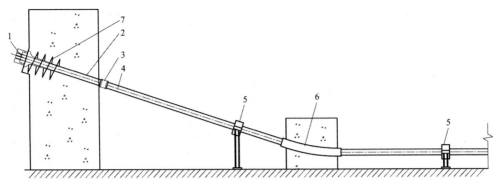

Figure 13.3 Components of an external cable assembly: (1) anchorage; (2) conduit; (3) sealing devices and joint; (4) external cable; (5) damping device; (6) steering device; (7) spiral reinforcing steel bar.

embedded in the end or middle diaphragms, or arranged in the specially made protruding anchor blocks. The steering device changes the direction of the cable, set in the middle diaphragms or specially made protruding structures. For long external cables, in order to avoid resonance when it is close to the natural frequency of the structure, damping devices are installed on the cables at a certain interval to change the natural frequency of the cables. When the free length of the cable is too long, positioning devices such as restraint supports are set.

The prestressing tendons in cables are manufactured to comply with the requirements of the Steel wire for prestressing of concrete (GB/T5223-2014) and *PAN-based carbon fiber* (GB/T 26752-2020), and the finished cables are fabricated to conform to *Hot-extruded PE protection paralleled high strength wire cable for cable-stayed bridge* (GB/T 18365-2018).

13.2.2 Anchorage system

The anchorages for external cables include cast-type anchorages (Fig. 13.4A and B) and steel–plate–type anchorages (Fig. 13.4C and D). The anchorage system is mainly composed of the anchor plate, anchor base plate, spiral reinforcing steel bar, protective cover, connecting pipe, outer sheath, and horn tube. For anchorages in which the external cables can be stretched many times, adjusting nuts are set outside the anchor plate (Fig. 13.4B and D).

The anchorages are required to bear reliable anchoring performance, sufficient bearing capacity, and good applicability. The external prestressing tendon—anchorage assembly needs to pass the performance test by fatigue loading stipulated in the *Anchorage, grip, and coupler for prestressing tendons* (GBT 14370-2015), and the efficiency coefficient and total elongation of assembly should satisfy Eqs. (3.2) and (3.3).

13.2.3 Steering devices

The steering devices are important components in an external prestressing concrete structure, by which the external cables are connected with the concrete structure at the middle regions. A steering gear is embedded in the steering device matching with the

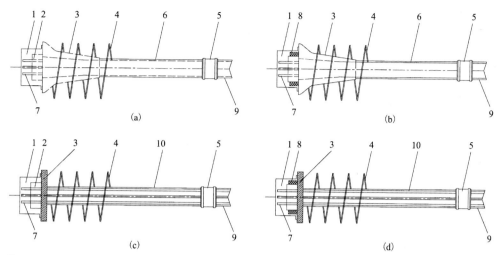

Figure 13.4 Structural diagram of typical anchorages for external cables: (1) protective cover; (2) anchor plate; (3) base plate; (4) spiral reinforcing steel bar; (5) connecting pipe; (6) isolation bushing; (7) reserved section for stretching; (8) nut; (9) outer sheath; (10) conduit.

type of external cable and the steering requirements. The steering devices not only fulfill the steering function for external cables, but also reliably transfer the force caused by the steering cables to the concrete structure.

The steering device is commonly made of steel or reinforced concrete. The steel steering device has the advantages of being simple in structure, small in section dimensions, light self-weight, and easy to construct, however special anticorrosion measures need to be taken. For a reinforced concrete steering device, although its section dimensions are large and heavy, it is integrated with the concrete structure which can be fabricated simultaneously with the main structure. Therefore, reinforced concrete steering devices are used widely in external prestressed concrete structures.

In terms of form, the steering devices can be divided into block type, beam type, and rib type, which can be further subdivided into transverse and vertical rib types, as shown in Fig. 13.5. The block types are relatively small, and can only bear the vertical component of cable force. The transverse rib types are usually arranged transversely above the bottom plate of the box girder which can bear the vertical and transverse components of cable force, and they are commonly used in the inclined and curved flexural structures. The vertical rib types are vertically arranged beside the box girder web, which can bear a large vertical component of cable force. If the steering device needs to bear a large bending moment, the beam types can be adopted.

The core component of the steering device is the steering gear, which is divided into centralized and decentralized types. The former refers to the centralized arrangement of prestressing tendons in the external cable (Fig. 13.6A and B), in which all tendons are concentrated close to the pressure side of the steering gear (concave side). For decentralized types, the prestressing tendons are scattered by the guide pipes, as shown in

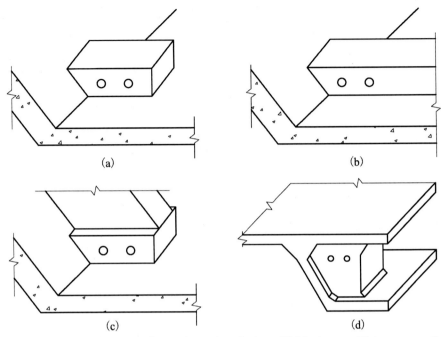

Figure 13.5 Structural diagram of typical steering devices: (1) block type; (2) transverse rib type; (3) vertical rib type; (4) beam type.

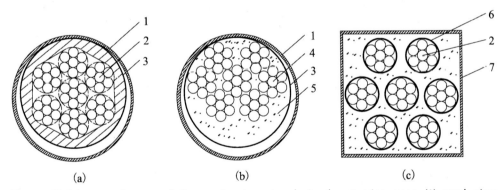

Figure 13.6 Layout diagram of the prestressing strands in the steering gear: (1) steel pipe; (2) unbonded strand; (3) HDPE sheath; (4) bonded strand; (5) cement mortar; (6) guide pipe; (7) accessory structure.

Fig. 13.6C. The seamless steel pipes should be used for cluster arrangement, and the seamless steel pipes or HDPE pipes should be used for scattered arrangements. The centralized-type steering gear is suitable for finished cables, while the decentralized-type steering gear can be used both for finished and nonfinished cables.

To avoid additional stress and stress concentration in the external cable at the steering positions during stretching and applying external loads, it is generally required that the bending angle of the external cable at each steering device be less than 15 degrees, and the radius of the steering gear be sufficiently large. For centralized-type steering gears, $R_{min} \geqslant 22D$ is required (R_{min} is the minimum radius of the steering gear and D is the diameter of the outer sheath, respectively). For decentralized-type steering gears, $R_{min} \geqslant 580\phi$ is required (ϕ is the maximum diameter of the steel wires in the strands). The minimum bending radius of the steering gear in the vicinity of the anchorage shall be 1000 mm greater than the above values.

13.2.4 Damping devices

Under the action of dynamic load, both the concrete structure and the external cables vibrate. When their frequencies are close to each other and close to the frequency of the dynamic load, resonance may occur, which can easily cause fatigue failure in the anchorages and external cables. The *Code for design of prestressed concrete members* (JGJ 369-2016) stipulates that the damping devices should be installed when the free length of the external cables between the anchorage and the steering device or between two adjacent steering devices exceeds 12 m.

A damping device is composed of positioning parts and vibration-isolating materials, as shown in Fig. 13.7. The positioning parts are made of steel and other materials with high stiffness, and one end is fixed on the concrete. The vibration-isolating material should be polychloroprene rubber which is placed between the external cable and the

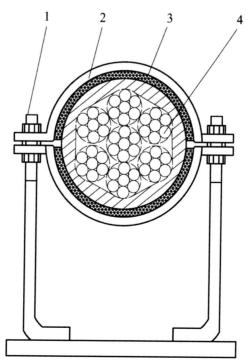

Figure 13.7 A typical damping device: (1) adjustable pull rod; (2) rubber cushion; (3) haver buckle; (4) external cable.

positioning device. For nonfinished cables without fillers, the vibration isolating material is set between the tendons and the outer sheath. Appropriate anticorrosion measures are taken for the damping devices. The damping devices are of repeatable installing and removing type for easy maintenance, and all replaceable parts can be easily installed and removed.

13.3 Calculation of stress in external tendons

13.3.1 Effective stress in external tendons in the service period

Referring to Eq. (12.9), the effective stress in external prestressing tendons in the service period can be calculated by

$$\sigma_{pe,ex}(x, t) = \sigma_{con,ex} - [\sigma_{l1}(x) + \sigma_{l2}(x) + \sigma_{l4}(x) + \sigma_{l5}(x, t) + \sigma_{l6}(x, t)] \qquad (13.1)$$

where $\sigma_{con,ex}$ = control stress in external tendons at jacking.

$\sigma_{l1}(x)$ = prestress loss due to friction.

$\sigma_{l2}(x)$ = prestress loss due to anchorage set.

$\sigma_{l4}(x)$ = prestress loss due to elastic shortening by stretching subsequent tendons.

$\sigma_{l5}(x, t)$ = prestress loss due to relaxation of tendon.

$\sigma_{l6}(x, t)$ = prestress loss due to shrinkage and creep of concrete.

$\sigma_{con,ex}$ should be greater than $0.4f_{ptk}$ but less than $0.6f_{ptk}$. When the prestress loss needs to be offset partially, it can be increased by $0.05f_{ptk}$.

When stretching the external cables, the frictional loss occurring within the pipes and steering devices can be calculated by

$$\sigma_{l1}(x) = \sum \sigma_{pi}\left[1 - e^{-(\mu_i\theta_i + k_i x_i)}\right] \qquad (13.2)$$

where σ_{pi} = stress in external tendons at the starting position of each anchorage or steering device when the tendons are stretched.

μ_i = frictional coefficient between the external tendons and the pipes or steering devices.

k_i = coefficient of wobble friction in the pipes or steering devices.

x_i = contact length between each pipe or steering device and the external tendons.

θ = sum of the angular change at each pipe or steering device.

μ, κ should be determined by in situ tests. If reliable data are not available, the values in Table 13.1 for external strands can be adopted.

Table 13.1 The values of μ and κ.

Pipe material and type of finished cable	μ	κ
Smooth strand goes through steel pipe	0.30 (0.20–0.30)	0.001 (0)
Smooth strand goes through HDPE pipe	0.13 (0.12–0.15)	0.002 (0)
Unbonded strand	0.09 (0.08–0.10)	0.004 (0)

The data outside and within parentheses in this table are reproduced from JGJ 369-2016 and JTG3362-2018 codes, respectively.

The prestress loss due to anchorage set can be obtained by

$$\sigma_{l2}(x) = \frac{\sum \Delta l}{l_{ex}} E_p \tag{13.3}$$

where Δl = retraction of the external tendons due to anchorage set.

l = effective length of the external tendons.

E_p = modulus of elasticity of external tendon.

When the external tendons are stretched in batches, the previously stressed tendons shorten when stretching the subsequent tendons, therefore, prestress loss $\sigma_{l4}(x)$ due to elastic shortening occurs in the previously stressed tendons.

The length change of the previous tendons by stretching the subsequent tendons shall be calculated according to the structural characteristics and the layout form of the external tendons. In simplified analysis, $\sigma_{l4}(x)$ can be calculated by Eq. (5.55) or (5.56).

The prestress loss of the external steel strands due to relaxation (σ_{l5}) can be calculated from Table 2.25 (Grade II relaxation), and the loss due to concrete shrinkage and creep (σ_{l6} and σ'_{l6}) can be calculated from Eqs. (5.79)–(5.80). For prestressing FRP, the prestress loss due to relaxation can be calculated from Eq. (2.38), and the prestress loss due to shrinkage and creep can be calculated by Eq. (12.7).

13.3.2 Stress change in external tendons under loading

For externally prestressed concrete structures with a short or medium span, the vertical deflection is low at service loads, and the secondary effect of prestressing can be neglected. In this case, the stress in external tendons in the service period can be given by

$$\sigma_{p,ex}(x) = \sigma_{pe,ex}(x) + \Delta\sigma_{p,ex} \tag{13.4}$$

where $\sigma_{pe,ex}(x)$ = effective stress in external tendons in the service period, given by Eq. (13.1).

$\Delta\sigma_{p,ex}$ = stress change in external tendons under loading.

$\Delta\sigma_{p,ex}$ can be derived by Eqs. (12.10) and (12.11), or by obtaining the tension change in the tendons from the structural analysis under loading. The JGJ 369-2016 code stipulates that $\sigma_{p,ex}$ should not exceed $0.6f_{ptk}$ for external prestressing steel strands.

The following is used to compute the stress change in external tendons in a simply supported beam subjected to the external loads, as shown in Fig. 13.8, where the tendons can freely slide along the steering devices.

Cutting off the external tendons at the middle results in

$$\delta_{11}\Delta N_p + \delta_{1p} = 0 \tag{13.5}$$

where ΔN_p = tension change in the external tendons due to external loads.

δ_{11} = displacement in the external tendons due to unit tension.

δ_{1p} = displacement in the external tendons due to external loads.

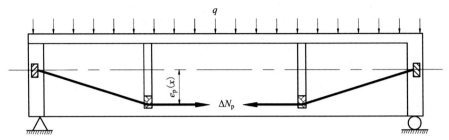

Figure 13.8 Calculation diagram of stress change in external tendon.

From which

$$\Delta N_p = -\frac{\delta_{1p}}{\delta_{11}} = \frac{\int_0^{l_{ex}} \dfrac{M(x)e_p(x)}{E_c I(x)} dx}{\int_0^{l_{ex}} \dfrac{e_p^2(x)}{E_c I(x)} dx + \dfrac{1}{A_{p,ex} E_p} l_{ex}} \tag{13.6}$$

where $M(x)$ = bending moment caused by the external loads in the section under consideration.

$e_p(x)$ = eccentricity of the external tendons.

l_{ex} = initial effective length of the external tendons between two anchor points.

$I(x)$ = moment of inertia of section converted to concrete,

$I(x) = I_n(x) + \alpha_{EP} A_{p,ex} e_p^2(x)$.

$I_n(x)$ = moment of inertia of the net concrete section.

$A_{p,ex}$ = cross-sectional area of the external tendons.

E_c = modulus of elasticity of concrete.

α_{EP} = modulus ratio of the tendon to concrete.

The stress change in the external tendons can be obtained from Eq. (13.6):

$$\Delta\sigma_{p,ex} = \frac{1}{A_{p,ex}} \cdot \frac{\int_0^{l_{ex}} \dfrac{M(x)e_p(x)}{E_c I(x)} dx}{\int_0^{l_{ex}} \dfrac{e_p^2(x)}{E_c I(x)} dx + \dfrac{1}{A_{p,ex} E_p} l_{ex}} \tag{13.7}$$

13.3.3 Ultimate stress in external tendons at flexural failure

For statically determinate and indeterminate externally prestressed concrete structures with numerous profiles and multiple broken lines, it is difficult to obtain the accurate value of the ultimate stress in external tendons at flexural failure. However, the ultimate stress should be obtained first in the design, so the approximate formulas are generally provided in the design codes based on theoretical analysis and experimental results.

The JTG 3362-2018 code stipulates that the ultimate stress in external tendons can be adopted as $\sigma_{pe,ex}(x)$, approximately given by Eq. (13.1) in the calculation of the bearing capacity.

The JGJ 369-2016 code stipulates that the ultimate stress in external tendons in the calculation of the bearing capacity may be taken as follows.

For flexural bearing capacity of the normal section of the simply supported beams

$$\sigma_{pu,ex} = \sigma_{pe,ex} + 100 \tag{13.8}$$

For flexural bearing capacity of the normal section of the continuous or cantilever beams

$$\sigma_{pu,ex} = \sigma_{pe,ex} + 50 \tag{13.9}$$

For bearing capacity of the oblique section

$$\sigma_{pu,ex} = \sigma_{pe,ex} + 100 \tag{13.10}$$

where $\sigma_{pu,ex}$ = design value of tensile stress of external tendons in the calculation of the bearing capacity, MPa.

$\sigma_{pe,ex}(x)$ = effective stress in external tendons in the service period, MPa, given by Eq. (13.1).

13.4 Bearing capacity of the externally prestressed concrete sections

For externally prestressed concrete structures, the flexural bearing capacity of the normal sections and the shear bearing capacity of the oblique sections should be verified in the design, as expressed in Eqs. (9.1) and (9.2). In calculating bearing capacity, the basic assumptions adopted are almost the same as those for bonded post-tensioned flexural members.

The following introduces the calculation of section bearing capacity of the externally prestressed flexural members with internal bonded tendons provided in the JTG 3362-2018 code, where the effective stress in external tendons given by Eq. (13.1) is used in the calculation of the bearing capacity.

13.4.1 Flexural bearing capacity of T-section when the flange is in the compression zone

When the flange is in the compression zone, a typical externally prestressed concrete T-section subjected to external bending moment is shown in Fig. 13.9.

If the depth of neutral axis falls within the flange, the following equation should be satisfied ($x \leqslant h'_f$):

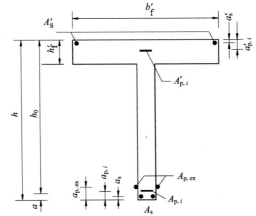

(a) Externally prestressed concrete Tee section

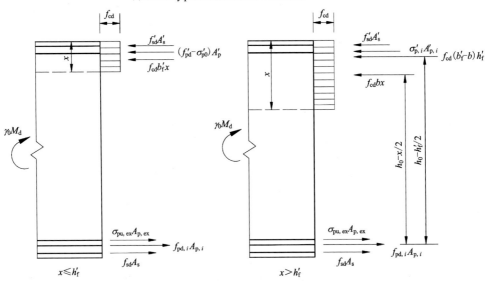

(b) Calculation diagram for bearing capacity

Figure 13.9 Typical externally prestressed concrete T-section.

$$f_{sd}A_s + f_{pd,i}A_{p,i} + \sigma_{pe,ex}A_{p,ex} \leqslant f_{cd}b'_f x + f'_{sd}A'_s + \left(f'_{pd,i} - \sigma'_{p0,i}\right)A'_{p,i} \tag{13.11}$$

Otherwise, the neutral axis falls outside the flange $(x > h'_f)$ if

$$f_{sd}A_s + f_{pd,i}A_{p,i} + \sigma_{pe,ex}A_{p,ex} > f_{cd}b'_f x + f'_{sd}A'_s + \left(f'_{pd,i} - \sigma'_{p0,i}\right)A'_{p,i} \tag{13.12}$$

where $\sigma_{pe,ex}(x) =$ effective stress in external tendons, given by Eq. (13.1).

$A_{p,ex} =$ cross-sectional area of external tendons.

f_{cd} = design strength of concrete in compression.

f_{sd} = design tensile strength of reinforcing steels.

f'_{sd} = design strength of reinforcing steels in compression.

$f_{pd,i}$ = design tensile strength of bonded tendons.

$f'_{pd,i}$ = design strength of bonded tendon in compression.

$\sigma'_{p0,i}$ = stress in bonded tendons in the compression zone corresponding to zero concrete strain at the same level.

$A_{p,i}$ = cross-sectional area of bonded tendons in the tension zone.

$A'_{p,i}$ = cross-sectional area of bonded tendons in the compression zone.

A_s = cross-sectional area of reinforcing steels in the tension zone.

A'_s = cross-sectional area of reinforcing steels in the compression zone.

h'_f = depth of flange in a T-section.

b'_f = width of the flange in a T-section.

13.4.1.1 When the neutral axis falls within the flange

When the neutral axis falls within the flange, the flexural bearing capacity of the T-section can be obtained from Fig. 13.9:

$$M_u = f_{cd}b'_f x\left(h_0 - \frac{x}{2}\right) + f'_{sd}A'_s\left(h_0 - a'_s\right) + \left(f'_{pd,i} - \sigma'_{p0,i}\right)A'_{p,i}\left(h_0 - a'_{p,i}\right) \qquad (13.13)$$

The depth of the neutral axis is given by

$$f_{sd}A_s + f_{pd,i}A_{p,i} + \sigma_{pe,ex}A_{p,ex} = f_{cd}b'_f x + f'_{sd}A'_s + \left(f'_{pd,i} - \sigma'_{p0,i}\right)A'_{p,i} \qquad (13.14)$$

And the depth of the neutral axis should satisfy

$$x \leqslant \xi_b h_0 \qquad (13.15)$$

To ensure that the strain in bonded tendons and reinforcing steels in the compression zone reaches a certain value at failure, the depth of the neutral axis should also satisfy

$$x \geqslant 2a' \qquad (13.16)$$

If only reinforcing steels are arranged in the compression zone, or $\sigma'_{p,i}$ is tensile (negative) when both the bonded tendons and reinforcing steels are arranged in the compression zone, the depth of the neutral axis should also satisfy

$$x \geqslant 2a'_s \qquad (13.17)$$

where M_u = flexural bearing capacity of the section under consideration.

h_0 = effective depth of the section, $h_0 = h - a$.

a = distance from the centroid of all reinforcements in the tension zone to the same side face, $a = \dfrac{a_s A_s + a_{p,i} A_{p,i} + a_{p,ex} A_{p,ex}}{A_s + A_p + A_{p,ex}}$.

d' = distance from the centroid of all reinforcements in the compression zone to the same side face, $d' = \dfrac{A_s' d_s' + A_{p,i}' d_{p,i}'}{A_s' + A_p'}$.

$a_{p,ex}$ = distance from the centroid of external tendons to the face of the tension zone.

$a_{p,i}$ = distance from the centroid of bonded tendons in the tension zone to the same side face.

$d_{p,i}'$ = distance from the centroid of bonded tendons in the compression zone to the same side face.

a_s = distance from the centroid of reinforcing steels in the tension zone to the same side face.

d_s' = distance from the centroid of reinforcing steels in the compression zone to the same side face.

13.4.1.2 When the neutral axis falls within the web

When the neutral axis falls within the web, the flexural bearing capacity of the T-section can be obtained from Fig. 13.9:

$$M_u = f_{cd} \left[bx \left(h_0 - \frac{x}{2} \right) + \left(b_f' - b \right) h_f' \left(h_0 - \frac{h_f'}{2} \right) \right] + f_{sd}' A_s' \left(h_0 - d_s' \right)$$

$$+ \left(f_{pd,i}' - \sigma_{p0,i}' \right) A_{p,i}' \left(h_0 - d_{p,i}' \right) \tag{13.18}$$

The depth of neutral axis is given by

$$f_{sd} A_s + f_{pd,i} A_{p,i} + \sigma_{pe,ex} A_{p,ex} = f_{cd} \left[bx + \left(b_f' - b \right) h_f' \right] + f_{sd}' A_s' + \left(f_{pd,i}' - \sigma_{p0,i}' \right) A_{p,i}' \tag{13.19}$$

And the depth of the neutral axis should satisfy Eqs. (13.15)–(13.17).

13.4.2 Flexural bearing capacity of the T-section when the flange is in the tension zone

When the flange is in the tension zone, a typical externally prestressed concrete T-section subjected to external bending moment is shown in Fig. 13.10.

From Fig. 13.10, the flexural bearing capacity of the T-section can be obtained by

$$M_u = f_{cd} bx \left(h_0 - \frac{x}{2} \right) + f_{sd}' A_s' \left(h_0 - d_s' \right) + \left(f_{pd,i}' - \sigma_{p0,i}' \right) A_{p,i}' \left(h_0 - d_{p,i}' \right) \tag{13.20}$$

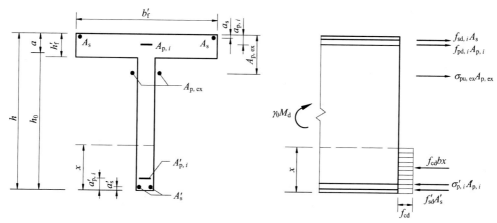

Figure 13.10 Typical externally prestressed concrete T-section.

The depth of neutral axis is given by

$$f_{sd}A_s + f_{pd,i}A_{p,i} + \sigma_{pe,ex}A_{p,ex} = f_{cd}bx + f'_{sd}A'_s + \left(f'_{pd,i} - \sigma'_{p0,i}\right)A'_{p,i} \quad (13.21)$$

When the longitudinal reinforcement in the compression zone is considered in Eq. (13.20) and Eq. (13.16) or (13.17) does not hold, the calculation of flexural capacity of the normal section shall comply with the following provisions.

If only bonded tendons and reinforcing steels are arranged in the compression zone and σ'_p is in compression:

$$M_u = f_{sd}A_s(h - a_s - a') + f_{pd,i}A_{p,i}\left(h - a_{p,i} - a'\right) + \sigma_{pe,ex}A_{p,ex}\left(h - a_{p,ex} - a'\right) \quad (13.22)$$

If only reinforcing steels are arranged in the compression zone, or σ'_p is tensile (negative) when both the tendons and reinforcing steels are arranged in the compression zone:

$$\begin{aligned}
M_u &= f_{pd,i}A_{p,i}\left(h - a_{p,i} - a'_s\right) + \sigma_{pe,ex}A_{p,ex}\left(h - a_{p,ex} - a'_s\right) \\
&+ f_{sd}A_s\left(h - a_s - a'_s\right) - \left(f'_{pd,i} - \sigma'_{p0,i}\right)A'_{p,i}\left(a'_{p,i} - a'_s\right)
\end{aligned} \quad (13.23)$$

The calculation of flexural bearing capacity of the rectangular-, box-, or I-section can be calculated by referring to the above approach on the T-section.

13.4.3 Shear bearing capacity of the oblique section

In an externally prestressed concrete structure, the bearing capacity of the oblique sections includes the shear bearing capacity and flexural bearing capacity which can be achieved by referring to the bonded post-tensioned structures in Sections 9.3 and 9.4.

A typical analytical model for shear bearing capacity for externally prestressed concrete structures is shown in Fig. 13.11, where the external tendons and bonded tendons in the tension zone and vertical bonded tendons are not drawn.

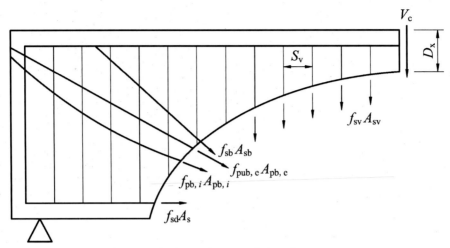

Figure 13.11 Calculation diagram of shear bearing capacity.

The shear bearing capacity of an externally prestressed structure with rectangular-, T-, or I-section can be calculated by

$$V_u = V_{cs} + V_{sb} + V_{pb,i} + V_{pb,ex} \tag{13.24}$$

$$V_{cs} = 0.45 \times 10^{-3} \alpha_1 \alpha_2 \alpha_3 b h_0 \sqrt{(2 + 0.6P)\sqrt{f_{cu,k}} \left(\rho_{sv} f_{sd,v} + 0.6 \rho_{pv} f_{pd,v}\right)} \tag{13.25}$$

$$V_{sb} = 0.75 f_{sd} \sum A_{sb} \sin \theta_s \tag{13.26}$$

$$V_{pb,i} = 0.75 f_{pb,i} \sum A_{pb,i} \sin \theta_{p,i} \tag{13.27}$$

$$V_{pb,ex} = 0.75 \sigma_{pe,ex} \sum A_{pb,ex} \sin \theta_{p,ex} \tag{13.28}$$

where V_{cs} = shear capacity of concrete provided by the concrete and the transverse stirrups intersecting with the oblique section.

V_{sb} = shear bearing capacity provided by the bent-up reinforcing steels intersecting with the oblique section.

$V_{pb,i}$ = shear bearing capacity provided by the bent-up bonded tendons intersecting with the oblique section.

$V_{pb,ex}$ = shear bearing capacity provided by the bent-up external tendons intersecting with the oblique section.

$A_{pb,i}$ = cross-sectional area of bent-up bonded tendons.

$\theta_{p,i}$ = angle between the bent-up bonded tendons and the beam's axis.

$\sigma_{pe,ex}$ = effective stress in external tendons.

$A_{pb,ex}$ = cross-sectional area of bent-up external tendons.

$\theta_{p,ex}$ = angle between the bent-up external tendons and the beam's axis, $\theta_{p,ex} \leqslant 45$ degrees.

θ_s = angle between the bent-up reinforcing steels and the beam's axis.

α_1 = coefficient reflecting the influence of the positive and negative bending moments.

α_2 = increasing coefficient by the prestressing.

α_3 = influence coefficient on compression flange.

$f_{cu,k}$ = characteristic value of cubic compressive strength of concrete.

ρ = ratio of longitudinal reinforcement in the tension zone in the oblique section, $\rho = \frac{A_s + A_{p,i}}{bh_0}$, $P = 100\rho$, $P = 2.5$ is taken if $P > 2.5$.

b = width of section at the level of the shear plane considered.

h_0 = effective depth.

ρ_{sv} = bent-up reinforcing steel ratio, $\rho_{sv} = A_{sv}/s_v b$.

ρ_{pv} = bent-up tendon ratio, $\rho_{pv} = A_{pv}/s_{pv} b$.

s_v = spacing between two adjacent stirrups.

s_{pv} = spacing between two adjacent vertical tendons.

$f_{sd,v}$ = design tensile strength of stirrup.

$f_{pd,v}$ = design tensile strength of vertical tendon.

Eqs. (13.24)–(13.28) are used to calculate the shear bearing capacity of the oblique section with a constant depth. For continuous beams and cantilever beams with variable section depth, the effect of additional shear stress should be considered.

To prevent the oblique section from inclined compression failure, or control the inclined cracking, limiting the sectional dimensions and increasing the concrete strength rather than increasing the quantity of reinforcement are adopted. The JTG 3362-2018 code stipulates that the dimensions of flexural rectangular-, T-, I-, or box-sections should satisfy:

$$V_d \leqslant 0.51 bh_0 \sqrt{f_{cu,k}} \qquad (13.29)$$

where V_d = maximum design value of shear force in the section considered.

13.5 Stress analysis in externally prestressed flexural sections

When the forces caused by the external tendons are treated as the external concentrated loads acting on the anchor points and steering devices, the section stress analysis of an externally prestressed concrete flexural member can be achieved similarly to that of a bonded prestressed member. The stress calculation in the uncracked section of the externally prestressed flexural members with bonded tendons provided in the JTG 3362-2018 code is introduced below.

It is assumed that the long-term strain change in reinforcing steels due to concrete creep and shrinkage is approximately equal to the long-term strain change in the nearby bonded tendons in the service period, and the effects due to the stress redistribution in the section and the tendon relaxation are neglected. The effective stresses in bonded tendons are calculated from Eqs. (7.31) and (7.32), and the effective stress in external tendons are obtained from Eq. (13.1).

For externally prestressed members, the internal bonded tendons are stretched and anchored first, then grouting mortar is used. When the external tendons are stretched, generally, the mortar has become hardened. Therefore, in section stress analysis, the properties of the net section (composed of concrete and reinforcing steels) should be used before grouting mortar and the properties of the converted section (composes of concrete, reinforcing steels, and bonded tendons) should be used later, respectively. For simplification, the properties of the net section are adopted in stress calculation due to the prestressing force. At service loads, the properties of the converted section are used.

Based on the above assumptions, the result provided by the prestressing tendons and reinforcing steels and its eccentricity in the section of an externally prestressed flexural member in the service period can be obtained from Fig. 13.12.

From Fig. 13.12, the generalized prestressing force and its eccentricity can be obtained:

$$N_{p,ex} = \sigma_{pe,ex}A_{p,ex} + \sigma'_{pe,ex}A'_{p,ex} + \sigma_{pe,i}A_{p,i} + \sigma'_{pe,i}A'_{p,i} - \sigma_{l6}A_s - \sigma'_{l6}A'_s \qquad (13.30)$$

$$e_{p,ex} = \frac{\sigma_{pe,ex}A_{p,ex}y_{p,ex} - \sigma'_{pe,ex}A'_{p,ex}y'_{p,ex} + \sigma_{pe,i}A_{p,i}y_{p,i} - \sigma'_{pe,i}A'_{p,i}y'_{p,i} - \sigma_{l6}A_s y_s + \sigma'_{l6}A'_s y'_s}{N_{p,ex}}$$

$$(13.31)$$

where $N_{p,ex} =$ generalized prestressing force.

$e_{p,ex} =$ eccentricity of $N_{p,ex}$.

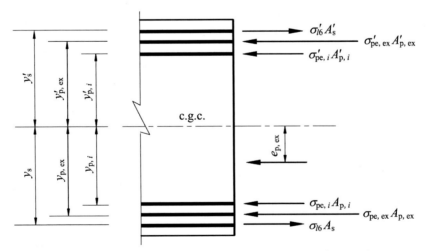

Figure 13.12 Diagram for calculation of the generalized prestressing force and its eccentricity.

$\sigma_{pe,ex}(x)$ = effective stress in external tendons in the tension zone after all losses.

$\sigma'_{pe,ex}$ = effective stress in external tendons in the compression zone after all losses.

$\sigma_{pe,i}$ = effective stress in bonded tendons in the tension zone after all losses.

$\sigma'_{pe,i}$ = effective stress in bonded tendons in the compression zone after all losses.

$A_{p,ex}$ = cross-sectional area of external tendons in the tension zone.

$A'_{p,ex}$ = cross-sectional area of external tendons in the compression zone.

$A_{p,i}$ = cross-sectional area of bonded tendons in the tension zone.

$A'_{p,i}$ = cross-sectional area of bonded tendons in the compression zone.

A_s = cross-sectional area of reinforcing steels in the tension zone.

A'_s = cross-sectional area of reinforcing steels in the compression zone.

$y_{p,ex}$ = distance from the centroid of the external tendons in the tension zone to the centroid of the net section.

$y'_{p,ex}$ = distance from the centroid of the external tendons in the compression zone to the centroid of the net section.

$y_{p,i}$ = distance from the centroid of the bonded tendons in the tension zone to the centroid of the net section.

$y'_{p,i}$ = distance from the centroid of the bonded tendons in the compression zone to the centroid of the net section.

y_s = distance from the centroid of the reinforcing steels in the tension zone to the centroid of the net section.

y'_s = distance from the centroid of the reinforcing steels in the compression zone to the centroid of the net section.

σ_{l6} = prestress loss of tendons in the tension zone due to concrete shrinkage and creep.

σ'_{l6} = prestress loss of tendons in the compression zone due to concrete shrinkage and creep.

The normal stress of concrete caused by the effective prestressing force in the service period is

$$\sigma_{pc} = \frac{N_{p,ex}}{A_n} \mp \frac{N_{p,ex}e_{p,ex}}{I_n}y_{cn} \tag{13.32}$$

The normal stress of concrete caused by the prestressing force and the design load in the service period is

$$\sigma_c = \sigma_{pc} + \sigma_{Lc}$$

$$= \left(\frac{N_{p,ex}}{A_n} \mp \frac{N_{p,ex}e_{p,ex}}{I_n}y_{cn} \right) + \left(\frac{N}{A_0} \pm \frac{M}{I_0}y_{c0} \right) \tag{13.33}$$

where N = design value of axial force.

M = design value of bending moment.

A_n = cross-sectional area of the net section.

I_n = moment of inertia of the net section.

A_0 = cross-sectional area of the converted section.

I_0 = moment of inertia of the converted section.

y_{cn} = distance from the level considered to the centroid of the net section.

y_{c0} = distance from the level considered to the centroid of the converted section.

In the service period, the maximum tensile stress in external tendons is given by Eq. (13.4). The maximum tensile stress in bonded tendons can be obtained by

$$\sigma_{p,i} = \sigma_{pe,i} - \alpha_{EP}\sigma_{c,p}$$

$$= \sigma_{pe,i} - \alpha_{EP}\left[\left(\frac{N_{p,ex}}{A_n} + \frac{N_{p,ex}e_{p,ex}}{I_n}y_{0p,i}\right) + \left(\frac{N}{A_0} - \frac{M}{I_0}y_{0p,i}\right)\right] \quad (13.34)$$

where $y_{0p,i}$ = distance from the outermost layer of the bonded tendons in the tension zone to the centroid of the converted section.

α_{EP} = modulus ratio of the tendon to concrete.

$\sigma_{c,p}$ = normal stress of concrete at the centroid of the outermost layer bonded tendons in the tension zone.

The maximum tensile stress in reinforcing steels can be obtained by

$$\sigma_s = \alpha_{ES}\sigma_{c,s} = -\sigma_{L6} + \alpha_{ES}\left[\left(-\frac{N_{p,ex}}{A_0} - \frac{N_{p,ex}e_{p,ex}}{I_0}y_{0s}\right) + \left(-\frac{N}{A_0} + \frac{M}{I_0}y_{0s}\right)\right]$$
$$(13.35)$$

where y_{0s} = distance from the outermost layer of the reinforcing steels in the tension zone to the centroid of the converted section.

α_{ES} = modulus ratio of the reinforcing steel to concrete.

$\sigma_{c,s}$ = normal stress of concrete at the centroid of the outermost layer reinforcing steel in the tension zone.

For highway flexural members in which cracking is not allowed in the service period, the JTG 3362-2018 code prescribes that the maximum normal compressive stress of concrete in the compression zone should satisfy Eq. (7.48), the maximum tensile stress in bonded tendons should satisfy Eq. (7.49), and the maximum tensile stress in external prestressing strands should satisfy:

$$\sigma_{p,ex} \leqslant 0.60 f_{ptk} \quad (13.36)$$

where f_{ptk} = characteristic tensile strength of prestressing strand.

In the service period, the calculation and limitation of concrete principal stress and shear stress in the section of externally prestressed concrete flexural members can refer to that of bonded post-tensioned flexural members.

13.6 Calculation and control of deflection and crack

Compared with the deflection calculation of post-tensioned bonded flexural members at service loads, there are two obvious characteristics for externally prestressed flexural members: first, the external concentrated forces acting on the anchor points and steering devices due to external tendons are calculated; for those members in which their deflections are sensitive to the secondary effect of prestressing, such as the long span members with a high ratio of span to section depth, the secondary effect of prestressing should be considered.

For externally prestressed members in which cracking is allowed, the crack width needs to be controlled. Once the tensile stress in the outermost layer reinforcing steels in the tension zone, or the stress change in outermost layer bonded tendons in the tension zone from the decompression state to the cracked state is obtained, the crack width can be derived as that in post-tensioned bonded members.

13.6.1 Deflection of the externally prestressed concrete flexural structures

The short-term deflection of an externally prestressed flexural member at service loads can be calculated by the bilinear method, as discussed in Chapter 8, whether the secondary effect of prestressing is taken into account or not. Due to the complexity of obtaining the external concentrated forces acting on the anchor points and steering devices due to external tendons and the difficulty in calculating the deflections considering the secondary effect of prestressing, the deflection calculation of an externally prestressed structure only can be fulfilled by a computer in most cases. The following introduces briefly the principles of deflection calculation recommended in the JGJ 369-2016 code.

When $M(x) \leqslant M_{cr}$

$$f_s = \int_0^L \frac{\overline{M(x)} \cdot M(x)}{B_1} dx \qquad (13.37)$$

When $M(x) > M_{cr}$

$$f_s = \int_0^L \overline{M(x)} \left[\frac{M_{cr}(x)}{B_1} + \frac{M(x) - M_{cr}(x)}{B_2} \right] dx \qquad (13.38)$$

where $M(x) =$ bending moment caused by the prestressing force and the design load.

$\overline{M(x)}$ = bending moment caused by the unit force acting on the considered position.
$M_{cr}(x)$ = cracking moment.
B_1 = short-term flexural rigidity of the uncracked section.
B_2 = short-term flexural rigidity of the cracked section.
L = effective span.

When the secondary effect of prestressing is neglected, the short-term flexural rigidity cab be obtained by the following.

For uncracked sections:

$$B_s = 0.85E_c I_0 \tag{13.39}$$

For cracked sections:

$$B_s = \frac{0.85E_c I_0}{k_{cr} + w(1 - k_{cr})} \tag{13.40}$$

$$w = \left(1.0 + \frac{0.21}{\alpha_E \rho}\right)\left(1 + 0.45\gamma_f\right) - 0.7 \tag{13.41}$$

$$\gamma_f = \frac{\left(b_f - b\right)h_f}{bh_0} \tag{13.42}$$

$$\rho = \frac{A_s + 0.2A_{p,ex}}{bh_0} \tag{13.43}$$

$$M_{cr} = \left(\sigma_{pc} + \gamma f_{tk}\right)W_0 \tag{13.44}$$

where M_k = bending moment caused by the characteristic combination of actions.

M_{cr} = cracking moment.

$k_{cr} = \frac{M_{cr}}{M_k}$, $k_{cr} = 1.0$ is taken if $k_{cr} > 1.0$.

ρ = reinforcement ratio.

E_c = modulus of elasticity of concrete.

I_0 = moment of inertia of the converted section.

$A_{p,ex}$ = cross-sectional area of external tendons in the tension zone.

b_f = width of the flange in the tension zone.

h_f = depth of the flange in the tension zone.

b = width of rectangular section or the web width of the T- or I-section.

h_0 = effective depth of the section.

α_E = modulus ratio of reinforcement to concrete.

If the ratio of span to overall section depth is greater than 12, the secondary effect of prestressing must be considered, and the short-term flexural rigidity should be determined by

$$B_s = \frac{\left(E_s A_s + E_p A_{p,ex}\right) h_0^2}{\varphi\left(0.15 - 0.4\dfrac{h_0}{e + \Delta}\right) + 0.2 + \dfrac{6\alpha_{EP}\rho}{1 + 3.5\gamma_f}} \tag{13.45}$$

$$e = \frac{M_s}{N_{p0}} + e_{p0,ex} \tag{13.46}$$

$$\Delta = \frac{k_1 M_k L^2 - k_2 \sigma_{pe,ex} A_{p,ex} e L^2}{E_c I_e} \tag{13.47}$$

where M_k = bending moment at midspan caused by the characteristic combination of actions.

N_{p0} = resultant provided by the external tendons and reinforcing steels when the normal stress of concrete at the point of resultant is zero.

E_p = modulus of elasticity of external tendon.

E_s = modulus of elasticity of reinforcing steel.

I_e = effective moment of inertia.

$A_{p,ex}$ = cross-sectional area of external tendons in the tension zone.

A_s = cross-sectional area of reinforcing steels in the tension zone.

h_0 = effective depth of the section.

L = effective span.

φ = nonuniform coefficient of stress in longitudinal reinforcement.

Δ = relative displacement of the external tendons in the section under consideration.

$e_{p0,ex}$ = eccentricity of the external tendon at the end of the member.

κ_1, κ_2 = coefficients related to the load form and support condition, checked in Table 13.2.

Table 13.2 The values of κ_1 and κ_2.

Steering device		Coefficient	
Number	**Location**	κ_1	κ_2
0	—	0.106	0.125
1	Midspan	0	0
2	One–third span	0.014	$\frac{\cos\alpha}{72} + \frac{L\sin\alpha}{216e_{p,ex}}$

α is the ratio of the middle horizontal length of external tendons to the entire length of the member.

The long-term deflection of externally prestressed flexural members can be calculated by Eqs. (8.36) and (8.37), in which the short-term flexural rigidity is given by Eqs. (13.39)−(13.47).

13.6.2 Crack width of the externally prestressed concrete flexural structures

The crack width of externally prestressed flexural members can be calculated by Eqs. (12.39)−(12.42), in which the longitudinal reinforcement ratio and equivalent stress of longitudinal tensile reinforcement may be taken as

$$\rho_{te} = \frac{A_s}{A_{te}} \tag{13.48}$$

$$\sigma_{sk} = \frac{M_k \pm M_{p2} - 1.03 N_{pe,ex}(z - e_p)}{(0.20 A_{p,ex} + A_s)z} \tag{13.49}$$

where ρ_{te} = longitudinal reinforcement ratio in the tension zone.

A_{te} = effective area of concrete in tension, $A_{te} = 0.5bh + (b_f - b)h_f$.

A_s = cross-sectional area of reinforcing steels in the tension zone.

σ_{sk} = equivalent stress of longitudinal reinforcement in the tension zone in the cracked section caused by the characteristic combination of actions.

b_f = width of the tensile flange.

h_f = depth of the tensile flange.

Under the characteristic combination of these actions, considering the influence of the long-term effect, the JGJ 369-2016 stipulates that the maximum crack width should not exceed the allowable values specified in Table 8.7 or Table 8.8.

Suggested readings

[1] JGJ 369: 2016. Prestressed concrete structure design code. Beijing: China Construction Industry Press; 2016.
[2] GB/T 18365: 2018. Hot-extruded PE protection paralleled high strength wire cable for cable-stayed bridge. Beijing: China Construction Industry Press; 2018.
[3] GB/T5223: 2014. Steel wire for prestressing of concrete. Beijing: China Construction Industry Press; 2014.
[4] GB/T 26752: 2020. PAN-based carbon fiber. Beijing: China Construction Industry Press; 2020.
[5] GB/T 14370: 2015. Anchorage, grip, and coupler for prestressing tendons. Beijing: China Standards Press; 2015.
[6] JTG 3362: 2018. Specifications for design of highway reinforced concrete and prestressed concrete bridges and culverts. Beijing: People's Communications Press; 2018.
[7] Hu Di. Basic principles of prestressed concrete structure design. 2nd ed. Beijing: China Railway Publishing House; 2019.

Index

Note: 'Page numbers followed by "*f*" indicate figures and "*t*" indicate tables.'

Printed in the United States
by Baker & Taylor Publisher Services